CONTROL OF VOLATILE ORGANIC COMPOUND EMISSIONS

CONTROL OF VOLATILE ORGANIC COMPOUND EMISSIONS

Conventional and Emerging Technologies

PAIGE HUNTER
S. TED OYAMA

A Wiley-Interscience Publication

JOHN WILEY & SONS, INC.

New York • Chichester • Weinheim • Brisbane • Singapore • Toronto

This book is printed on acid-free paper. ♾

Published simultaneously in Canada.

For ordering and customer service, call 1-800-CALL-WILEY.

Library of Congress Cataloging-in-Publication Data:

Oyama, S. Ted, 1955-
Control of volatile organic compound emissions : conventional and emerging technologies / S. Ted Oyama, Paige Hunter.
p. cm.
Includes bibliographical references and index.
ISBN 0-471-33369-7 (alk. paper)
1. Volatile organic compounds—Environmental aspects. 2. Air—Purification—Equipment and supplies. 3. Air—Purification—Technological innovations. I. Title. II. Hunter, Paige.
TD885.5.O74 O94 2000
628.5′3 21—dc21 99-043060

Printed in the United States of America.

10 9 8 7 6 5 4 3 2 1

To my parents—who reared three daughters to believe that we could do anything we set our minds to,
and for our husbands—who patiently tolerate that trait.
Paige Hunter

To Hideko—who inspired me to do better
and to Monika and Leo—who made it worthwhile.
S. Ted Oyama

CONTENTS

PREFACE

Release of airborne contaminants into the environment continues to be one of the major problems of the early twenty-first century industrial world. As the demand for reduced emissions increases, the market for innovative control technologies also improves. This book explores conventional and emerging technologies for treating volatile organic compound (VOC)–containing streams. Standard methods of air pollution control are well covered in a number of publications, and the attention here is on techniques specifically applied to VOCs. These include condensation, adsorption, absorption/scrubbing, thermal incineration, catalytic incineration, flaring, biofiltration, odor modification, and air dilution. A critical examination of the range of applicability and economics of these techniques is presented. Emerging methods discussed include membrane filtration, ultraviolet oxidation, and corona destruction. Particular emphasis is on catalytic ozonation, a novel technology for treating gaseous streams. This technology is suitable for retrofitting on existing process or control systems, an important consideration for most industries.

This book begins in Chapter 1 with a discussion of the major types of air pollutants and the definition of VOCs. Chapter 2 then summarizes and compares the major methods of air pollution control. Chapters 3–9 describe conventional methods in detail, while Chapter 10 introduces emerging technologies. Chapter 11 presents the basic properties and methods for producing and handling ozone, and Chapter 12 discusses the science and technology of catalytic ozone destruction of VOCs. Methods for determining estimated cost and applicability for each technology are included. These calculations are intended to assist in the preliminary control technology assessment. Equipment vendors and referenced material should be consulted to confirm the results before a technology determination is finalized.

S. Ted Oyama
Paige Hunter

ACKNOWLEDGMENTS

It is with pleasure that we express our appreciation to colleagues, friends, and family who have supported us in this project. In particular, James McKenna did an excellent job preparing artwork for Chapters 3 through 10. Hideko Oyama endured long periods of absence as the project was being completed and cheerfully encouraged it. Mark Hunter has gone far beyond the usual "patient and understanding" husband, doing whatever was needed, but, more importantly, demonstrating endless tolerance and providing continuous reassurance. Mark Furgurson has been a constant source of encouragement, taking on additional responsibilities so that Paige could have more time to concentrate on writing and attending classes. Yasuhiro Iwasawa was very supportive during Ted's sabbatical leave in his laboratory, when most of the chapters on ozone were written. But this book would never have been possible without Don Shepherd and the other members of the Roanoke Regional Office of the Virginia Department of Air Pollution Control, who instilled in Paige a deep commitment to air quality and took the time to teach her about pollution control systems and regulations. Finally, we acknowledge Michel Boudart, who has always been a source of inspiration for Ted in all matters academic and personal.

1

THE PROBLEM OF VOLATILE ORGANIC COMPOUNDS

1.1 SCOPE

Volatile organic compounds (VOCs) constitute a broad range of odorous and toxic substances that include hydrocarbons, olefins, aromatics, and various oxygen-, nitrogen-, sulfur-, and halogen- containing molecules. The effects of these substances on human health range from a simple nuisance to a serious hazard, and regulations have been enacted to limit their emissions. VOCs represent an ubiquitous problem in indoor and outdoor air pollution. They are emitted from virtually all industrial sources in one form or another. In the atmosphere, they combine with oxides of nitrogen and sunlight to form ozone, an oxidant that has harmful effects on plant and animal life.

1.2 TYPES OF AIR POLLUTANTS

There are many types of pollutants in ambient air, ranging from solid to liquid to gaseous. The solid and liquid substances may be in the form of fine particles or aerosols. Classification of these pollutants may be based on origin, composition, chemical properties, physiologic effects, physical location, or even legislation. There is significant overlap between many of the categories and no unique method of categorization. For example, pollutants may be described as (1) *natural* or *man-made*, (2) *particulate* or *gaseous*, and (3) *primary* or *secondary* [1]. Primary contaminants are emitted to the atmosphere from a process, whereas secondary contaminants arise from chemical reactions in the atmosphere. A number of other common classifications of pollutants are summarized in the following sections.

This chapter concentrates on definitions based on U.S. law. The Environmental Protection Agency's (EPA's) Office of Air and Radiation (OAR) Internet home page (http://www.epa.gov/docs/oar/oarhome.html) provides a useful overview and current updates on all aspects of air pollution technology and legislation. Regulations dealing with these compounds differ from state to state in the United States [2, 3] and from country to country [3, 4]. Although there are, of course, some considerable divergences, there are many similarities in the levels regulated. These are based on epidemiologic studies.

Further information on U.S. state or territory regulations is available from individual state offices. A listing of addresses and contacts is provided in EPA Form 8700-12 [5] as well as on the Internet (http://www.epa.gov/oar/oarrsrc.html). Rules concerning international aspects have been outlined (http://www.epa.gov/oar/caa/caa115.txt). Links to international efforts in pollution prevention are reported in the EPA's home page and reproduced in Appendix 1.1.

1.3 CRITERIA AND NONCRITERIA POLLUTANTS

U.S. federal law distinguishes between two basic categories of air contaminants. These categories are the *criteria* pollutants and the *noncriteria* pollutants [6].

Criteria air pollutants are those for which specific concentration limits have been set to distinguish between acceptable air quality and poor or unhealthy air quality. These limits have been defined by the EPA and are known as the National Ambient Air Quality Standards (NAAQS). They represent the concentration over a specific period of time that is generally regarded as safe for the majority of the population [6]. The criteria air pollutants consist of four gases and two solids (Table 1.1).

The standards for ozone and particulate matter are undergoing revision [7], but recent legal action in the U.S. Court of Appeals for the District of Columbia has challenged the new standards. As of the writing of this book, they are on hold pending appeal by the EPA. The standard for ozone was proposed to be reduced from 0.12 ppmv for a 1-h average to 0.08 ppmv for an 8-h average, with the average fourth highest concentration over a 3-year period determining whether an area was out of compliance [8, 9]. The 1-h ozone exposure standard was planned to be gradually phased out. The standard for particulate matter was updated to $PM_{2.5}$, but would not have been enforced until 2002 at the earliest, after a monitoring program and a review process was implemented. The existing PM_{10} standards would have continued to apply until states file mandatory SIPs (state implementation plans) within 3 years.

Table 1.1 also presents values of permitted industrial concentrations for the same substances. These are regulated by the Occupational Safety and Health Administration (OSHA) and determined by the American Conference of Governmental Industrial Hygienists (ACGIH). The permitted industrial

TABLE 1.1 Air Quality Standards

	National Air Quality Standards		Allowed Industrial Concentrations	
Contaminant	Concentration	Duration/h	Concentration	Duration/h
Ozone[b]	0.12 ppmv	1	0.30 ppmv	15 min
	0.08 ppmv	8	0.08 ppmv	8
Carbon monoxide	35 ppmv	1	400 ppmv	15 min
	9 ppmv	8	50 ppmv	8
Nitrogen dioxide	0.05 ppmv	Annual	5 ppmv	15 min
			3 ppmv	8
Sulfur dioxide	0.14 ppmv	24	5 ppmv	15 min
	0.03 ppmv	Annual	2 ppmv	8
Particulate matter[b]	150 μg m^{-3}	24	—[a]	—[a]
PM_{10}	50 μg m^{-3}	Annual	—[a]	—[a]
$PM_{2.5}$	65 μg m^{-3}	24	—[a]	—[a]
	15 μg m^{-3}	Annual	—[a]	—[a]
Lead	1.5 μg m^{-3}	3 months	450 μg m^{-3}	15 min
			150 μg m^{-3}	8

[a] Standards exist for specific types of particles.
[b] Standards for 0.08 ppmv ozone and $PM_{2.5}$ are being legally contested.

concentrations are much higher than the NAAQS values, and this is for two reasons: (1) a work period is only 40 h out of a total of 168 h in a week, so exposures are limited; and (2) the workforce does not include members of the population most susceptible to air pollution (infants and elderly). Unlike air quality standards, industrial standards are designed to protect just the workers [10].

Noncriteria pollutants are the remaining substances. These can fall into two other categories, *hazardous* or *toxic*, as defined by legislation or regulation. In general, the hazardous air pollutants (HAPs) cause a variety of general symptoms (irritation, asphyxia, and so forth), whereas the toxic substances are associated with specific physiologic responses (toxicity) [6]. HAPs constitute 172 individual compounds and 17 categories of substances as classified in the 1990 Clean Air Act Amendments. A downloadable listing is available on the Internet (http://www.epa.gov/oar/caa/caa112.txt).

A full listing of substances regulated by the EPA is provided at the end of this book [11]. This includes the criteria pollutants, NSPS-regulated substances, HAPs, Accidental Release substances, Ozone Depleting Substances (ODS), and VOCs. The New Source Performance Standards (NSPS) are regulations pertaining to newly constructed facilities (a description can be found at http://www.epa.gov/oar/caa/caa111.txt). Regulations on ozone-depleting substances are discussed at (http://www.epa.gov/ozone/title6/usregs.html).

1.4 PARTICULATE MATTER

Particulate matter (PM) in the atmosphere consists of nongaseous particles in the size range of 0.01 to over 100 μm [10]. (The term PM_{10} in the NAAQS refers to particulate matter smaller than 10 μm, and $PM_{2.5}$ refers to particles smaller than 2.5 μm.) The particles may be solid or liquid; those suspended in a gaseous medium are also known as *aerosols*. Particulates are difficult to categorize on the basis of fundamental properties such as their settling rates, opacity, optical activity, structure, surface-to-volume ratio, reaction activity, physiologic action, and so forth [1]. Their particle size, shape, and density, however, are probably the most important [12]. Particles larger than 100 μm settle too quickly and are not included in the definition. Popular descriptive terms such as *smoke, dust*, and *mist*, which are based on the mode of formation, are convenient and useful [1]. Figure 1.1 shows typical particulate matter found in the atmosphere [13].

There are many methods of dealing with particulate matter that utilize different principles for separation [12, 14]. These include gravity settling chambers [15], cyclones [16], fabric filter (baghouse) dust collectors [17, 18], and electrostatic precipitators [19]. The differences result from the nature of the forces applied to the particles in order to collect them.

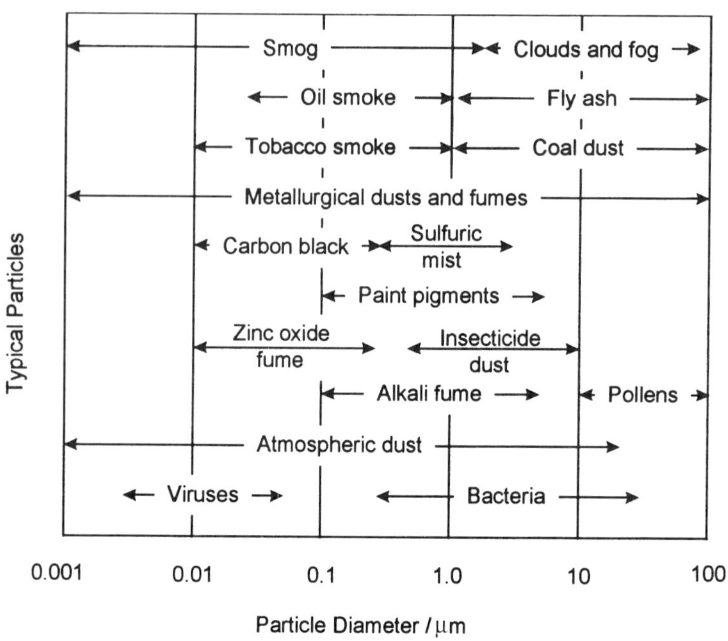

Figure 1.1 Sizes of airborne particles. (Adapted from Lapple [13].)

1.5 VOLATILE ORGANIC COMPOUNDS

Volatile organic compounds (VOCs) are a class of substances in which organic carbon is bonded to hydrogen or to other elements. As an approximate rule, VOCs may be defined as organic liquids or solids whose room temperature vapor pressures are greater than about 0.01 psia (0.0007 atm, 70 Pa) and whose atmospheric boiling points are up to about 500°F (260°C, 533 K) [10]. There are two main classes of VOCs [20]: hydrocarbons and compounds (Table 1.2).

Most organic compounds with less than 12 carbon atoms are VOCs [10], and this includes most HAPs (Appendix 1.2). There are a number of VOCs (e.g., ethylene, propane), however, that are not HAPs. Strictly speaking, VOCs do not include particulate matter. Certain aerosols are formed from condensed droplets of VOCs, however, and their treatment can involve similar methods.

The U.S. EPA defines a VOC as any compound of carbon that participates in atmospheric photochemical reactions. This definition excludes pure carbon itself, carbonate carbon, solid carbidic compounds in which the carbon is bonded to a metal, CO, and CO_2 [10]. A list of VOCs is provided in Appendix 1.2. This list is based on a tabulation of compounds in an EPA database [21] and, although updated here, should not be considered complete. The original tabulation contained some compounds that are no longer considered VOCs (methane, ethane, acetone). The EPA has developed a further list of substances (subject to revisions) that are excluded from the VOC definition because of their low photochemical reactivity (Table 1.3) [11].

Control of volatile organic compound emissions is important because they are a principal component in the chemical and physical atmospheric reactions that form ozone and other photochemical oxidants. The EPA [22] reports that overall emissions of VOCs in the United States increased from 1900 to 1970, peaking in the early 1970s, but have decreased steadily until 1992, with a slight increase in 1993 and 1994 (Fig. 1.2). Considerable changes occurred over the years in the types and amounts of VOCs emitted. For example, in 1900, emissions from all

TABLE 1.2 Classes of Volatile Organic Compounds

Hydrocarbons	Examples	Compounds	Examples
Paraffins	Propane, butane, octane	Alcohols	Methanol, ethanol
Olefins	Ethylene, butadiene	Aldehydes	Formaldehyde
Acetylenes	Acetylene	Ketones	Methyl ethylketone
Aromatics	Toluene, benzopyrene	Acids	Formic acid
		Hydroperoxides	Peroxyacylnitrite (PAN)
		Halides	Vinyl chloride, bromobenzene
		S compounds	Dimethyl sulfide
		N compounds	Trimethyl amine

TABLE 1.3 Substances Excluded from the VOC Definition[a]

Methane, ethane	*t*-Butyl acetate
Acetone	Methylene chloride (dichloromethane)
1,1,1-Trichloroethane (methyl chloroform)	Trichlorofluoromethane (CFC-11)
Dichlorodifluoromethane (CFC-12)	Chlorodifluoromethane (CFC-22)
1,1,1-Trichloro-2,2,2-trifluoroethane (CFC-113)	Trifluoromethane (CFC-23)
1,2-Dichloro-1,1,2,2-tetrafluoroethane (CFC-114)	Chloropentafluoroethane (CFC-115)
1,1,1-Trifluoro-2,2-dichloroethane (HCFC-123)	1,1,1,2-Tetrafluoroethane (HFC-134a)
1,1-Dichloro-1-fluoroethane (HCFC-141b)	1-Chloro-1,1-difluoroethane (HCFC-142b)
2-Chloro-1,1,1,2-tetrafluoroethane (HCFC-124)	
1,1,2,2-Tetrafluoroethane (HFC-134)	Pentafluoroethane (HFC-125)
1,1-Difluoroethane (HFC-152a)	1,1,1-Trifluoroethane (HFC-143a)
Parachlorobenzotrifluoride	Volatile methyl siloxanes

[a] Also excluded are perfluorocarbon compounds that fall into these classes: (1) Cyclic, branched, or linear, completely fluorinated alkanes; (2) Cyclic, branched, or linear, completely fluorinated ethers with no unsaturations; (3) Cyclic, branched, or linear, completely fluorinated tertiary amines with no unsaturations; and (4) Sulfur-containing perfluorocarbons with no unsaturations, with S bonds only to C or F.

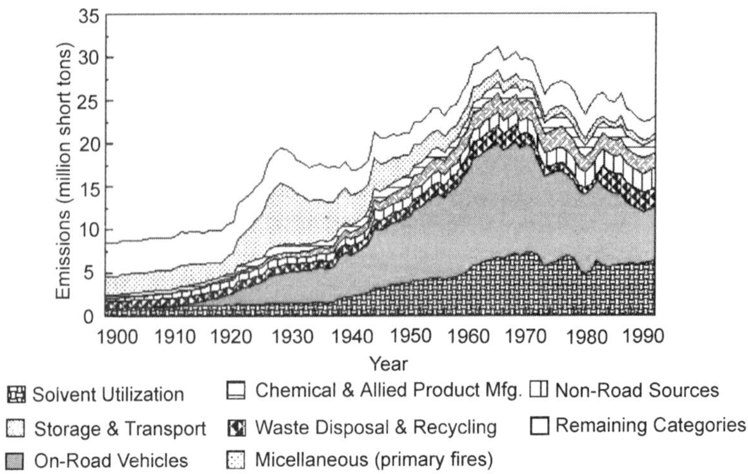

Figure 1.2 Trend in volatile organic compound emissions. (Reproduced from the EPA [22].)

fuel combustion sources represented 68% of the total national VOC emissions, but by 1994 this contribution was 4%. These decreases occurred despite population growth and were due to increased controls on the burning of fossil fuels for utilities and industry. A breakdown of the VOC emissions in the United States by category is provided for 1994 (Fig. 1.3).

VOC emissions from transportation sources increased 159% from 1940 to 1970 as a result of wider use of automobiles and aircraft. After 1970, the Federal Motor Vehicle Control Program (FMVCP) initiative resulted in a drastic decline

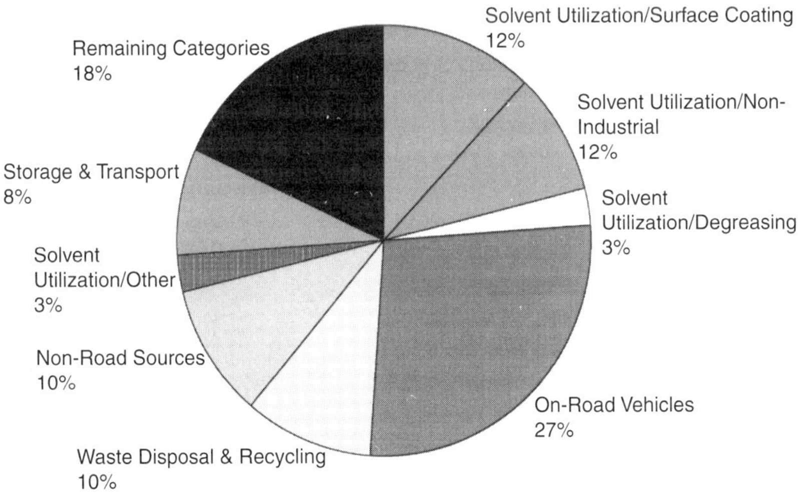

Figure 1.3 Volatile organic compound 1994 emission estimates by source category. (Reproduced from the EPA [22].)

in vehicle emissions, even with the enormous increase in vehicle motor traffic. From 1970 to 1994, on-road vehicle emissions dropped approximately 51% as a result of inspection/maintenance, reformulated fuels programs, and other control programs. VOC generation tracks large population and industrial centers [22]. The distribution in the United States is heavily concentrated in the coastal areas (Fig. 1.4).

1.6 INDUSTRIAL POLLUTANTS

As mentioned earlier, VOC release from industrial sources is universal. These releases vary from industry to industry and stem not only from designed elements like stacks and vents but also from the myriads of activities related to production, storage, and handling. Included in this category are releases from transportation vehicles. Because the number and types of sources are spread so widely geographically, the control of emissions poses considerable regulatory and technological challenges. Table 1.4 shows the number and diversity of the types of emission sources and the broad range of VOCs emitted from each source category [3].

The problem of VOC release is large. Despite successful control in a number of areas, the amounts of VOCs emitted are still substantial. In 1992, the estimated release of VOCs was 22.7 million metric tons (10.3 billion lb) [23]. The main sources were industry (44%) and transportation (35%). A National Research Council study concluded that VOC emissions were likely to have been underestimated by a factor of two to four [24].

Waste treatment and disposal facilities in the United States alone accounted for 2 million metric tons of VOC emissions per year [25]. These facilities emit VOCs

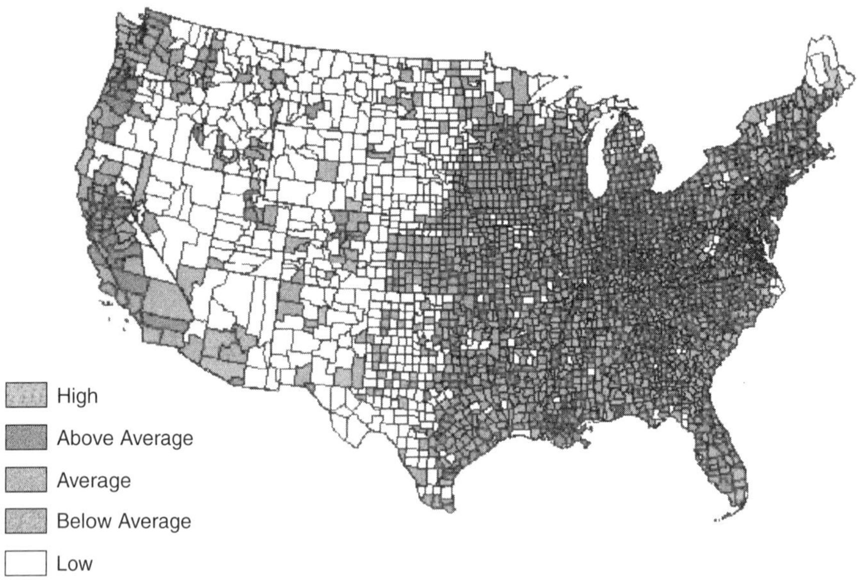

Figure 1.4 Density map of 1994 county level VOC emission estimates. (Reproduced from the EPA [22].)

from storage tanks and landfills. The types and quantities of emissions depend on their origin and have been the subject of a number of investigations. Emissions from stationary sources [26] such as electric utilities, coke ovens, municipal incinerators, and mobile sources [27] have been discussed in the literature.

1.7 INDOOR AIR POLLUTANTS

There are also many sources and types of pollutants in nonindustrial indoor air [28, 29]. Good coverage of the topic is found on the Internet (http://www.epa.gov/iaq/). Indoor air pollutants include tobacco smoke, radon, emissions from building materials, combustion products from inside buildings, pollutants from outside buildings, emissions from consumer products, and biological pollutants [30]. Biological pollutants include organisms like bacteria, molds, pollen, and viruses, which may breed in wet areas or stagnant water in ducts or humidifiers. One indoor bacterium, *Legionella*, has caused both Legionnaire's disease and Pontiac fever. Radon is a colorless, odorless radioactive gas that seeps into homes from the earth and rock beneath them, from well water, and from building materials. It is a serious health concern because it is a cause of cancer. A description of this topic is provided on the Internet (http://www.epa.gov/iaq/radon/).

Excluding radon and the biologic pollutants, the majority of the contaminants found in indoor air can be classified as VOCs [31]. Because these include a

TABLE 1.4 Typical Pollutants Related to Emissions from Industrial Sources

Substance	Mining	Coal Prod./ Comb.	Oil Prod./ Comb.	Energy Prod.	Metal	Chemical	Pulp/paper	Food	Solvent Use	Wast Treatment	Mobile Sources	Other[a]
Particulate	•	•	•	•	•	•	•	•	•	•	•	•
Hydrocarbons	•	•	•	•	•	•	•	•	•	•	•	•
Chlorinated VOCs				•	•	•	•	•	•	•	•	•
Aromatics	•	•	•	•	•	•	•	•	•	•	•	•
Phenols	•		•	•			•	•	•			•
Aldehydes			•	•		•	•	•	•	•	•	•
Organic acids						•		•				
Dioxins/furans	•			•	•					•	•	
Mercaptans		•	•			•	•	•	•	•		•
Trichloroethylene												
Amines				•		•		•	•	•		•
Odors	•	•				•	•	•	•			•

[a] Other sources include printing industries, dry cleaning operations, and textile manufacturing.

large number of individual compounds, it is common to characterize organic content by an indicator known as the *total volatile organic compounds* (TVOCs). This quantity can be easily obtained through chemical analysis (e.g., with an integrating flame ionization detector) and is valuable because it has been related to human responses to pollution [32, 33]. For example, TVOC < 0.20 mg m^{-3} is found to be comfortable, TVOC in the range 0.2–3.0 mg m^{-3} is found to be irritating, TVOC in the range 3.0–25 mg m^{-3} is acutely uncomfortable, and TVOC > 25 mg m^{-3} is toxic.

Tobacco smoke is a special component of indoor air pollution. It is a complex mixture of thousands of volatile, semivolatile, and particulate organic and inorganic compounds formed from the incomplete combustion of tobacco leaves. *Environmental tobacco smoke* (ETS) is smoke that is emitted from the smoldering end of cigarettes between puffs. Smoking in indoor spaces increases the contamination level in CO, NO_2, VOCs, and particulate matter to concentrations above the levels allowed by NAAQS standards [31].

Two special effects of indoor air pollution are the so-called *sick building syndrome* (SBS) and *building-related illness* (BRI) [28]. SBS is used to describe situations in which building occupants experience acute health and comfort effects that appear to be linked to time spent in a building but for which no specific illness or cause can be identified. Complaints include headache; eye, nose, or throat irritation; dry cough; dry or itchy skin; dizziness and nausea; difficulty in concentrating; fatigue; and sensitivity to odors. The symptoms generally subside once the building is left. In contrast, BRI is associated with diagnosable illnesses that can be attributed directly to airborne building contaminants. Complaints include cough, chest tightness, fever and chills, and muscle aches.

1.8 ODOROUS POLLUTANTS

Odorous waste gases are a special kind of air pollutant. Even extremely small amounts of odorant can be perceived by humans. It is estimated that only 10^8 or 10^9 molecules of odorant vapor in the nose is enough to trigger detection [34]. To put this in perspective, 1 μg of ethyl mercaptan in air constitutes approximately 10^{16} molecules, 10^7 or 10^8 times the amount necessary for detection [34].

Table 1.5 lists common classes of odorous pollutants and their lower odor threshold [34]. *Threshold* refers to the minimum concentration that produces an olfactory response or sensation. It is usually determined subjectively, by a panel of people, 50% of whom must correctly detect an odor [35]. Most odorous substances are VOCs.

Human acuity to odor requires stringent levels of control. This usually necessitates waste reduction efficiencies of 95–100%, which are substantially higher than needed for most gaseous emission [34].

TABLE 1.5 Properties of Odorous Air Pollutants

Category/Class	Systemic Name	Formula	Odor	Threshold[a]/ppm
		Sulfur Compounds		
Sulfides	Hydrogen sulfide	H_2S	Rotten eggs	4.7×10^{-3}
Mercaptans	Methyl mercaptan	CH_3SH	Decayed cabbage	2.0×10^{-5}
	Allyl mercaptan	$CH_2{=}CHCH_2SH$	Garlic	3.0×10^{-3}
	Benzyl mercaptan	$C_6H_5CH_2SH$	Unpleasant	2.6×10^{-3}
Thioethers	Dimethyl sulfide	$(CH_3)_2S$	Decayed cabbage	3.0×10^{-3}
	Diethyl sulfide	$(C_2H_5)_2S$	Foul, garlic	4.8×10^{-3}
		Nitrogen Compounds		
Inorganic	Ammonia	NH_3	Pungent	4.7×10^{-1}
Aliphatic amines	Dimethylamine	$(CH_3)_2NH$	Fishy	4.7×10^{-2}
Heterocyclic amines	Pyridine	C_5H_5N	Empyreumatic	3.0×10^{-3}
	3-Methylindole	C_9H_9N	Fecal	5.0×10^{-2}
Cyanides	Hydrogen cyanide	HCN	Bitter almonds	9.0×10^{-1}
		Hydrocarbons		
Aliphatics	2-Butene	$CH_3CH{=}CHCH_3$	House gas	2.4×10^{1}
	2-Methylpropene	$CH_2{=}C(CH_3)_2$	House gas	2.0×10^{1}
		Oxygenates		
Phenol	Phenol	C_6H_5OH	Empyreumatic	4.7×10^{-2}
Aldehydes	Formaldehyde	HCHO	Pungent	1.0
	Acetaldehyde	CH_3CHO	Lachrymogenous	6.6×10^{-2}
Organic acids	Butanoic acid	C_3H_7COOH	Rancid	1.0×10^{-3}
		Inorganic		
Chlorine	Chlorine	Cl_2	Pungent	3.1×10^{-1}
Ozone	Ozone	O_3	Cut hay	5.1×10^{-1}
Sulfur oxides	Sulfur dioxide	SO_2	Pungent	4.7×10^{-1}

[a] Lower limit.

Adapted from the National Academy of Sciences [34].

Industrial plants are common sources of odorous emissions. Surveys of odor-control technology cover various industrial enterprises [34, 36]. Specifically described are particular conditions in plants for livestock and poultry production [37], sewage treatment, rubber processing, steel manufacturing, metal casting, pulp and paper processing, petroleum refining, and chemical production [38].

1.9 EFFECTS OF POLLUTANTS

The effects of HAPs and VOCs have been extensively studied and documented, although some of the findings remain controversial [10, 31, 39]. Air pollution causes damage to the entire environment. This includes adverse effects on human and animal health, harm to crops, vegetation, and forests, as well as damage to physical materials.

Chemical agents have different effects on living animals. Depending on dose and route of exposure, they can attack specific organs or the entire body. Air pollution acts most directly on the skin, exposed mucous membranes in the eyes and nose, and the lungs. From there, effects may be concentrated in other organs or systems. *Hepatotoxic* agents (e.g., carbon tetrachloride) affect the liver, *nephrotoxic* agents (e.g., halogenated hydrocarbons) affect the kidneys, *hematopoietic* agents (e.g., aniline, toluidine, phenol) affect blood, *neurotoxic* agents (e.g., methanol, benzene) affect the nervous system, and *anesthetic* agents (e.g., acetylene hydrocarbons, olefins, ethers, alcohols) affect the brain [40]. Air pollutants differ in the time over which they act. Criteria pollutants generally have a faster action than HAPs, acting over minutes to months versus years, although the long-term effects of exposure to lead, a criteria pollutant, are well known.

Because of the deleterious effects of high concentrations of chemicals on humans, various limits have been set for maximum exposure. For example, the ACGIH has suggested threshold limit values (TLVs) (formerly known as maximum allowable concentrations [MACs]) for permissible air concentrations of a given chemical to which an individual can be repeatedly exposed for 8 h per day, 5 days per week [41]. Typical examples are given in Table 1.6.

OSHA and the American Society of Heating Refrigerating and Air-Conditioning Engineers, Inc. (ASHRAE) have both set one-tenth of the TLV as guidelines for indoor air contaminants. On the basis of a scientific study [42], the American

TABLE 1.6 Threshold Limit Values (TLVs) of Typical VOCs

Name	Formula	TLVs	
		ppm	mg m^{-3}
Benzene	C_6H_6	10	30
Cyclohexane	C_6H_{12}	300	1050
Toluene	C_7H_8	100	375
Styrene	C_8H_8	50	215
Acetone	C_3H_6O	750	1800
Carbon tetrachloride	CCl_4	5	32
Dichloromethane	CH_2Cl_2	100	350
Tetrachloroethylene	C_2Cl_4	50	340
p-Chlorotoluene	C_7H_7Cl	50	260

Industrial Hygiene Association (AIHA) has adopted a guideline of 5 mg m^{-3} for TVOCs.

Airborne pollutants can also cause widespread injury to plant life. The effects of acid rain on forests have been well publicized. *Phytotoxic* chemicals such as peroxyacylnitrate have devastated agricultural crops in large areas surrounding metropolitan centers. Ozone has the greatest effect among common pollutants and has been documented to decrease yields of agricultural crops in the United States by 5–10% [43]. As with animals, plant damage depends greatly on particular conditions. Leaves are the structure most affected because they represent high surface area bodies designed for gas exchange with the atmosphere. Injury can also, however, occur in the roots, stems, and reproductive organelles. Effects range from *chlorosis* (chlorophyll destruction) to *necrosis* (tissue death) [31]. One of the most serious effects is a general weakening of the plant, resulting in susceptibility to other diseases or to insect infestations.

Gaseous and particulate pollutants have a significant deleterious effect on materials. Particularly affected are metals, building stones, paints, textiles, fabric dyes, rubber, leather, and paper [31]. Effects may be due to physical abrasion from particulate matter carried by wind and rain to chemical and photochemical reactions resulting from the adsorption of gases on the surfaces of solids. Chemical reactions change the structure of materials, resulting in discoloration, corrosion, or complete erosion.

1.10 LEGISLATION

Laws in the United States arise from acts originated by Congress and signed by the president. Laws pertaining to air pollution have their basis in the Clean Air Act of 1963, as amended in 1970, 1977, and 1990. A simple explanation of the legislation can be found in the Plain English Guide to the Clean Air Act (http://www.epa.gov/oar/oaqps/peg_caa/pegcaain.html), and the full text is available at http://www.epa.gov/oar/caa/contents.html. Regulations are well discussed in a number of publications [1, 10, 26] and are not treated in depth here. Note is made of the exponentially increasing number of regulations (Fig. 5).

The EAP is charged with producing detailed regulations that specify how the laws are to be applied. Its Internet home page address is http://www.epa.gov/. These regulations undergo examination in public hearings, review by the Office of Management and Budget, and litigation by concerned parties before becoming enforceable.

Uniform standards that regulate the release of HAPs from specific industries are being developed by the EPA as Maximum Achievable Control Technology (MACT) standards. These supplement uniform standards (National Emission Standards for Hazardous Air Pollutants [NESHAPs]) already in place for benzene, beryllium, vinyl chloride, arsenic, and mercury [44]. In addition, compounds such as total reduced sulfur (TRS) and fluorides are regulated for specific industries.

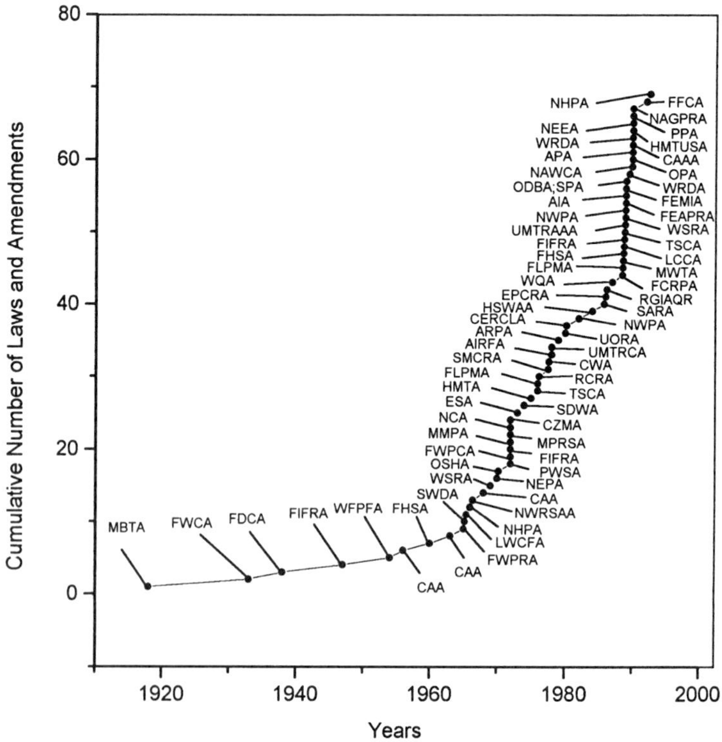

Figure 1.5

Some of the regulations, such as the New Source Performance Standards (NSPS) and the NESHAPs apply nationally. Other regulations are administered by individual states in the form of State Implementation Plans (SIPs). These plans, which are subject to review by the EPA, include local regulations and often the national standards. Because these standards constitute a minimum level of protection below which no location in the United States is allowed to operate, it is not possible for a region to attract polluting industries on the basis of lax laws.

1.11 UNITS AND CONVERSION FACTORS

Concentrations of pollutants are expressed in units of mass per unit volume (such as $\mu g\ m^{-3}$, micrograms per cubic meter) or in terms of a volumetric ratio (ppm, volume of contaminant per million volumes of air) [31]. The conversion between mass units and volumetric ratios at standard temperature and pressure is

$$\mu g\ m^{-3} = \text{ppm} \times \text{MW} \div 0.02445 = \text{ppb} \times \text{MW} \div 24.45$$

$$mg\ m^{-3} = \text{ppm} \times \text{MW} \div 24.45 = \text{ppb} \times \text{MW} \div 24450$$

where

$\mu g\ m^{-3}$ = micrograms per cubic meter
$mg\ m^{-3}$ = milligrams per cubic meter
ppm = parts per million by volume
MW = molecular weight of contaminant ($g\ mol^{-1}$)
ppb = parts per billion

REFERENCES

1. J.C. Mycock; J.D. McKenna; L. Theodore, *Handbook of Air Pollution Control Engineering and Technology*, Lewis Publishers, Boca Raton, **1995**.
2. K.D. Bailey, "Legislative and Legal Considerations," in *Toxic Air Pollution Handbook*, D.R. Patrick, Ed., Van Nostrand Reinhold, New York, **1994**, Ch. 16, pp. 291–312.
3. *Hazardous Air Pollutants, The London Workshop*, Organization for Economic Cooperation and Development, Paris, France, **1995**.
4. S.D. Lee; T. Schneider, "Control of Air Toxics in Other Countries," in *Toxic Air Pollution Handbook*, D.R. Patrick, Ed., Van Nostrand Reinhold, New York, **1994**, Ch. 27, pp. 505–513.
5. EPA Form 8700-12, *Notification of Regulated Waste Activity*, Rev. Nov. 30, **1993**. Lists state and territory regulatory office addresses and telephones. The report is available from the Office of Solid Waste (5301), U.S. EPA, Washington, D.C., 20460.
6. R.D. Griffin, "The Atmosphere and Its Contaminants," in *Principles of Air Quality Management*, Lewis Publishers, Boca Raton, **1994**, Ch. 1, pp. 1–22.
7. W.J. Clinton, Address, July 16, **1998**, in US EPA Office of Air & Radiation, available at http://ttnwww.rtpnc.epa.gov/naaqsfin/impfac.htm.
8. L.R. Raber, "EPA's Air Standards," *Chem. Eng. News*, April 14, **1997**, p. 10.
9. U.S. EPA Office of Air & Radiation, available at http://ttnwww.rtpnc.epa.gov/naaqsfin/o3pm.htm.
10. N. de Nevers, *Air Pollution Control Engineering*, McGraw-Hill, New York, **1995**.
11. D.K. Berry and S.R. Sabatini of Environmental Quality Management, Cincinatti, Ohio, compiled this table in **1996**. It was adapted and reproduced with their permission.
12. W. Licht, *Air Pollution Control Engineering, Basic Calculations for Particulate Collection*, Marcel Dekker, New York, **1988**.
13. C.E. Lapple, *Stanford Res. Inst. J.*, **1961**, 5, 94.
14. P.N. Cheremisinoff; R.A. Young, *Air Pollution Control and Design Handbook, Part 1*, Marcel Dekker, New York, **1977**.
15. P.N. Cheremisinoff; V. Van Brunt, "Gravity Settling Chambers," in *Air Pollution Control and Design Handbook, Part 1*, P.N. Cheremisinoff; R.A. Young, Eds., Marcel Dekker, New York, **1977**, Ch. 9, pp. 263–280.
16. M.V. Vatia; P.N. Cheremisinoff, "Cyclones," in *Air Pollution Control and Design Handbook, Part 1*, P.N. Cheremisinoff; R.A. Young, Eds., Marcel Dekker, New York, **1977**, Ch. 10, pp. 281–316.

17. M.G. Kennedy; P.N. Cheremisinoff, "Fabric Filter Dust Collectors," in *Air Pollution Control and Design Handbook, Part 1*, P.N. Cheremisinoff; R.A. Young, Eds., New York, Marcel Dekker, **1977**, Ch. 11, pp. 317–370.
18. L. Bergmann, "European Baghouse Design," in *Air Pollution Control and Design Handbook, Part 1*, P.N. Cheremisinoff; R.A. Young, Eds., Marcel Dekker: New York, **1977**, Ch. 12, pp. 371–384.
19. R.W. Ziminski, "Electrostatic Precipitators," in *Air Pollution Control and Design Handbook, Part 1*, P.N. Cheremisinoff; R.A. Young, Eds., Marcel Dekker, New York, **1977**, Ch. 13, pp. 385–398.
20. A. Stern, Ed., *Air Pollution*, Vol. III, Academic Press, New York, **1968**.
21. J.J. Shah; H.B. Singh, *Environ. Sci. Technol.*, **1988**, 22, **1381**.
22. U.S. Environmental Protection Agency, *National Air Quality and Emissions Trends Report 1900–1994*, Report EPA-454/R-95-011, U.S. EPA, Research Triangle Park, NC, 27711, Oct. **1995**.
23. U.S. Environmental Protection Agency, *National Air Quality and Emissions Trends Report 1992*, U.S. EPA, Research Triangle Park, NC, Oct. **1993**.
24. National Research Council, *Rethinking the Ozone Problem in Urban and Regional Air Pollution*, National Research Council, National Academy Press, Washington, D.C., **1991**.
25. T.T. Shen; C.E. Schmidt; T.R. Card, *Assessment and Control of VOC Emissions from Waste Treatment and Disposal Facilities*, Van Nostrand Reinhold, New York, **1993**.
26. D.R. Patrick, "Special Sources of Toxic Air Pollutants," in *Toxic Air Pollution Handbook*, D.R. Patrick, Ed., Van Nostrand Reinhold, New York, **1994**, Ch. 24, pp. 461–468.
27. M.P. Walsh, "Mobile Sources and Air Toxics," in *Toxic Air Pollution Handbook*, D.R. Patrick, Ed., Van Nostrand Reinhold, New York, **1994**, Ch. 25, pp. 469–488.
28. A.L. Hines; T.K. Ghosh; S.K. Loyalka; R.C. Warder, Jr., *Indoor Air Quality & Control*, Prentice Hall, Englewood Cliffs, NJ, **1993**.
29. L. Morawska; N.D. Bofinger; M. Maroni, *Indoor Air: An Integrated Approach*, Elsevier Science, Oxford, England, **1995**.
30. G.T. Wolff, "Air Pollution," in *Kirk Othmer, Encyclopedia of Chemical Technology*, 4th Ed., J.L. Kroschwitz, M. Howe-Grant, Eds. Vol. 1, John Wiley, New York, **1991**, p. 743.
31. R.D. Griffin, *Principles of Air Quality Management*, Lewis Publishers, Boca Raton, **1994**.
32. L. Mølhave, "Volatile Organic Compounds, Indoor Air Quality and Health," *Indoor Air*, **1991**, 4, 357.
33. L. Mølhave, "Interpretation and Limitations of the Concept 'Total Volatile Organic Compounds' (TVOC) as an Indicator of Human Responses to Exposures of VOC in Indoor Air," *Indoor Air* **1992**, 2, 65.
34. Committee on Odors from Stationary and Mobile Sources, "Methods of Controlling Odors," in *Odors from Stationary and Mobile Sources*, National Academy of Sciences, Washington, DC, **1979**, Ch. 6, pp. 179–242.
35. W.H. Prokop, "Odors," in *Air Pollution Engineering Manual*, A.J. Buoniore; W.T. Davis, Eds., Air & Waste Management Association, Van Nostrand Reinhold, New York, **1992**, Ch. 5, pp. 147–154.

36. F.H.H. Valentin; A.A. North, Eds., *Odour Control–A Concise Guide*, Warren Spring Laboratory, Hertfodshire, England, **1980**.

37. Committee on Odors from Stationary and Mobile Sources, "Control of Agricultural Odors," in *Odors from Stationary and Mobile Sources*, National Academy of Sciences, Washington, DC, **1979**, Ch. 7, pp. 243–293.

38. Committee on Odors from Stationary and Mobile Sources, "Control of Industrial Odors," in *Odors from Stationary and Mobile Sources*, National Academy of Sciences, Washington, DC, **1979**, Ch. 8, pp. 294–363.

39. A.L. Hines; T.K. Ghosh; S.K. Loyalka; R.C. Warder, Jr., "Investigation of Co-Sorption of Gases and Vapors as a Means to Enhance Indoor Air Quality," Report No. GRI-90/0194, NTIS Document No. PB91-178806, Gas Research Institute, Chicago, **1991**.

40. G.D. Clayton; F.E. Clayton, *Patty's Industrial Hygiene and Toxicology*, Vols. 1 and 2, John Wiley & Sons, New York, **1986**,

41. N.I. Sax, *Dangerous Properties of Industrial Materials*, 6th Ed., Van Nostrand Reinhold, New York, **1983**.

42. L. Mølhave; B. Bach; O.F. Pederson, "Human Reactions During Exposure to Low Concentrations of Organic Gases and Vapors Known as Normal Indoor Air Pollutants," in *Proc. 3rd Int. Conf. Indoor Air Quality and Climate*, Stockholm, Sweden, **1984**, Vol. 3, p. 431.

43. W.W. Heck, *Multiple Air Pollutants and Forest and Crop Damage in the US*, World Resources Institute, Yale University Press, New Haven, CT, **1989**.

44. Bureau of National Affairs, *Hazardous Emissions*, BNA, Washington, DC, **1996**, pp. 125–171.

2

EXISTING TECHNOLOGIES FOR VOLATILE ORGANIC COMPOUND ELIMINATION

2.1 REDUCTION AT THE SOURCE

Various strategies exist to control volatile organic compound (VOC) release. The most desirable is to improve processes so that emissions are minimized at the outset. This is the idea behind "environmentally conscious manufacturing" [1], "green engineering" [2], and "benign by design" [3] initiatives in various industries. These initiatives may be applied in two ways: improvements or redesign of processes. *Improvements* include altering unit operations, such as increasing reactor yields, raising separation levels, or simply cutting down on fugitive emissions. *Redesign* involves more fundamental changes in processes, such as switching from an organic phase to an aqueous phase, converting from stoichiometric to catalytic chemistry, or going from batch to continuous operation. In all these cases, the concept is to eliminate emissions at the point of creation and avoid "end of the pipe" treatment.

An effective technology for waste minimization that is extremely broad in scope is catalysis. Its use in a wide variety of environmental applications has been reviewed elsewhere [4]. For example, catalysis can be used to manufacture environmentally safer products like hydrofluorocarbons as replacements for the problematic chlorofluorocarbons (CFCs). It can be used as an alternative to hazardous and toxic chemicals like HCN, HF, HCl, and phosgene in a variety of processes. It can be used to improve yields, reduce side-products, and produce unique molecules in a variety of chemical and energy applications.

There are many situations in which it is impractical or impossible to avoid production of some waste. This can be due to inherent limitations in selectivity of reactions or to unavoidable inefficiencies in separations. Both kinetic and thermodynamic factors may be involved in these cases. Airborne contaminants may also be generated by the nature of the process itself. For example, manufacturing, painting, cooking, dry-cleaning, and animal rendering operations represent situations in which activities are carried out in relatively open spaces with release of volatile species into the immediate environment. Such operations usually require ventilation to sweep away any noxious substance. Treatment of this air stream containing low concentrations of contaminants now becomes the objective. Again, in this situation catalysis can be very effective [4]. Catalysis is used for the elimination of pollutants from fixed sources like power plants, mobile sources like vehicles, and increasingly in everyday environments like offices, homes, and retail outlets.

2.2 METHODS OF VOC CONTROL

A number of options are available for the elimination of VOCs [5]. These are described in a number of books and monographs [6–15] and are summarized in Table 2.1.

The use of these methods depends on the precise nature of the VOCs, the concentration levels, and the conditions (temperature, pressure, flow rate). These have been described in a general manner in a number of articles and books. They are briefly described here and are covered in more detail in subsequent chapters, with particular emphasis on their applications to VOCs.

2.2.1 Condensation

Condensation is the liquefaction of condensable contaminants by the use of low temperatures. Specifically, the compounds to be removed from the gaseous phase are cooled to a temperature at which their partial pressure (fugacity) in the gas stream exceeds their dew point so that they transform to liquids. Conventional

TABLE 2.1 Methods of VOC Abatement

Methods	References
Condensation	6, 16
Adsorption	15, 17–19
Absorption/scrubbing	6, 15, 20, 21
Thermal incineration	9, 22, 23
Catalytic incineration	9, 24
Flaring	25, 26
Biological treatment	27–29
Odor modification	5, 30
Air dilution	6, 31, 32
Membrane separation	33
Ultraviolet treatment	34
Ozone catalytic oxidation	Present study

condensers are shell and tube heat exchangers [35], used because they contain a relatively large amount of surface area for heat transfer but are fairly compact. Critical parameters that establish the requirements and efficiency of the condenser include the overall heat transfer coefficient (a function of the hot and cool stream composition, flow characteristics, and the construction material), the difference in temperature between the streams, and the operating pressure [36].

Condensation is most effective for compounds with high boiling points [37]. If a significant portion of the organic material is composed of compounds that solidify at the condenser operating temperature, however, a standard shell and tube design condenser will not be effective, as these will quickly foul the heat transfer area, plugging the condenser. For those cases, a contact condenser in which the hot and cool streams come directly into contact with each other would be more appropriate.

2.2.2 Adsorption

Adsorption refers to the trapping of pollutants on a high-surface area material. The process is typically used to remove contaminants in fairly low concentrations from a gas stream. The pollutants are adsorbed onto the surface or interstitial areas of a material such as activated carbon or a molecular sieve by physical or chemical attraction. Once the carbon or adsorbent material is saturated, that is, it can no longer adsorb any more pollutant, the material is regenerated, typically by introducing steam to drive off the pollutants [38]. The vent stream is condensed, decanted if necessary to separate the aqueous layer from the organic layer, and, in some cases, further treated. For example, one of the layers may require distillation to further separate the constituents from the aqueous phase or from each other [39].

Because of the subsequent separation required to reuse the solvents, carbon adsorption is not well suited to streams that have a large number of VOCs, nor does it work effectively on streams that have constituents that are difficult to separate. In those circumstances, as well as when the adsorbed species has a large molecular weight and is therefore difficult to desorb, or in cases of low flows [40], the carbon can be used and then disposed of. This creates a similar dilemma to that of incineration, though, in that it treats a symptom (VOC emissions) but causes other problems such as solid waste disposal.

2.2.3 Absorption/Scrubbing

Absorption is a physical process consisting of the dissolution of a pollutant in a liquid. In absorbers (or scrubbers), the vapor stream is introduced into a chamber where it is intimately mixed with the liquid. The amount of pollutant remaining in the gas stream as it leaves the scrubber is governed by Henry's law, which establishes the amounts of a component in equilibrium in the gas and liquid streams for dilute mixtures [41]. With countercurrent gas and water flows, the mass transfer between the vapor and liquid phases is maximized. If the vapor

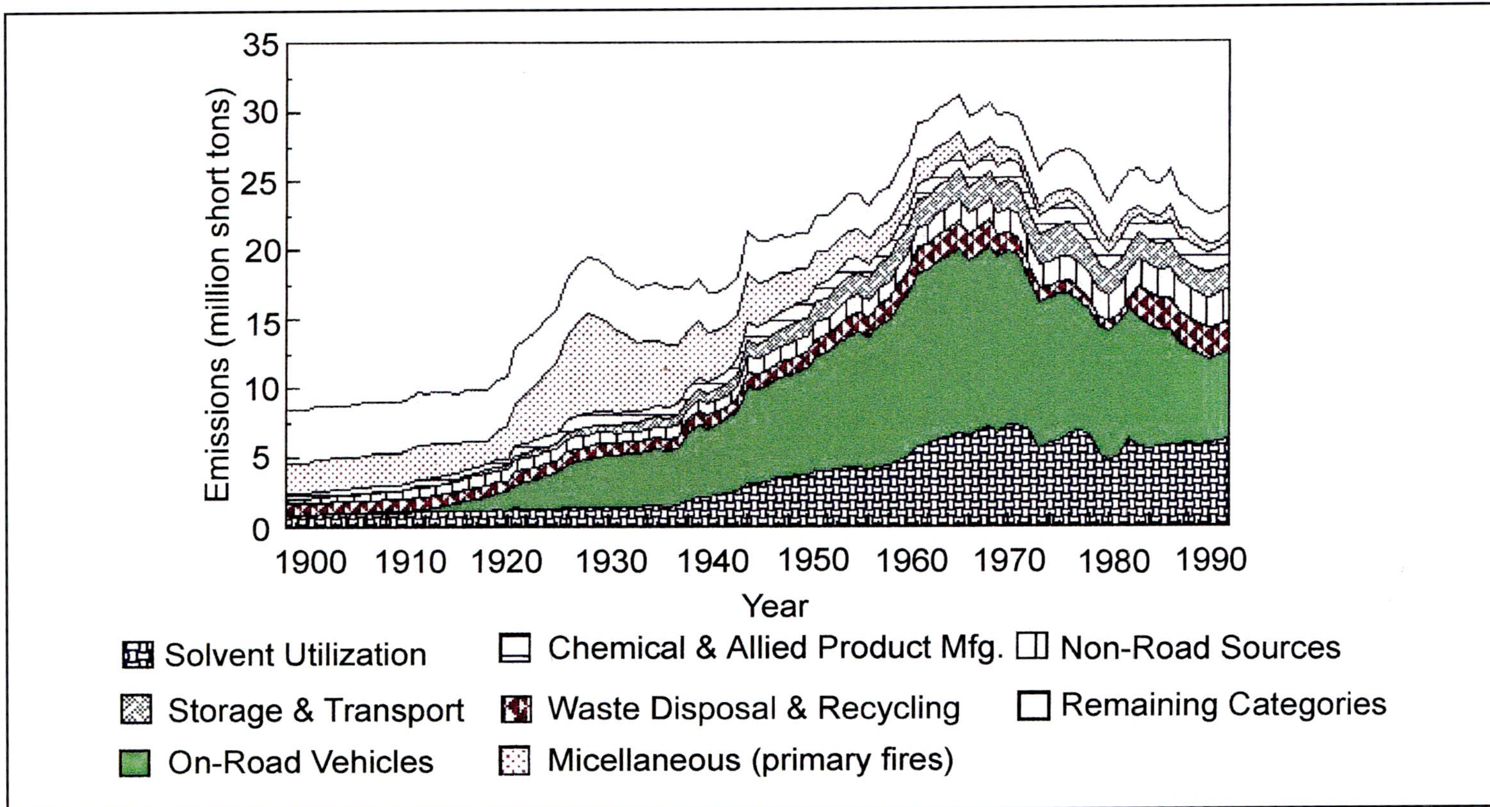

Figure 1.2 Trend in volatile organic compound emissions. (Reproduced from the EPA [22].)

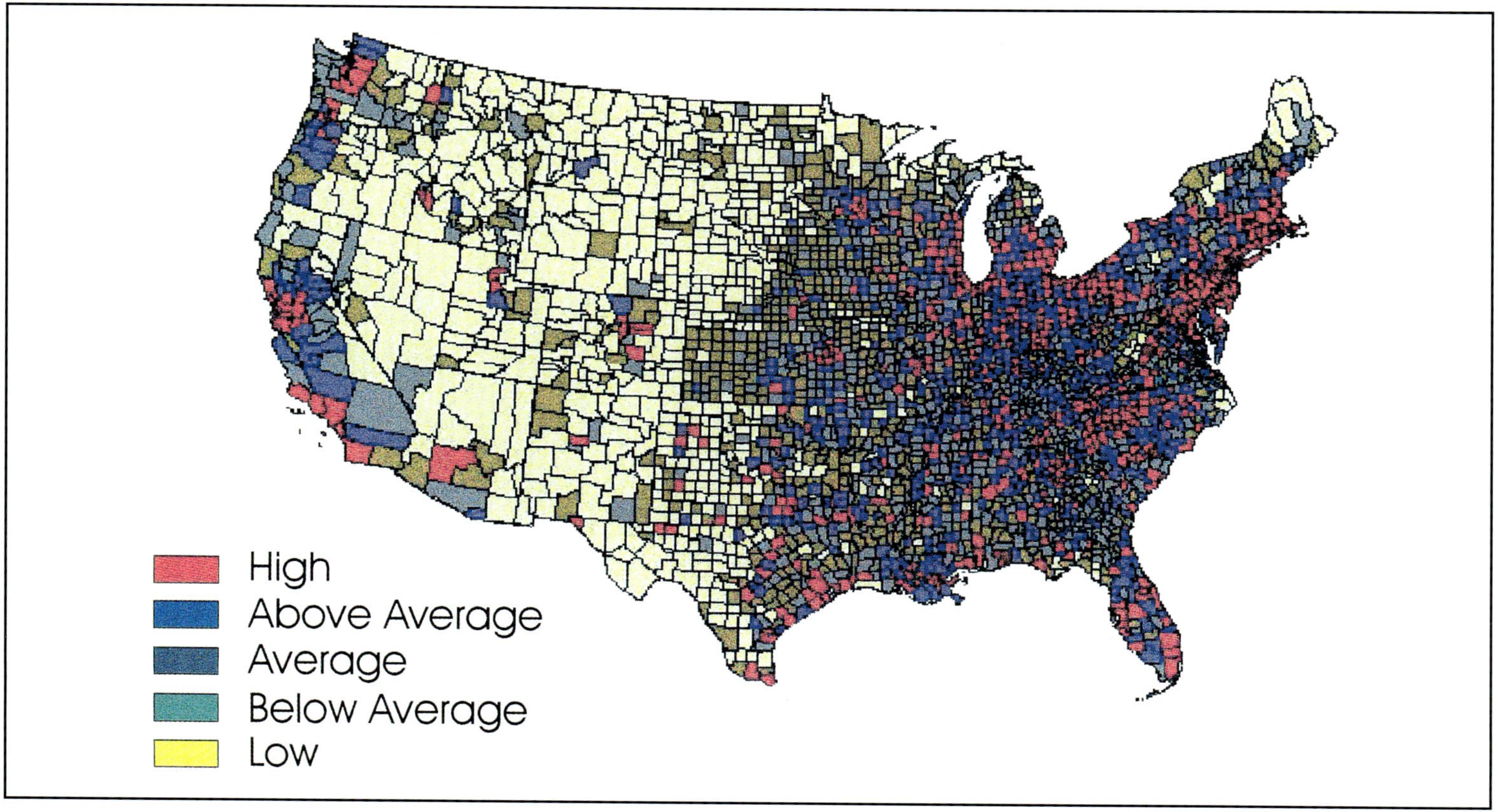

Figure 1.4 Density map of 1994 county level VOC emission estimates. (Reproduced from the EPA [22].)

and liquid have adequate contact, the gaseous components that are soluble in the liquid are absorbed, where they may react or be removed by discharge of the liquid. Some additional removal is accomplished by the evaporative cooling that occurs in the scrubber; some components condense out and are removed as the liquid is removed. Absorption, however, is not effective on constituents that are not soluble in the liquid medium. Therefore, it, too, is not well suited to vapor streams that have a large variety of constituents.

There is a variation of absorption known as *wet scrubbing* in which the gas-phase component undergoes a chemical reaction [42] with a component in the liquid phase. The reactions can be of an acid–base nature or an oxidation type. The reaction of the liquid-phase compound results in a lower concentration of the liquid-phase compound than would otherwise exist at equilibrium. Thus, the driving force for removal of the gas-phase component is higher than in simple adsorption. One drawback, however, is that the scrubbing agent is consumed in the process, and hence the method is more costly than adsorption where the liquid phase can be reused. Also, in some cases, the process generates a by-product that must be separated and disposed [43].

2.2.4 Thermal Incineration

Incineration is basically the burning of combustible wastes [44]. For many years, thermal incinerators have been considered to be one of the most effective control devices for VOCs. They are relatively easy to operate, not likely to foul, and do not require additional processes to remove the pollutant. They are sometimes coupled with recuperative heat exchangers for energy efficiency or, more recently, with more efficient regenerative heat exchangers. Operating costs can be further minimized by ensuring that the gas stream is as concentrated as possible so that energy is not expended to heat excess air. The concentration is usually limited to approximately 25% of the lower explosive limit (LEL) by fire insurance regulations [45].

Incinerators are ideal candidates for gas streams, which contain a variety of pollutants and from which any solvent that is recovered is not viable for reuse. Under proper conditions, the design efficiencies are typically in excess of 98% and depend primarily on the gaseous components combusted, the residence time, the amount of mixing, and the combustion temperature [46]. Costs associated with incineration processes are typically higher than those associated with solvent recovery processes like carbon adsorption or condensation due to the cost of the natural gas and increased solvent cost. In addition, incineration can be perceived as a "band-aid" approach to VOC control because it treats the major problem of VOC emissions but generates smaller problems like NO_x and CO_2 generation.

2.2.5 Catalytic Incineration

Catalytic incineration, or combustion, refers to the burning of combustible wastes with the aid of a catalyst [47–50]. The catalyst opens up a different reaction pathway with a lower activation energy than gas-phase combustion and allows oxidation to proceed at a lower temperature. Advantages of catalytic combustion are lower energy requirements and lower production of NO_X. Caution must be exercised with catalytic units, however, to avoid the introduction of particulate matter or metal fumes, which can foul the catalyst [24]. The catalyst is usually a noble metal like platinum or palladium dispersed on an inert oxide support. To avoid pressure drop problems, the support is usually deposited on the surface of a monolith carrier.

2.2.6 Flaring

Flaring is another combustion technique that involves the direct burning of an organic stream in an open flame. It is carried out at the tip of a burner, usually with the addition of extra fuel and, sometimes, steam [26]. There is no combustion chamber, so flame temperature and proper mixing of the combustible mixture with air are critical design aspects [51]. Flaring can tolerate large variations in flow rate, so it is commonly employed in transient situations such as startup and shutdown. It is routinely used in refining as a method of disposing of VOCs in that industry.

2.2.7 Biological Treatment

Biological treatment consists of using microorganisms to biologically degrade contaminants [34]. Biological purification is a relatively new technology in the field of air pollution control, although the same concept has been used for years to treat wastewater streams. Microorganisms are grown on a substrate *(biofiltration)* or are suspended in a scrubber liquor *(bio-scrubbers)* [52]. The scrubber liquid or substrate may also serve as an adsorbent for the pollutants so that the microbes have a constant food supply, even if the process is not operating. In less variable processes, the microbes are contained in compost, peat, or a similar soil, and the air stream is passed through the bed for treatment. As in water treatment operations, the microorganisms break down the large organic molecules into smaller, less harmful molecules, consuming a portion of the molecules for energy to sustain the microbial activity. Although this technology is still not well accepted in the United States, there are several hundred installations operating worldwide [53]. Like adsorption, biopurification is a good alternative for streams that have relatively low concentrations of organic constituents and is most effective for systems containing alcohols, ethers, aldehydes, and ketones.

Biofiltration offers excellent VOC removal, even from dilute gas streams with relatively low maintenance and operating costs. Care must be exercised, however, to remove particulates, ensure that adequate moisture is present, avoid gas channeling, and maintain a fairly constant temperature [29]. As with wastewater treatment systems, the microbes are susceptible to significant changes in the nature or concentration of pollutants and can be killed by sudden changes. Therefore, highly variable processes, or batch systems, in which products change daily or weekly are not well suited to biofiltration. In addition, design of a biofiltration system generally requires a pilot unit to determine the treatability of the gas stream and the residence time required for effective removal.

2.2.8 Odor Modification

Odor modification involves the addition of a chemical agent to an odorous substance to improve the nature of the emitted smell [30, 54]. The concept is to pair gases in proportions that render the mixture odorless or nearly odorless [55, 56]. The action of odor cancellation is attributed to (1) temporary inactivation of the smelling nerves, (2) odor compensation, or (3) chemical combination. Examples of pairs of compensating odors are ethyl mercaptan and eucalyptol, skatole and coumarin, and butyric acid and oil of juniper [30].

Odor masking also involves incorporation of a malodor into an overall mixture that is pleasant. For example, major constituents of sewage odor are indole and skatole. They are also, however, components of oil of jasmine. By adding an odorant that has the attributes of jasmine, but is lacking the indole and skatole components, the overall mixture can be made to be made to have a pleasant floral aroma [30].

Odor modification is commercially applied in chemical plants, waste treatment facilities, lagoons, and refineries. The odor modifier can be added directly to the processed product as in cooking operations in a rendering plant, or digestion in a kraft sulfate plant, or can be mixed with gaseous effluents as a final step in gas cleanup [30].

2.2.9 Dilution

Dilution for gaseous wastes refers to their dispersion in an excess amount of air [31]. The subject has been reviewed elsewhere [57]. The wastes may originate from evaporative and fugitive emissions, but also commonly from industrial smokestacks [10]. The release of such streams is governed in the United States by the EPA through the Primary and Secondary National Ambient Air Quality Standards.

Considerable work has been carried out in the design of chimney stacks [58]. The contaminated gas stream rising from a stack is known as a *plume*, if continuous, and a *puff*, if intermittent. An important quantity is the plume rise distance, which is affected by such parameters as the stack diameter, the exit velocity at the stack, the temperature difference between exit gas and the ambient air, the wind speed, and the atmospheric stability class [58].

The release of gas streams from stacks is intimately tied to geographical and meteorological conditions. These include effects due to mountains, valleys, vegetation, urban heat islands, ocean breezes, jet streams, and global circulation cells [59]. Much effort has been expended in the area of modelling [60, 61]. The most common dispersion model is the gaussian plume model, which assumes that horizontal convection is balanced by vertical and transverse turbulent diffusion. A more sophisticated box model divides the air region into a number of boxes or volumes and solves mass continuity equations by numerical techniques. An important factor in these models is the atmospheric lifetime as determined by chemical reactions and physical removal processes [62].

2.2.10 Membrane Separation

Membrane technology involves the use of semipermeable membranes to separate VOCs from a process stream [33]. The technology has been used in water purification and has been adapted for use with gaseous media. Basically, the separation is based on preferential diffusion of VOCs across thin polymer layers wrapped around a perforated central tube. The driving force is a pressure differential between the inner and outer portions of the tube. The method is effective for recovering VOCs such as chlorinated hydrocarbons, chlorofluorocarbons, and hydrofluorocarbons which have been traditionally difficult to separate because of their high volatility.

2.2.11 Ultraviolet Treatment

Ultraviolet treatment is another emerging technology in which ultraviolet light is used to oxidize organic compounds. The oxidation reaction is promoted by the photocatalytic effect of the ultraviolet light and is conducted in the presence of a catalyst to activate the oxygen into highly reactive species, including ozone. After reaction, the air stream is scrubbed to perform additional oxidation and dried, and remaining VOC compounds are adsorbed in activated carbon. The technique has been reported to be up to 99% effective in the destruction of organic compounds [34, 63]. Problems are activity at high flow rates and the low photoefficiencies of the catalytic process, which results in high costs [64].

2.2.12 Catalytic Ozone Oxidation

Ozone catalytic oxidation technology is already being commercialized in a number of household applications such as the deodorization of refrigerators and toilets and the purification of room air. A contaminated gas stream is mixed with ozone and passed over a catalyst bed to carry out a low-temperature catalytic combustion reaction. The process utilizes the oxidizing power of ozone together with the activity and selectivity of oxidation catalysts to convert VOCs to CO_2 and H_2O. The process works at high and low VOC concentrations and at high flow rates. Because of the use of ozone, the air stream does not require appreciable heating. Improvements in generation equipment have decreased the cost of the ozone substantially, and this promises to become a broadly used technology.

2.3 COMPARISON OF TECHNOLOGIES

Figures 2.1 and 2.2 illustrate the ranges of concentrations and flow rates that are generally suitable for different technologies. These, in combination with required control efficiency and the availability of utilities, labor, and capital, form the basis for a control technology choice. Usually when a VOC control is desired or required, one immediate and easy solution is incineration. Incineration is a very effective, flexible control technology that allows for changes in pollutants, concentration, and flow. In addition, thermal incineration devices are available today with greater than 95% heat recovery, which minimizes the amount of supplemental fuel required. Catalytic incineration is also capable of high removal efficiencies with minimal natural gas requirements, but is more susceptible to fouling from particulates and metal fumes.

The other technique that is recognized as capable of high efficiencies (greater than 99% in some cases [J. Hoskins, personal communication, 1985–1988]) is adsorption. Adsorption has additional advantages, as well, because the pollutants can be recovered for reuse. Not all VOCs are suitable for adsorption, and a source

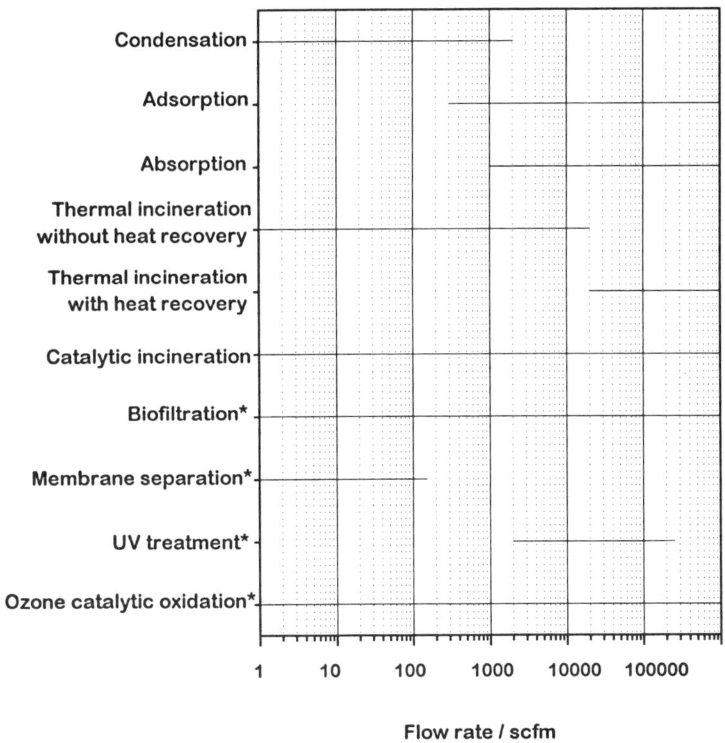

Figure 2.1 Range of flow rates for various VOC treatment technologies. (Adapted from Mukhopadhyay and Moretti [14].)

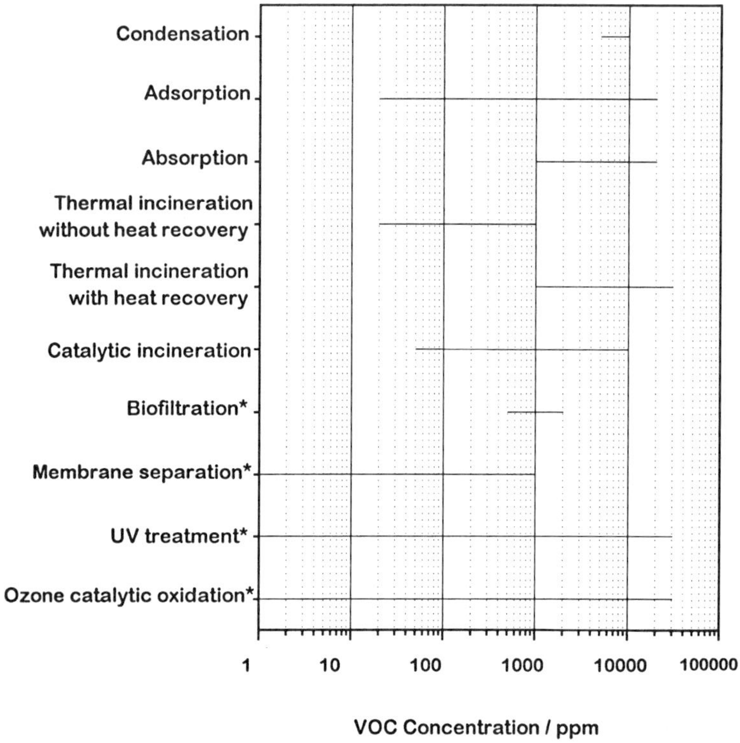

Figure 2.2 Range of concentrations for various VOC treatment technologies. (Adapted from Mukhopadhyay and Moretti [14].)

of steam or heat must be available for regenerative systems. Nonregenerative systems result in additional waste for disposal.

Biofiltration has also been reported to be capable of efficiencies greater than 90% [65], but are susceptible to sudden changes in temperature, concentration, and humidity. Condensation is generally considered to be less effective than the other technologies, but is capable of approximately 90% efficiency when refrigerants are used to cool the gas stream [66].

Membrane separation, ultraviolet treatment, and ozone catalytic oxidation are all emerging technologies that should be further investigated to determine their suitability to the situation, as well as to assess the efficiency of commercial units. This book provides an in-depth coverage of the topic of catalytic ozone oxidation of VOCs. As is discussed, this is a versatile technology that is effective for both large-scale industrial pollution abatement and low-concentration indoor air treatment. Already, household appliances fitted with catalytic ozone devices have been commercialized for the elimination of odors in toilets and refrigerators.

It is not likely that any of these technologies will be exactly suited to a given process. The choice of a control strategy is almost always a choice of lesser evils. One option that can be very beneficial in the long run, however, is a process change to reduce or eliminate organic compounds, that is, *reduction at the source*. Operational changes may have initial costs greater than the expenses associated with control equipment and are frequently ruled out on that basis. Savings in operating costs, though, can frequently pay for the changes in a relatively short period of time. If new equipment is needed, the updated equipment usually also has better process control to improve quality and efficiency as well as reduce emissions. When coupled with the fact that fewer man hours can be dedicated to complying with permits, regulations, and laws, process changes are frequently the optimum solution. Although they can be difficult initially, the rewards make operational changes well worth serious consideration as a control strategy.

REFERENCES

1. M.P. Dudukovic; P.L. Mills, "Special Issue on Symposium on Catalytic Reaction Engineering for Environmentally Benign Processes," *Ind. Eng. Chem. Res.*, **1994**, 33, 2885–3290.
2. P.T. Anastas; T.C. Williamson, Eds., *Green Chemistry: Designing Chemistry for the Environment*, ACS Symposium Series No. 626, American Chemical Society, Washington, DC, **1996**.
3. P.T. Anastas; C.A. Farris, *Benign by Design—Alternative Synthetic Design for Pollution Prevention*, ACS Symposium Series 577, American Chemical Society, Washington, DC, **1993**.
4. A.T. Bell; L.E. Manzer; N.Y. Chen; V.W. Weekman; L.L. Hegedus; C.J. Pereira, "Protecting the Environment through Catalysis," *Chem. Eng. Prog.*, Feb. **1995**, p. 26.
5. W.H. Prokop, "Odors," in Air Pollution Engineering Manual, A.J. Buonicore; W.T. Davis, Eds., Van Nostrand Reinhold, New York, **1992**, Ch. 5, p. 147.
6. Air & Waste Management Association, *Emerging Solutions to VOC & Air Toxics Control*, Proc. Specialty Conf. San Diego, Feb. 26–28, **1997**, AWMA, Pittsburgh, **1997**.
7. H.J. Rafson, Ed., *Odor and VOC Control Handbook*, McGraw-Hill, New York, **1998**.
8. N. de Nevers, *Air Pollution Control Engineering*, McGraw-Hill, New York, **1995**.
9. R.M. Heck; R.J. Farrauto, *Catalytic Air Pollution Control*, Van Nostrand Reinhold, New York, **1995**.
10. J.C. Mycock; J.D. McKenna; L. Theodore, *Handbook of Air Pollution Control Engineering and Technology*, Lewis Publishers, Boca Raton, **1995**.
11. C.D. Cooper; F.C. Alley, *Air Pollution Control, A Design Approach*, 2nd Ed., Waveland Press, Prospects Heights, IL, **1994**.
12. R.D. Griffin, *Principles of Air Quality Management*, Lewis Publishers, Boca Raton, **1994**, Ch. 5, pp. 113–158.
13. D.R. Patrick, Ed., *Toxic Air Pollution Handbook*, Van Nostrand Reinhold, New York, **1994**.

14. N. Mukhopadhyay; E.C. Moretti, *Current and Potential Future Industrial Practices for Reducing and Controlling Volatile Organic Compounds*, Center for Waste Reduction Technologies, American Institute of Chemical Engineers, New York, **1993**.
15. A.L. Hines; T.K. Ghosh; S.K. Loyalka; R.C. Warder, Jr., *Indoor Air Quality & Control*, Prentice Hall, Englewood Cliffs, NJ, **1993**.
16. A.J. Buonicore, "Control of Gaseous Pollutants-Condensation," in *Air Pollution Engineering Manual*, A.J. Buonicore; W.T. Davis, Eds., Van Nostrand Reinhold, New York, **1992**, Ch. 2, p. 31.
17. T.E. Cannon, "Carbon Adsorption Applications," in *Air Pollution Control and Design Handbook*, Part 1, P.N. Cheremisinoff; R.A. Young, Eds., Marcel Dekker, New York, **1977**, Ch. 15, p. 421.
18. M.H. Stenzel, "Remove Organics by Activated Carbon Adsorption," *Chem. Eng. Prog.*, **1993**, 89(4), 36.
19. M.J. Ruhl, "Recover VOCs via Adsorption on Activated Carbon," *Chem. Eng. Prog.*, **1993**, 89(7), 37.
20. J.W. MacDonald, "Packed Wet Scrubbers," in *Air Pollution Control and Design Handbook*, Part 2, P.N. Cheremisinoff; R.A. Young, Eds., Marcel Dekker, New York, **1977**, Ch. 25, p. 699.
21. N.D. Phillips, "Countercurrent Contact Wet Scrubbing," in *Air Pollution Control and Design Handbook*, Part 2, P.N. Cheremisinoff; R.A. Young, Eds., Marcel Dekker, New York, **1977**, Ch. 26, p. 721.
22. R.D. Ross, "Thermal Incineration," in *Air Pollution Control and Design Handbook*, Part 1, P.N. Cheremisinoff; R.A. Young, Eds., Marcel Dekker, New York, **1977**, Ch. 17, p. 449.
23. A. Garg, "Specify Better Low-NO_x Burners for Furnaces," *Chem. Eng. Prog.*, **1994**, 90(1), 46.
24. T.H. Snape, "Catalytic Incineration," in *Air Pollution Control and Design Handbook*, Part 1, P.N. Cheremisinoff; R.A. Young, Eds., Marcel Dekker, New York, **1977**, Ch. 21, p. 543.
25. R.Y. Purcell; G.S. Shareef, *Control Technologies for Hazardous Air Pollutants*, US EPA, Research Triangle Park, NC, **1986**, pp. 25–26, 60–63.
26. W.M. Vatavuk, "Flares," in *OAQPS Control Cost Manual*, 4th Ed. (and subsequent revisions), US EPA, Research Triangle Park, NC, **1990**, Ch. 7, p. 1–45.
27. G. Leson; A.M. Winer, "Biofiltration: An Innovative Air Pollution Control Technology for VOC Emissions," *J. Air Waste Manag. Assoc.*, **1991**, 41, 1045.
28. H. Bohn, "Consider Biofiltration for Decontaminating Gases," *Chem. Eng. Prog.*, **1992**, 88, 34.
29. R.F. Michelsen, "Biofiltration," in *Handbook of Air Pollution Control Engineering and Technology*, J.C. Mycock; J.D. McKenna; L. Theodore, Eds., Lewis Publishers, Boca Raton, **1995**, Ch. 21, pp. 375–394.
30. T.M. Hellman, "Odor Control by Odor Modification," in *Air Pollution Control and Design Handbook*, Part 2, P.N. Cheremisinoff; R.A. Young, Eds., Marcel Dekker, New York, **1977**, Ch. 44, p. 1005.
31. J.T. Yeh, "Modeling Atmospheric Dispersion of Pollutants," in *Air Pollution Control and Design Handbook*, Part 1, P.N. Cheremisinoff; R.A. Young, Eds., Marcel Dekker, New York, **1977**, Ch. 6, pp. 167–178.

32. J.B. Christian, "Estimate the Effects of Air Emissions with this Process Screening Model," *Chem. Eng. Prog.*, June **1995**, p. 59.
33. R.W. Baker; J. Kaschemekat; J.G. Wijmans; V.L. Simmons (Membrane Technology and Research, Inc.), *Membrane Vapor Separation Systems for the Recovery of VOCs.* Presentation at the 85th Annual AWMA Meeting and Exhibition, Kansas City, MO, June **1992**.
34. G. Gschwandtner; S. Fairchild, *Controlling Odorous Emissions from Iron Foundries*, US EPA, Research Triangle Park, NC, **1992**, pp.17–18.
35. G. Hounsell, "Condensers," in *Handbook of Air Pollution Control Engineering and Technology*, J.C. Mycock; J.D. McKenna; L. Theodore, Eds., Lewis Publishers, Boca Raton, **1995**, Ch. 12, pp. 223–256.
36. R.Y. Purcell; G.S. Shareef, *Control Technologies for Hazardous Air Pollutants*, US EPA, Research Triangle Park, NC, **1986**, pp. 27, 75–80.
37. A.J. Buonicore, "Condensation," in *Air Pollution Engineering Manual*, A.J. Buonicore; W.T. Davis, Eds., Van Nostrand Reinhold, New York, **1992**, pp. 52–58.
38. A. Turk, "Adsorption," in *Air Pollution, Third Edition, Volume IV: Engineering Control of Air Pollution*, A.C. Stern, Ed., Academic Press, New York, **1977**, Ch. 8, pp. 329–363.
39. L. Theodore; A.J. Buonicore, *Air Pollution Control Equipment, Volume II: Gases*, CRC Press, Inc., Boca Raton, **1988**, Ch. 10, pp. 43–77.
40. S. Calvert; H.M. Englund, Eds., *Handbook of Air Pollution Technology*, John Wiley & Sons, New York, **1984**, Ch. 3, pp. 162–185.
41. R.M. Bethea, *Air Pollution Control Technology: An Engineering Point of View*, Van Nostrand Reinhold, New York, **1978**, Ch. 12, pp. 331–365.
42. U. von Stockar; C.R. Wilke, "Absorption," in *Kirk-Othmer Encyclopedia of Chemical Technology*, 4th Ed., Vol. I, J.L. Kroschwitz; M. Howe-Grant, Eds., John Wiley & Sons, New York, **1991**, pp. 38–93.
43. T. Godish, *Air Quality*, 2nd Ed., Lewis Publishers, Chelsea, MI, **1991**, pp. 327–328.
44. S.M. Weiss, "Afterburners," in *Air Pollution Engineering Manual*, J.A. Danielson, Ed., US EPA, Research Triangle Park, NC, **1973**, pp. 171–183.
45. W.M. Vatavuk, "Thermal and Catalytic Incinerators," in *OAQPS Control Cost Manual*, 4th Ed., US EPA, Research Triangle Park, NC, **1990**, Ch. 3, pp. 1–66.
46. R.L. Burglund, "Exhaust Control, Industrial," in *Kirk-Othmer Encyclopedia of Chemical Technology*, 4th Ed., Vol. 9, J.I. Kroschwitz; M. Howe-Grant, Eds., John Wiley & Sons, New York, **1994**, pp.1024–1027.
47. J.J. Spivey, "Complete Catalytic Oxidation of Volatile Organics," *Ind. Eng. Chem. Res.*, **1987**, 26, 2165.
48. M.F.M. Zwinkels; S.G. Järås; P.G. Menon; T.A. Griffin, "Catalytic Materials for High-Temperature Combustion," *Catal. Rev. Sci. Eng.*, **1993**, 35, 319.
49. L.D. Pfefferle; W.C. Pfefferle, "Catalysis in Combustion," *Catal. Rev.-Sci. Eng.*, **1987**, 29, 219.
50. R. Prasad; L.A. Kennedy; E. Ruckenstein, "Catalytic Combustion," *Catal. Rev.-Sci. Eng.*, **1984**, 26, 1.
51. A.J. Buonicore, "Incineration," in *Air Pollution Engineering Manual*, A.J. Buonicore; W.T. Davis, Eds., Van Nostrand Reinhold, New York, **1992**, pp. 58–70.

52. M.J. Suess; K. Grefen; D.W. Reinisch, *Ambient Air Pollutants from Industrial Sources: A Reference Handbook*, World Health Organization Regional Office for Europe, Elsevier, New York, **1985**, pp. 708–713.
53. G. Lesen; A.M. Winer, "Biofiltration: An Innovative Air Pollution Control Technology for VOC Emissions," *J. Air Waste Manag. Assoc.*, **1991**, 41(8), 1045–1054.
54. T.W. Planker, "Masking and Odor Neutralization," in *Odor and VOC Control Handbook*, H.J. Rafson, Ed., McGraw-Hill, New York, **1998**, pp. 8.18–8.24.
55. P.N. Chermisinoff; R.A. Young, *Industrial Odor Technology Assessment*, Ann Arbor Science Publishers, Ann Arbor, MI, **1975**.
56. H. Zwaardemaker, *Die Physiologie des Gerüchs*, Englemann, Leipzig, **1895**.
57. K.T. Valsaraj, "Emission Modeling-Metereology, Topography, and Dispersion," in H.J. Rafson, Ed., *Odor and VOC Control Handbook*, McGraw-Hill, New York, **1998**, pp. 5.1–5.25.
58. R.T. Fellman; P.N. Cheremisinoff, "Design of Chimneys and Stacks for Pollution Control," in *Air Pollution Control and Design Handbook*, Part 1, P.N. Cheremisinoff; R.A. Young, Eds., Marcel Dekker, New York, **1977**, Ch. 7, pp. 179–214.
59. R.D. Griffin, "Metereology, Dispersion and Modeling," in *Principles of Air Quality Management*, Lewis Publishers, Boca Raton, **1994**, Ch. 5, pp. 113–158.
60. N. de Nevers, "Air Pollutant Concentration Models," in *Air Pollution Control Engineering*, McGraw-Hill, New York, **1995**, Ch. 6, pp. 105–140.
61. J.F. Lape, "Air Dispersion and Deposition Models," in *Toxic Air Pollution Handbook*, D.R. Patrick, Ed., Van Nostrand Reinhold, New York, **1994**, Ch. 11, pp. 226–240.
62. M.P. Ligocki, "Atmospheric Transformation and Removal of Air Toxics," in *Toxic Air Pollution Handbook*, D.R. Patrick, Ed., Van Nostrand Reinhold, New York, **1994**, Ch. 12, pp. 241–248.
63. T.J. Overcamp; G.R. DeHollander; H. Chang; C.P.L. Grady, *A Biologically Enhanced Scrubber for Volatile Organic Compounds*, Paper No. 91–180.12, AWMA Annual Meeting, Vancouver, British Columbia, **1991**.
64. E. Wilson, "TiO_2 Appears Inefficient for Water Treatment," *Chem. Eng. News*, July 1, **1996**, p. 29.
65. *VDI-Handbuch Reinhalten der Luft*, VDI Richtlinien 3477, Biological Waste Gas/Waste Air Purification, Biofilters, Band 6, Dec. **1991**.
66. G.S. Shareef; W.J. Barbour; S.K. Lynch; W.R. Pelt; W.M. Vatavuk, "Refrigerated Condensers," in *OAQPS Control Cost Manual*, 4th Ed. (and subsequent amendments), US EPA, Research Triangle Park, NC, **1990**, Ch. 8, pp. 1–50.

3

CONDENSATION

3.1 TECHNOLOGY DESCRIPTION

Condensation is a process used to separate compounds by either cooling the gas stream or pressurizing a gas stream to conditions in which one or more of the components separates from the mixture as a liquid. This occurs when the partial pressure (fugacity) of the constituent in the gas phase equals or exceeds the vapor pressure of the pure component in liquid form under the same conditions [1]. The efficiency of removing volatile organic compounds (VOCs) from a gas stream by condensation depends on the vapor pressure of the component to be separated, the pressure and temperature that can be achieved during condensation, and the concentration of the constituent in the gas stream [2]. A typical flow diagram of a condensation system for air pollution control is shown in Figure 3.1.

Water-cooled condensers are commonplace in industry [3]; they are used to prevent the escape of raw materials from the process, to liquefy components separated by distillation columns, and to liquefy steam to reuse it as boiler feedwater. Condensation is generally assumed to be considerably less efficient than other technologies for removing pollutants from the low-concentration gas streams encountered in air pollution control [2]. An alternative coolant [4] such as calcium chloride, freons, or methylene chloride may, however, yield condensing systems with 90% [2, 5] or better efficiency. The increased efficiency, which results from outlet gas temperatures of 194 to 277 K (−79° to 4°C, −110° to 40°F) [6], comes at the cost of increased maintenance and system complexity, though. A refrigerant must be selected carefully to ensure that potential leakage of the coolant does not result in a more severe environmental threat than that posed by the original gas stream. The potential danger from refrigerant leaks and

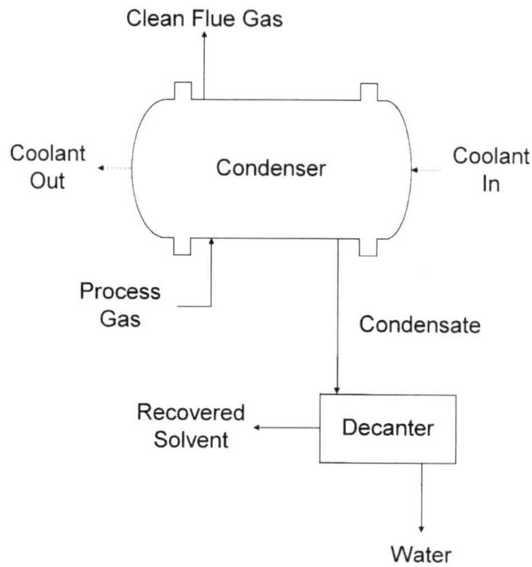

Figure 3.1 Typical condensation process flow.

the energy consumed by the refrigeration cycle should be weighed against the toxicity and impact of the untreated gaseous stream.

Condensers, particularly water-cooled condensers, are frequently used in combination with other control technologies [7]. In carbon adsorption systems, condensers are used to liquefy the concentrated plume of steam and pollutant that are removed during the desorption stage. They are also used for pretreatment of a gas stream, such as in a rendering plant where as much as 95% of the gas volume may be reduced by condensation [8]. Because cooling of the gas stream is inherent in a condensing system, the system reduces the volume and pollutant loading of a gas stream before it enters a final "polishing" control device, such as an absorber, adsorber, or incinerator. This enables the final control device to be smaller and to achieve lower emissions. Condensers may be used to reduce the loading to a level below the lower explosive limit (LEL) of the gas stream. When placed in front of a destructive control device such as an incinerator or absorber, a condenser provides for partial recovery of raw materials before the final destruction necessary to achieve low emission rates. Such a system, however, requires that the gas stream be cooled, only to be heated up again to reach the combustion temperature. Condensation systems are more effective on compounds that have low vapor pressures at the achievable outlet gas temperature; typically higher molecular weight compounds are most suited to recovery by condensation. Table 3.1 lists several applications of condenser systems for VOC control [9–11].

3.2 TYPES OF CONDENSERS

There are two primary types of condensers, *contact* and *noncontact* [12, 13]. The most common is the noncontact, or surface, condenser [14], in which heat

TABLE 3.1 Typical Industrial Applications of Condensers

Storage tanks
Petroleum refining
Phthalic anhydride manufacturing
Ammonia manufacturing
Dry cleaning
Styrene manufacturing
Rendering plants
Vapor degreasers
Kraft pulp mills

Data adapted from refs. 9–11.

is transferred across a barrier (usually a metal pipe or plate) that separates the gas stream from the coolant. Examples of noncontact condensers include shell-and-tube heat exchangers (Fig. 3.2), and plate-and-frame heat exchangers (Fig. 3.3). In shell-and-tube heat exchangers, the coolant normally flows through the tubes [16]; the gas stream flows through the shell outside the tubes, and the condensible constituents form a film on the outside surface of the tubes. If there is little moisture in the gas stream, noncontact condensing systems can provide for pure product recovery [17] when only one condensible compound is present in the gas stream. Even with moisture in the gas stream, the recovery is fairly simple: the organic is separated from the water either by decanting (if it is not soluble in the water) or by distillation [18].

In a contact condensing system, the coolant is sprayed directly into the gas stream to obtain the maximum heat transfer. These systems, while they are usually

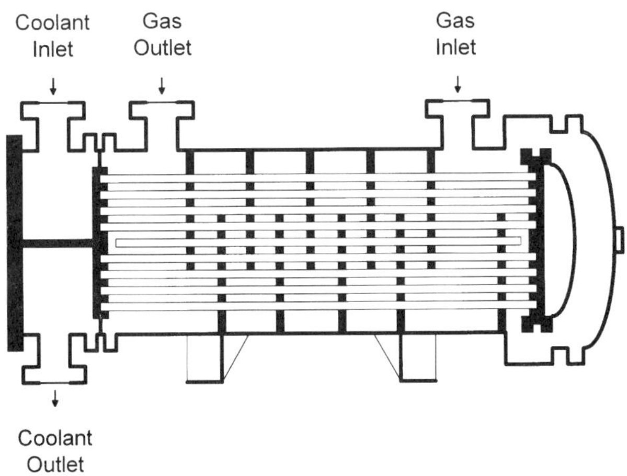

Figure 3.2 Shell-and-tube condenser. (Modified from Walsh and Murray [8].)

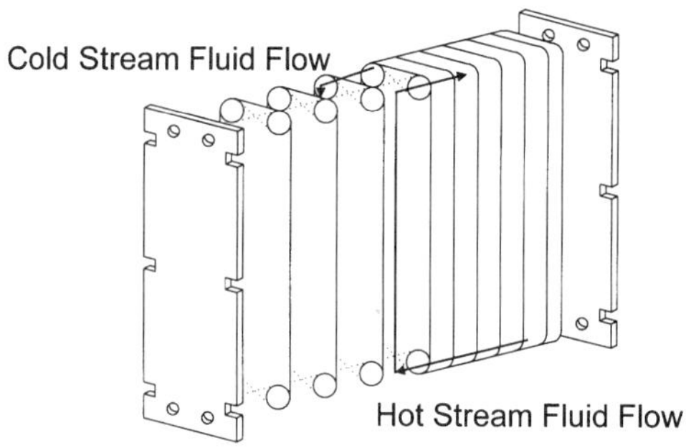

Figure 3.3 Plate and frame heat exchanger. (Modified from Polaris Corporation [15].)

less complicated and less expensive [19], create additional difficulties because the coolant must be separated from the condensed vapor if either one is to be reused. If the condensate is not reused, the liquid stream must be discarded, usually through a wastewater treatment plant. Thus, an air pollution problem is simply converted to a wastewater problem [16]. Contact condensers, however, also have the advantage that the liquid absorbs any soluble compounds, acting as both a condenser and an absorber [20, 21], at least until the liquid becomes saturated with the pollutant.

Contact condensers come in a variety of designs [16], including simple spray chambers [8] (Fig. 3.4) and venturi scrubbers [8] (in which the velocity of the gas is accelerated to increase the contact; Fig. 3.5). These types of vessels are also effective particulate collection devices [22, 23], an important consideration if the gas stream also contains particulate that might foul a shell-and-tube condenser.

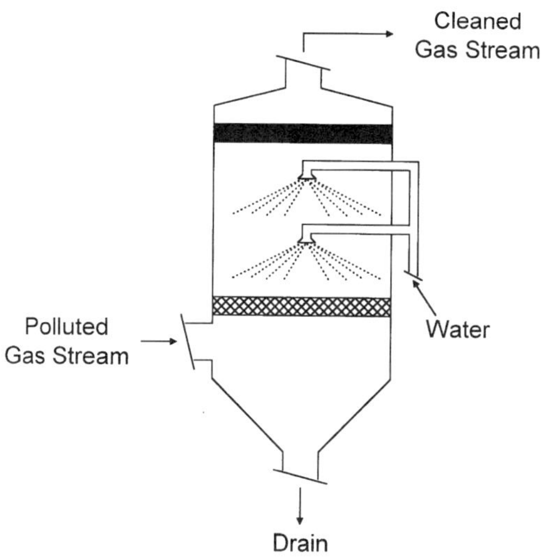

Figure 3.4 Spray chamber. (Modified from Walsh and Murray [8].)

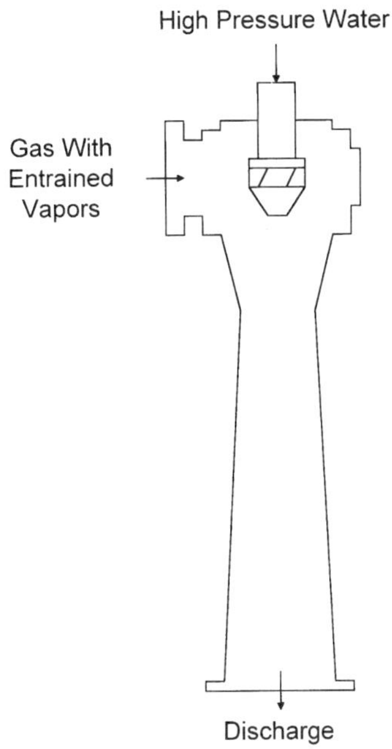

Figure 3.5 Venturi scrubber (or condenser). (Modified from Walsh and Murray [8].)

Some contact condensing systems cool the condensate further than normally required and use the chilled condensate as the recirculating coolant [24]. Although this method is only effective with pollutants that have a relatively low vapor pressure, it provides a good solution to the wastewater disposal problem encountered with miscible coolant and condensate systems.

3.3 CONDENSER SYSTEM DESIGN

Choosing and specifying a condensation system for air pollution control involves several considerations. In addition to evaluating whether the technology is appropriate, the designer must also consider safety, efficiency, economics, and maintenance requirements. An important aspect of the safety considerations is designing the system to fail safe. Assuming that every condenser will eventually leak, at least for a short while, the designer should determine whether the coolant should leak into the gas stream or vise versa. The stream under the most pressure will inevitably make its way into the other stream once the vessel begins to leak, so the condenser should be designed for pressures that ensure that the leakage is safe and easy to detect.

The efficiency of the condensation system depends on adequate heat transfer and the concentration of the pollutant in the gas stream. The gas stream should therefore be as concentrated as possible, because additional air consumes cooling capacity [25] that would otherwise be used for condensing the pollutant. Excess air also dilutes the concentration of the pollutant, making it more difficult to achieve the partial pressure necessary for removal. As with the other control devices, unnecessary air flow should be eliminated [13] by reducing the airflow at pickup points if possible, properly maintaining the seals in the ductwork, eliminating unnecessary pickup points, and maintaining valves and valve seats in excellent condition. The concentration in the ductwork should never exceed fire safety levels. This is of particular importance when storage tanks are vented to a condenser because the exhaust stream from the tank is saturated with solvent vapor.

Because of the strong influence of adequate heat transfer on the overall operation of the condenser, ensuring an adequate supply of coolant at as low a temperature as possible and minimizing potential fouling are important aspects of the system design. It is important to ensure that adequate heat transfer surface is available. Overdesigning the heat transfer area, however, can actually be detrimental to the condenser operation [3] if the overdesign results in less air velocity and less liquid washing per unit of condensing surface so that more fouling occurs.

A rough estimate of the surface area required for condensation in a simple countercurrent heat exchanger can be obtained by determining the total amount of heat, q ($J\ h^{-1}$ or $Btu\ h^{-1}$) that must be transferred out of the gas stream to achieve the desired outlet temperature. The specific heat of the vapor, the specific heat of the liquids, and the latent heat of vaporization of the condensible compounds must be known. The required surface area is then estimated by the equation [4]

$$A_c = q/U\Delta T_{lm} \tag{3.1}$$

where

A_c = Amount of heat transfer area required, m^2 (or ft^2)
q = Amount of heat to be transferred, $J\ h^{-1}$ (or $Btu\ h^{-1}$)
U = Overall heat transfer coefficient, $J\ m^{-2}\ h^{-1}\ K^{-1}$ (or $Btu\ ft^{-2}\ h^{-1}\ R^{-1}$)
ΔT_{lm} = Log–mean temperature difference, K (or °R)

The log–mean temperature difference is calculated by the formula

$$\Delta T_{lm} = \frac{\Delta T_1 - \Delta T_2}{\ln(\Delta T_1/\Delta T_2)} \tag{3.2}$$

where

ΔT_1 = Temperature difference between the hot fluid and the cold fluid at the entrance of the hot fluid to the heat exchanger, K (or °R)
ΔT_2 = Temperature difference between the hot fluid and the cold fluid at the entrance of the cold fluid to the heat exchanger, K (or °R)

Some typical overall heat transfer coefficients are listed in Table 3.2. These values are intended as a general guide only, not as the sole source of information

TABLE 3.2 Typical Overall Heat Transfer Coefficients

Vapor	Coolant	U ($J\ K^{-1}\ m^{-2}\ s^{-1}$)	U (Btu $°F^{-1}\ ft^{-2}\ h^{-1}$)
Ammonia	Water	260–880	150–500
Alcohol	Water	80–350	45–200
Methanol	Water	440–880	250–500
Kerosene	Water	50–110	30–65
Light organics (viscosity <0.5 cP; benzene, toluene, acetone, ethanol, gasoline)	Water	130–260	75–150
Heavy organics	Heavy organics	50–200	30–110
Heavy organics	Light organics	150–300	85–170
Heavy organics	Water	25–370	15–210
Light organics	Heavy organics	50–200	30–110
Light organics	Light organics	200–400	110–230
High boiling hydrocarbons	Water	35–90	20–50
Low boiling hydrocarbons	Water	140–350	80–200
Organic solvents (with large amount of noncondensibles)	Water or brine	35–140	20–80
Naphtha	Water	90–130	50–75
Vegetable oil vapors	Water	35–90	20–50
Steam	Water	350–1400	200–800
Gases (in general)	Water	4–90	2–50
Light organics	Brine	70–20	40–10
Gases	Air	4–15	2–8

Data are from refs. 3, 11, 20, 31–34.

for heat transfer. A more rigorous estimation such as that given by Knudson et al. [26], Bejan [27], Sherwood and Reed [28], McAdams [29], or Jakob [30] should be used for design. Heat transfer equipment vendors are also excellent sources for precise information regarding the heat transfer coefficient for a given situation.

Adequate provisions must also be made for subcooling the condensed liquid [8] so that the vapor is not simply re-evaporated from the warm liquid once the liquid reaches a storage tank or open vessel. This may be accomplished by maintaining a pool of condensate covering a small portion of the heat transfer surface. If this method is used, adjustments in the amount of heat transfer area should be made in the design of the condenser to ensure that adequate heat transfer area exists above the liquid pool to achieve the desired removal rate. Alternatively, a separate liquid-to-liquid heat exchanger or subcooler can be installed to maintain the desired condensate temperature. Separate subcoolers are typically much smaller than the condenser. The instrumentation that should be considered for condenser systems is listed in Table 3.3.

TABLE 3.3 Recommended Instrumentation

Gas stream temperature and pressure (inlet and outlet)
Pollutant concentration
Fan speed
Coolant flow and temperature (inlet and outlet)
Condensate flow
Refrigerant pressure (inlet and outlet), if applicable

3.4 ECONOMIC ESTIMATES

The cost of a standard shell-and-tube heat exchanger can be estimated with the method published by Cowie [35]. If necessary, the cost of auxiliary equipment such as pumps and fans can be estimated with the equations given by Vatavuk [4, 6]. The amount of heat transfer area per meter length of the exchanger is given in Table 3.4. Figures 3.6 and 3.7 give two components of the heat exchanger cost: a component that represents the cost of the heat exchanger per meter of length and the additional cost that must be added to account for tube length.

These factors are based on the cost of a standard split ring floating head heat exchanger made of a carbon steel shell and 19.05 mm (3/4 inch) tubes. Additional allowance must be made for other materials of construction, heat

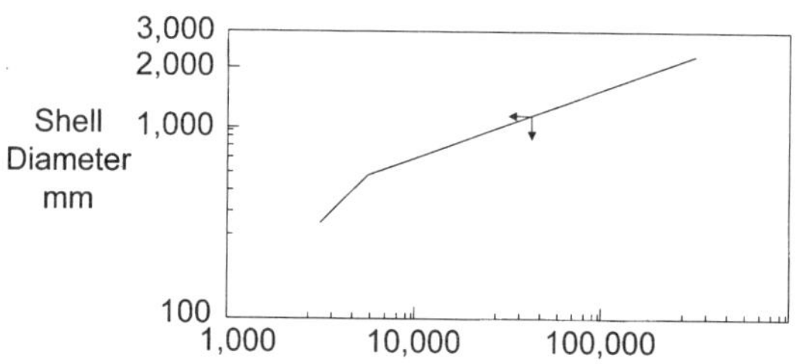

Figure 3.6 Cost of heat exchanger per meter of length in dollars. (Redrawn from Cowie [35].)

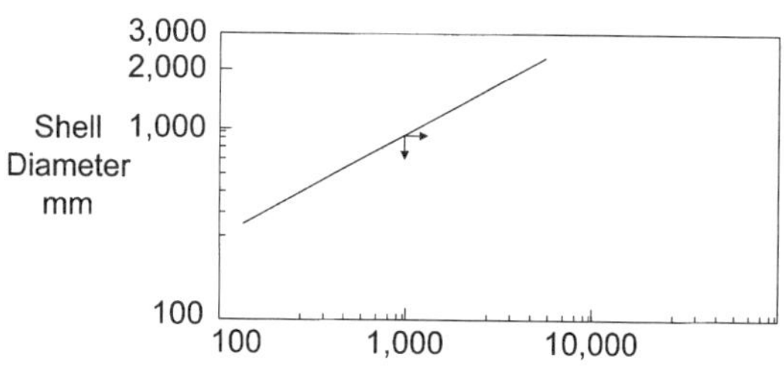

Figure 3.7 Additional cost of heat exchanger per meter of length in dollars. (Redrawn from Cowie [35].)

TABLE 3.4 Heat Exchange Area Per Unit Length of Exchanger

Shell Diameter		Area/Tube Length	
OD,[a] mm	in	m^2/m	ft^2/ft
203	8	1.67	5.48
254	10	3.1	10
305	12	4.54	14.9
356	14	5.6	18
406	16	7.89	25.9
457	18	10.6	34.9
508	20	15	49
560	22	18.2	59.7
610	24	22.2	72.8
635	25	24.3	79.6
688	27	28.0	91.9
737	29	33.1	109
787	31	38.2	125
838	33	43.5	143
890	35	49.5	162
940	37	55.3	182
991	39	62.2	204
1067	42	72.7	238
1143	45	84	276
1219	48	96.2	316
1295	51	108	353
1372	54	121	396
1524	60	149	490

[a] OD, Shell diameter is outside diameter. Modified from Cowie [35].

exchanger design, temperatures above 343°C, or different tube or baffle configurations [35]. The costs should be scaled to current values with the methods given by Vatavuk [4]. In addition to the cost of the heat exchanger and associated equipment, the installation costs must be included in the estimate. Installation costs associated with heat exchangers are typically 30–60% of the cost of the equipment alone [33].

3.5 POTENTIAL PROBLEMS

Fouling of the heat transfer area is one of the most prevalent concerns with noncontact heat exchangers [12]. *Fouling* refers to the build up of a layer of foreign material on the heat transfer surface [3] that reduces the rate of heat

TABLE 3.5 Cost Ratio of Various Materials of Construction Relative to Carbon Steel

304 Stainless steel	7
316 Stainless steel	10
Copper	7
Monel	10
Nickel	12
Hastelloy C	40
Inconel	13

Data are from Peters and Timmerhaus [33].

transfer. This may occur due to particulates in the incoming gas stream, contaminants in the cooling stream, organic growth in the cooling stream, or a variety of other factors. When river water is used as a coolant, special care must be taken to ensure that seeds from living organisms do not reach the heat exchanger tubes. For example, in the case of river mussels, the condenser tubes provide a relatively warm spot and the mussels can multiply quickly, choking off the cooling water flow [H.R. Dickson, personal communication, 1985–1988].

Corrosion is also a consideration in condensation equipment [18]. Appropriate materials are critical for maintaining the structural integrity of the condenser and ensuring that leaks do not occur. The durability of several materials of construction in corrosive atmospheres are given by Krisher and Siebert [36]. Table 3.5 lists the relative costs of several materials. Although a titanium or hastelloy condenser has a much higher initial cost, it may easily pay for itself in maintenance, lost production, and replacement costs in a corrosive environment.

Effective removal of the pollutant from the gas stream depends on the movement of gas molecules to the cold surface [37]. If the gas is cooled by more than 40 or 50 K (75° to 95°F), there is the possibility that fog will form in the condenser. This happens when the rate of heat transfer exceeds the rate of mass transfer, thereby cooling the gas before the gas molecules disperse to the cold surface and condense. The resulting condensate fog is comprised of relatively small droplets (approximately 1μm in diameter or smaller), which are difficult to collect [7].

3.6 INSTALLATION, OPERATION, AND MAINTENANCE CONSIDERATIONS

It is important to minimize thermal and mechanical stress and vibration [3] to reduce the opportunity for cracks and leaks [12]. Due to the shape and connections on a shell-and-tube condenser, it is very tempting to install it wherever it can be squeezed in and to force piping into place around it. It is, however, critical that the piping be designed and constructed to fit properly with adequate support for

TABLE 3.6 Advantages and Disadvantages of Condensation Systems

Advantages	Disadvantages
Allows for recovery and concentration of the pollutant	Effectiveness is pollutant specific, depends on vapor pressure of compound
No additional pollutants formed	Potential corrosion of ductwork due to moisture in gas stream
	Generally less efficient than other alternatives
	Potential for wastewater problems [39]

the condenser so that the connections and condenser shell are not stressed unnecessarily. To avoid unnecessary thermal stress, the condenser should be warmed up slowly and uniformly by starting the coolant flow, then allowing the warm vapor in gradually [3]. Adequate venting is also key to avoiding thermal stress during start up because it minimizes any cold spots [38].

Ample space must be maintained near the end of a shell-and-tube condenser to provide access and allow the tubes to be pulled from the shell for cleaning. Such routine maintenance should include a thorough cleaning of tubes on the inside and outside surfaces. Any cleaning methods must be used with care to avoid loosening seals between the tube and the tube sheet; steaming individual tubes is of particular concern [14]. During the outage, any tubes that have developed thin spots or leaks should be plugged to prevent loss of coolant or vapor. Careful documentation of tube failures will assist later in assessing erosion or stress problems in tubes. Similarly, proper instrumentation is essential to identifying malfunctions of a condenser. If measurements are made once the condenser is operational, but before fouling occurs, conditions can later be compared with the initial conditions to determine if the condenser is working properly or to identify problems. A summary of the advantages and disadvantages of condensation is given in Table 3.6.

REFERENCES

1. G.T. Joseph; D.S. Beachler, *APTI Course 415: Control of Gaseous Emissions, Student Manual*, US EPA, Research Triangle Park, NC, **1981**, Ch. 6.
2. R.Y. Purcell; G.S. Shareef, *Control Technologies for Hazardous Air Pollutants*, US EPA, Research Triangle Park, NC, **1986**, pp. 24, 27, 75–80.
3. L. Theodore; A.J. Buonicore, *Air Pollution Control Equipment; Volume II: Gases*, CRC Press, Boca Raton, **1988**, Ch. 9, pp. 31–40.
4. W.M. Vatavuk, *OAQPS Control Cost Manual*, 4th Ed. (and subsequent amendments), US EPA, Research Triangle Park, NC, **1990**, Ch. 6.
5. R.Y. Purcell; G.S. Shareef, *Evaluation of Control Technologies for Hazardous Air Pollutants*, EPA, Research Triangle Park, NC, **1986**.

6. W.M. Vatavuk, *Estimating Costs of Air Pollution Control*, Lewis Publishers, Chelsea, MI, **1990**, pp. 172–179.
7. B.B. Crocker, "Air Pollution Control Methods," in *Kirk-Othmer Encyclopedia of Chemical Technology*, 4th Ed., Vol. 1, J.I. Kroschwitz; M. Howe-Grant, Eds., John Wiley & Sons, New York, **1991**, pp. 749–825.
8. R.T. Walsh; R.C. Murray, "Vapor Condensers," in *Air Pollution Engineering Manual*, J.A. Danielson, Ed., US EPA, Research Triangle Park , NC, **1973**, pp. 198–207.
9. A.C. Stern; R.W. Boubel; D.B. Turner; D.L. Fox, *Fundamentals of Air Pollution*, 2nd Ed., Academic Press, Orlando, **1984**, pp. 433–436.
10. B. Henderson-Sellers, *Pollution of Our Atmosphere*, Adam Hilger, Ltd., Bristol, England, Britain, **1984**, pp. 177–181.
11. R.T. Walsh; P.G. Talens; W.C. Rogers, "Reduction of Inedible Animal Matter," in *Air Pollution Engineering Manual*, J.A. Danielson, Ed., US EPA, Research Triangle Park, NC, **1973**, pp. 815–829.
12. A.J. Buonicore, "Condensation," in *Air Pollution Engineering Manual*, A.J. Buonicore; W.T. Davis, Eds., Van Nostrand Reinhold, New York, **1992**, pp. 52–58.
13. G. Hounsell, "Condensers," in *Handbook of Air Pollution Control Engineering and Technology*, J.C. Mycock; J.D. McKenna; L. Theodore, Lewis Publishers, Chelsea, MI, **1995**, pp. 223–256.
14. J.J. Spivey, "Recovery of Volatile Organics From Small Industrial Sources," *Environ. Prog.*, **1988**, 7(1), 31.
15. Polaris Corporation, http://www.engin.umich.edu/equipment/exchange/platefra.html, August **1996**.
16. R.D. Griffin, *Principles of Air Quality Management*, Lewis Publishers, Ann Arbor, MI, **1994**, pp. 187–191.
17. H.W. Parker, *Air Pollution*, Prentice-Hall, Englewood Cliffs, NJ, **1977**, pp. 258–260.
18. S.J. Priebe, *Informal Report —The Recovery of Volatile Organic Compounds*, Idaho National Engineering Laboratory, Idaho Falls, **1989**, pp. 5–11.
19. W.F. Connery, "Condensers," in *Air Pollution Control Equipment: Selection, Design, Operation, and Maintenance*, L. Theodore; A.J. Buonicore, Eds., ETS, Inc., Roanoke, VA, **1992**, Ch. 6, pp. 153–169.
20. A.J. Buonicore; L. Theodore; R.E. McKinney; G. Tchobanoglous, "Waste Management" in *Perry's Chemical Engineers' Handbook*, 6th Ed., D.W. Green; J.O. Maloney, Eds., McGraw-Hill, New York, **1984**, Sect. 26, pp. 30–31.
21. J. Joseph; D. Beachler; J. Jahnke; M. Leslie; J. Henry, *APTI Course 415: Control of Gaseous Emissions: Instructor's Guide*, US EPA, Research Triangle Park, NC, **1981**, Ch. 16.
22. D.L. Harmon; L.E. Sparks, "Conclusions from EPA Scubber Research and Development," in *Plant Inspection Workshop—Techniques for Evaluating Performance of Air Pollution Control Equipment: Operation and Maintenance of Wet Scrubbers*, compiled by PEDCo Environmental, Inc., US EPA, Research Triangle Park, NC, **1981**, pp. 77–101.
23. C.E. Lapple, "Dust and Mist Collection," in *McGraw-Hill Encyclopedia of Environmental Science and Engineering*, 3rd Ed., S.P. Parker; R.A. Corbitt, Eds. in Chief, McGraw-Hill, New York, **1992**, pp. 130–132.

24. B.B. Crocker; K.B. Schnelle, Jr, "Control of Gases by Absorption, Adsorption, and Condensation," in *Handbook of Air Pollution Technology*, S. Calvert; H.M. Englund, Eds., John Wiley & Sons, New York, **1984**, Sect. 4, pp. 186–189.
25. A.C. Stern; R.W. Boubel; D.B. Turner; D.L. Fox, *Fundamentals of Air Pollution*, Academic Press, New York, **1973**, pp. 419–422.
26. J.G. Knudson; K.J. Bell; A.D. Holt; H.C. Hottel; A.F. Sarofim; F.C. Standiford; D. Stuhlberg; V.W. Uhl, "Heat Transmission" in *Perry's Chemical Engineers' Handbook*, D.W. Green; J.O. Maloney, Eds., McGraw-Hill, New York, **1984**, Sect. 10.
27. A. Bejan, *Heat Transfer*, John Wiley & Sons, New York, **1993**, pp. 444–478.
28. T.K. Sherwood; C.E. Reed, *Applied Mathematics in Chemical Engineering*, McGraw-Hill, New York, **1939**, p. 68.
29. H.M. McAdams, *Heat Transmission*, 2nd Ed., McGraw-Hill, New York, **1942**, p. 289.
30. M. Jakob, *Heat Transfer*, Vol. II, Wiley, New York, **1957**, pp. 356–365.
31. Kern, *Process Heat Transfer*, McGraw-Hill, New York, **1950**.
32. R.N. Robinson, *Chemical Engineering Reference Manual*, 4th Ed., Professional Publications, Belmont, CA, **1988**, Ch. 7.
33. M.S. Peters; K.D. Timmerhaus, *Plant Design and Economics for Chemical Engineers*, 3rd Ed., McGraw-Hill, New York, **1980**, pp. 645–652.
34. J.P. Holman, *Heat Transfer*, McGraw-Hill, New York, **1986**, p. 527.
35. R.C. Cowie, "Costing of Shell-and-Tube Heat Exchangers" in *Hemisphere Handbook of Heat Exchanger Design*, G.F. Hewitt, Ed., Hemisphere Publishing, New York, **1990**, Sect. 4.8.2, pp. 1–5.
36. A.S. Krisher; O.W. Siebert, "Materials of Construction," in *Perry's Chemical Engineers' Handbook*, 6th Ed., D.W. Green; J.O. Maloney, Eds., McGraw-Hill, New York, **1984**, Sect. 23, pp. 15–39.
37. H.E. Hesketh, *Air Pollution Control*, Ann Arbor Science Publishers, Ann Arbor, MI, **1979**, pp. 137–141.
38. K.J. Bell, "Operational Problems in Process Condensers," in *Condensation and Condenser Design*, J. Taborek; J. Rose; I. Tanasawa, Eds., ASME, New York, **1993**, pp. 55–65.
39. J.H. Siegell, "Exploring VOC Control Options," *Chem. Eng.*, **1996**, 103(6), pp. 92–96.

4

ADSORPTION

4.1 TECHNOLOGY DESCRIPTION

Adsorption is a technique commonly used to remove relatively low concentrations of pollutants from a gas stream by trapping them on a solid with a large surface area. The solid is typically an activated carbon or a crystalline material with high internal porosity whose surface holds (adsorbs) the pollutant through intermolecular forces. The pore structure includes macropores (diameters larger than 50 nm), micropores (diameters less than 2 nm), and mesopores (transitional sizes between micropores and macropores) [1]. As shown in Figure 4.1 [2], for micropores, adsorbed molecules can actually block access to internal surfaces. Diffusion and mass transfer rates govern the movement of the pollutant from the air stream to the exterior surface of the solid and then from the exterior into the pore structure. The efficiency of removing VOCs from a gas stream by adsorption depends on the type and concentration of pollutant, the pressure, temperature, and humidity of the gas stream, as well as the pore structure, quantity of active sites, and nature of adsorbent material.

There are two types of adsorption. The first is physical adsorption, where the pollutant molecules are held in place in the pores by relatively weak physical attraction forces of a dipole or van der Waals nature. These forces are the same intermolecular attractive forces that cause condensation of gases [3] and are typically of low energy (1–15 kcal/g-mol or 4–50 kJ/g-mol). The second type of adsorption is chemical adsorption (or chemisorption), in which chemical bonding forces are also present. The bonds formed by chemical adsorption are much stronger (10–100 kcal/g-mol or 40–400 kJ/g-mol) and the process of desorption is therefore more energy intensive. The process of removing mercury from a gas stream by using a carbon impregnated with sulfur is an example of

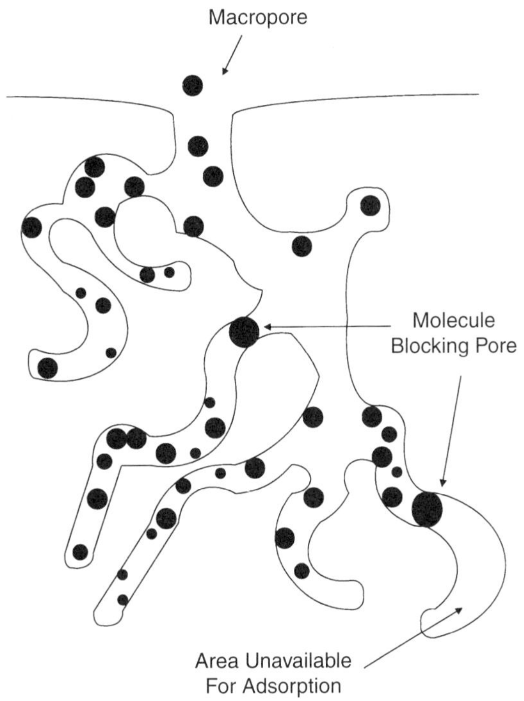

Figure 4.1 Pore structure. (Redrawn from Joseph and Bachler [2].)

chemisorption [4]. Table 4.1 summarizes the characteristics of physisorption and chemisorption.

Adsorption systems like the one shown in Figure 4.2 [2] have been used since 1920 [6] to recover solvents in industries that use large quantities of organic liquids, such as synthetic fiber manufacturing, pharmaceuticals, munitions, and printing and coating. Although incineration is also capable of very high efficiencies, one primary advantage of adsorption systems is the ability to recover pollutants for sale or reuse. This advantage is frequently a governing factor in the cost effectiveness evaluation because the cost of solvent can be a significant portion of the manufacturing cost for some processes. One example is cellulose acetate manufacturing [J. Hoskins, personal communication, 1985–1988].

Another advantage of adsorption systems is that they work well with a wide range of adsorbate concentrations. If designed and operated properly, the removal efficiencies, which can exceed 98% [7], result in low outlet concentrations consistently, even with inlet concentrations exceeding 10,000 ppm.

Once the pollutant (or adsorbate) is adsorbed onto the surface of the solid (or adsorbent), the pollutant is held in place by van der Waals forces or by capillary condensation [3, 8–10]. Eventually, the pores are full or entry of additional molecules is blocked, and the solid is unable to adsorb any more pollutant. At this point, the solid is said to be "saturated" and is removed from service. It may

TABLE 4.1 Summary of Types of Adsorption

	Physisorption	Chemisorption
Adsorbent	All solids	Some solids
Adsorbate	All gases below critical point	Some gases that are chemically reactive
Temperature range	Low temperature	Generally high temperature
Heat of adsorption	Low, 1–15 kcal/g-mol	High, 10–100 kcal/g-mol
Rate	Very rapid	
Activation energy	Low	Ranges from low for nonactivated adsorption to high for activated adsorption
Coverage	Multiple layers	Monolayer or less
Reversibility	Reversible	Frequently not reversible
Uses	Determination of pore sizes and surface area	Determination of surface concentration, estimates of active center area, and adsorption/desorption rates

Data adapted from Carberry [5].

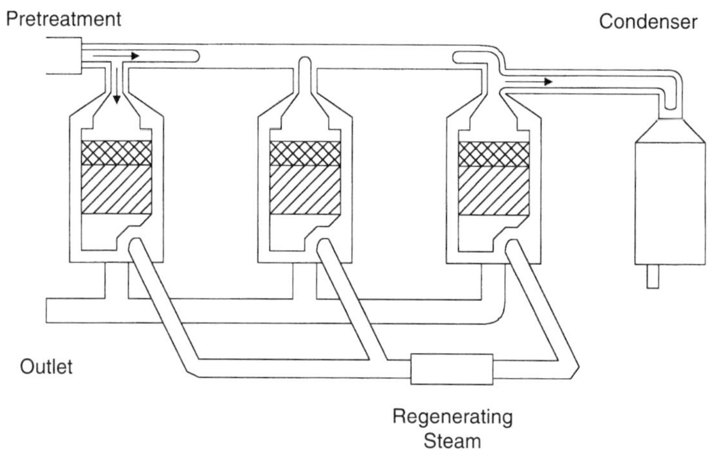

Figure 4.2 Typical process flow diagram of an adsorption system. (Redrawn from Joseph and Beachler [2].)

be discarded or sent back to the manufacturer to be "cleaned out," or it may be *regenerated* in place. *Regeneration* or *desorption* is the process of using heat or lower pressure to reverse the adsorption process. Direct steam injection is the most widely used method of providing heat for regeneration [11]. Heating coils, hot air, microwave radiation, and infrared radiation, however, have also been used or investigated [12, 13]. A technique called *pressure swing adsorption* [14] may also be used, in which the adsorption cycle is under high pressure and the desorption is accomplished by operating the adsorber at a lower pressure [15].

Although it may seem somewhat pointless to adsorb the pollutant onto the surface of a solid only to drive it off again, the amount of air present during the desorption process is considerably less than that in the original gas stream, and a vapor stream with a much higher concentration is produced. The vapor stream is usually condensed to recover the pollutant [16]. If the pollutant is not water soluble, the condensed liquid stream can be decanted to produce a stream of relatively pure organic liquid. If the pollutant is water soluble, the condensed liquid stream is typically distilled from the condensed steam to recover the organic liquid [17], or the entire stream is sent to a wastewater treatment plant for disposal.

Adsorbers are frequently paired with other control systems, such as condensers used after the desorption cycle to cool and liquify the steam/pollutant mixture. Condensers may also be used before the adsorption equipment to reduce the concentration of a pollutant to a safer level, such as one that is below the *lower explosive limit* of the gas stream or to a level at which the heat of adsorption is not sufficient to start a fire in the adsorber. Adsorption is an exothermic process, and the heat that results from adsorption of large concentrations of a pollutant can be dangerous if not properly dissipated or controlled with relative humidity [18]. Having water present in the bed or as humidity in the gas stream provides a heat sink and allows better control of the exothermic process. Water must be used with caution, however, because it competes to some degree with the pollutant for sites on the adsorbent. This effect is much less evident with activated carbon (a nonpolar adsorbent) than with other adsorbents such as zeolites and aluminosilicates (which are polar).

Adsorption is also used in conjunction with incinerators, providing a means of concentrating the gas stream before destruction and lowering the overall amount of energy required for combustion. For example, a pollutant in dilute concentrations can be adsorbed onto carbon and then desorbed with hot air. By the adsorption/desorption cycle, the concentration of pollutant in the air can be increased significantly, decreasing the volume of air that must be heated to the combustion temperature [19].

Adsorption is well suited for organic compounds that have molecular weights in the range of 40–130 [20, 21] and boiling points in the range of approximately 290–420 K (21°–150°C, 70°–300°F) [9]. These compounds have high enough boiling points to be adsorbed (compounds that are gases at room temperature are not adsorbed to a substantial amount) [22] and low enough boiling points to be removed from the adsorbent. Adsorption is similar to the process of distillation [3] in that heavier compounds are retained better than lower boiling compounds and remain in the carbon, much as they would remain on the lower trays of a distillation column. By this process, the heavier, higher boiling compounds replace the lighter weight, lower boiling compounds that are adsorbed. The higher boiling compounds are also more difficult to displace during regeneration, so they tend to accumulate in the bed. The presence of such compounds occupies active space in the pores and decreases the efficiency of the carbon bed. This is particularly a problem in processes where lubricating oils are used; even very small concentrations of lubricant can significantly affect the performance of a carbon bed [J. Hoskins, personal communication, 1985–1988].

It is generally accepted that adsorption in the carbon bed occurs only within a fairly small zone, called the *mass transfer zone* (MTZ) [12, 23]. The size of the MTZ depends on the velocity of air through the carbon bed, the pollutant characteristics, the temperature and pressure, and the humidity. Air is usually fed down through the carbon in the adsorber to avoid fluidizing the bed, to avoid creating channels for the air to bypass the carbon, and to avoid abrading the carbon any more than necessary. As vapor first enters the top surface of the carbon, the majority of the pollutant is removed from the gas stream before the gas travels very far into the bed. Mass transfer may actually only occur in the top 3 or 4 in of carbon [8]. In the active adsorption zone, an equilibrium is reached between the gaseous molecules that impinge and are retained on the surface and those that are in the gas phase. There is a constant process of adsorption and desorption as molecules move.

The equilibrium ratio of the amount of pollutant adsorbed to the partial pressure of pollutant in the gas stream at a constant temperature is represented by an *isotherm*. Isotherms of some typical pollutants are shown in Figure 4.3 [24, 25].

Figure 4.4 [24] depicts the change that occurs in the pollutant concentration at various depths in the bed as time progresses. At the beginning of the adsorption cycle (curve "a"), the bed immediately starts adsorbing pollutant, and the concentration declines rapidly over a very shallow portion of the bed (the *critical bed depth*, L_c). As the top layers become saturated, the concentration of pollutant in

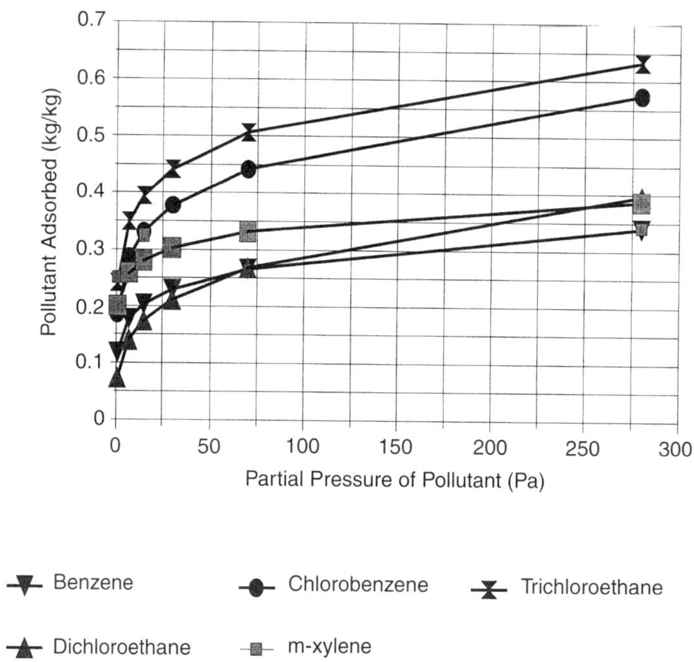

Figure 4.3 Isotherms of typical pollutants at 298 K (25°C, 77°F) [24, 25].

those layers becomes the same as the incoming gas. But there is relatively fresh carbon that has not been exposed to high concentrations of pollutant below the saturated layer. The MTZ (represented by T_c in the diagram) shifts down, and the adsorption continues (the same process is depicted in Fig. 4.5 [12]). By this process, the MTZ works its way down the bed until its leading edge comes to the bottom of the carbon bed (curve "b" in Fig. 4.4). This point, at which the concentration of organic in the exhaust stream begins to rise rapidly, is referred to as the *breakthrough point*, that is, the point at which the MTZ is exiting the carbon bed, and (virtually) complete adsorption of the pollutant is no longer occurring.

The MTZ concept makes it evident that, at any given time, only a small portion of the carbon is actually being used in the adsorption process; there is carbon that has not yet encountered the MTZ and carbon that is saturated, but only a few inches of the carbon may actually be in use. Because stationary carbon beds can be several feet deep [26], a substantial amount of carbon is needed to maintain a reasonable adsorption cycle time, but it is not in active use.

The length of the MTZ is determined by the pollutant characteristics, inlet concentration, and velocity through the bed. The amount of time that the bed can adsorb before reaching the "breakthrough" point can be calculated from the following equation [24]:

$$t_p = \left[\frac{w_p w_c}{C_m Q} - \frac{\rho Q}{K} \ln\left(\frac{C_m}{C_t}\right)\right] \tag{4.1}$$

where

t_p = breakthrough time, min
w_p = capacity of carbon to adsorb pollutant, kg pollutant/kg carbon (lb pollutant/lb carbon)
w_c = weight of carbon, kg (lb)
C_m = inlet concentration, kg/m^3 (lb/ft^3)
Q = actual flowrate, m^3/min (ft^3/min)
ρ = density of activated carbon, kg/m^3 (lb/ft^3)
K = pseudo first-order reaction rate constant derived experimentally
C_t = breakthrough concentration, kg/m^3 (lb/ft^3)

Operation of the carbon bed can be optimized with the amount of carbon, type of carbon, carbon replacement schedule, and bed depth to diameter ratio as variables. For large installations, collection of experimental data and pilot studies are essential to optimize these variables.

4.2 TYPES OF ADSORPTION SYSTEMS

The conventional adsorber vessel (Fig. 4.6) is a fairly long cylinder that can be installed in either a vertical or a horizontal position [22]. The bed of adsorbent material is several feet thick and is suspended on a screen. Vapor-laden air flows down through the bed. Regeneration steam is frequently introduced from the

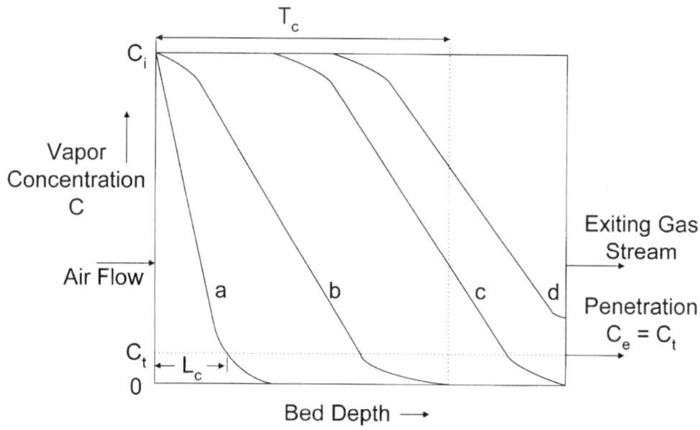

Figure 4.4 Progress of adsorption mass transfer rate through the bed. (Adapted from Dorling [24].)

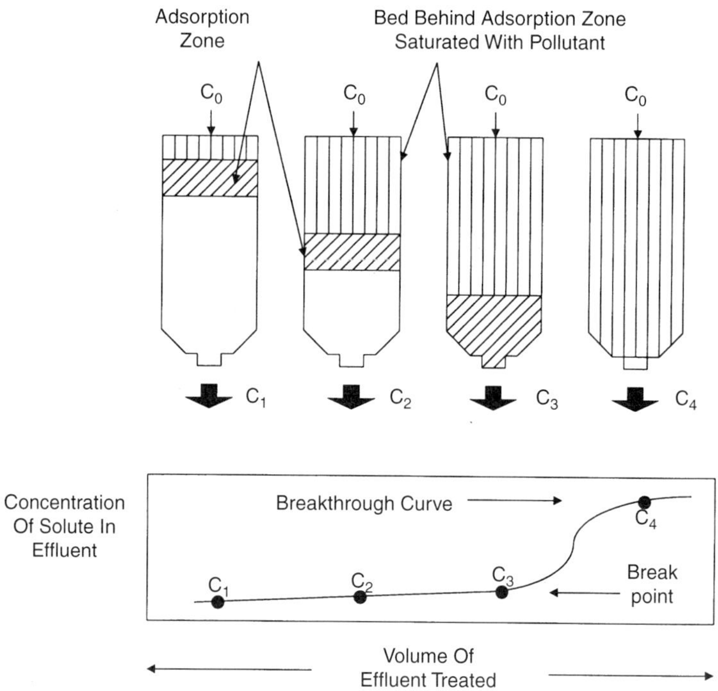

Figure 4.5 Flow of the mass transfer zone through the bed [12].

bottom of the vessel. Several other types of adsorber vessels have been developed to maximize contact of the adsorbent with the gas stream and to minimize the amount of carbon in the system.

Moving bed adsorbers are similar to fixed bed adsorbers, but the carbon is usually contained in a hollow cylinder or a disk-shaped packed bed (see Fig. 4.7)

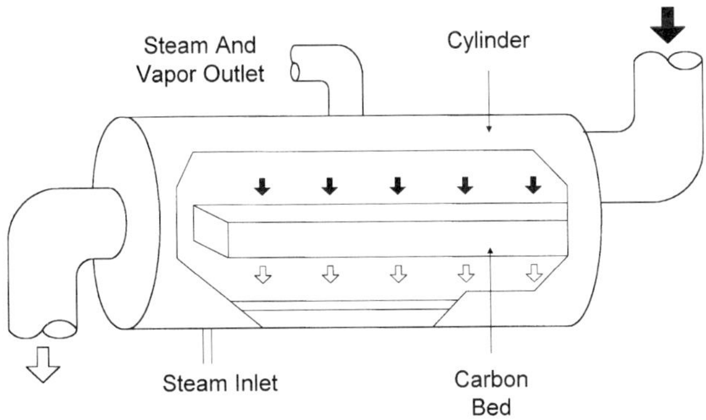

Figure 4.6 Conventional adsorber bed. (Adapted from Joseph and Beachler [2].)

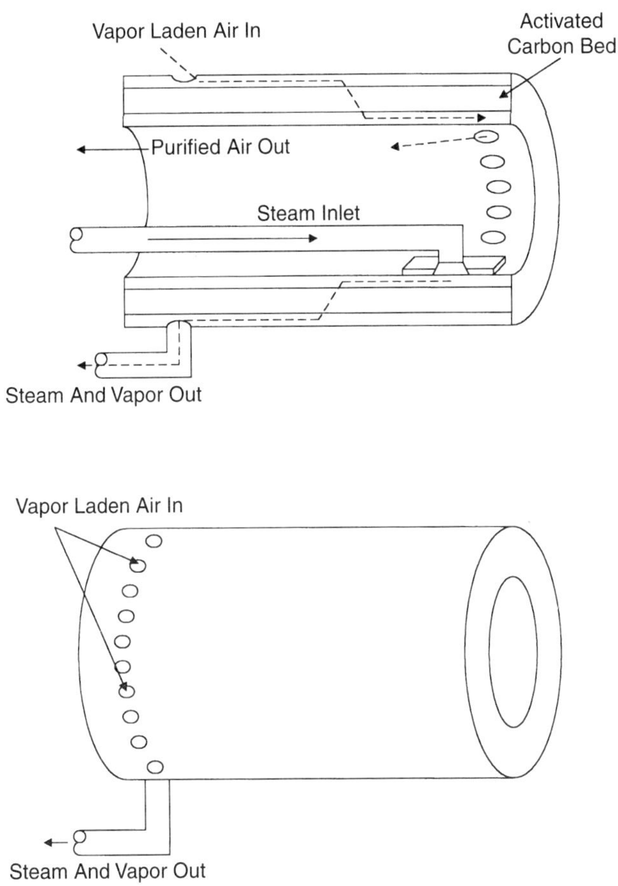

Figure 4.7 Moving bed adsorber. (Redrawn from Joseph and Beachler [2].)

[2, 27]. While one section of the carbon is adsorbing, another section is being steamed (Fig. 4.7).

Another type of adsorber designed to minimize the amount of carbon is the fluidized bed adsorber (Fig. 4.8 [10]). In a fluidized bed adsorber, the gas flows up through the carbon and suspends it. The carbon, which is in intimate contact with the gas stream, adsorbs the pollutants and works its way down the column and is replaced by recycled carbon introduced at the top of the adsorber. A regenerative section is also included in the fluidized bed; it may be at the top of the fluidized

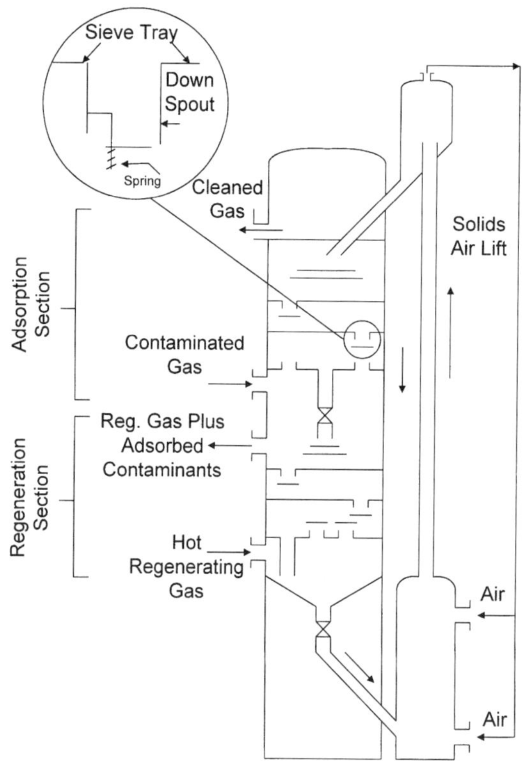

Figure 4.8 Fluidized bed adsorber. (Adapted from Buonicore and Theodore [10].)

TABLE 4.2 Instrumentation to Consider for Adsorption Systems

Gas stream temperature and pressure (inlet and outlet)
Fan speed
Outlet concentration
Bed temperature
Humidity

bed so that carbon is regenerated immediately before entering the adsorbing section [29], or it may be at the bottom of the adsorption section [3]. Table 4.2 lists the instrumentation that should be considered for adsorption systems.

4.3 ADSORPTION SYSTEM DESIGN

For continuous processes in which regenerative adsorption is used, several adsorbers are installed in parallel so that one or more are always available for adsorption while used ones are regenerating (see Fig. 4.9). It is also advantageous to operate the system in a "cooling" mode [29] in which the adsorber that has just finished steaming is not put directly on line, but takes the exhaust gas from an active adsorber. Routing the exhaust gas from the primary adsorbing bed through the "cooling" bed accomplishes two things. First, the relatively cool air serves to remove the heat from the steamed bed and dries it out so that sites are available for organic vapor adsorption instead of being occupied by water molecules. Second, as the primary adsorbing bed becomes saturated and the exhaust concentration starts to rise, the exhaust gas is still being routed through the "cooling" bed. Because the cooling bed by this point in the cycle is cooled and almost dry, it provides a clean bed to adsorb any residual organics before they are exhausted. Once the adsorbing bed reaches the breakthrough point (the point at which the concentration of organic compounds in the exhaust begins to increase

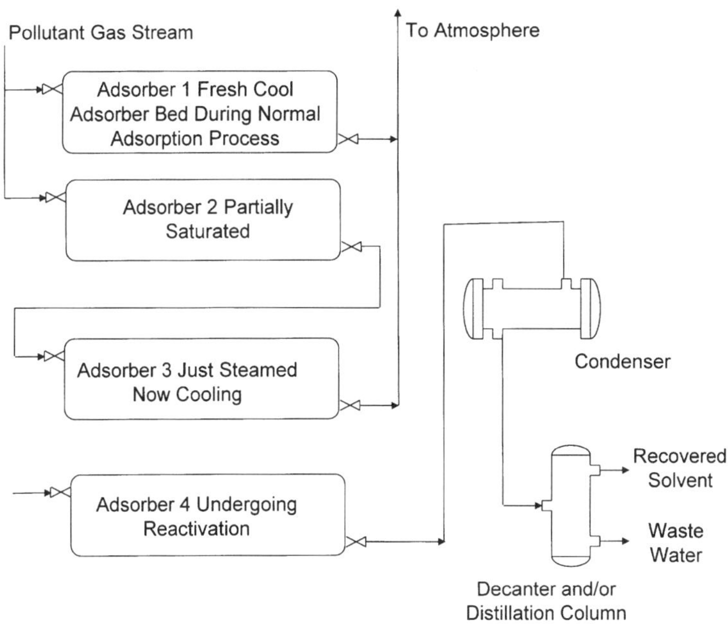

Figure 4.9 Schematic diagram of the cooling cycle. (Adapted from Crocker and Schnelle [19].)

dramatically), it is taken off line and steamed. Similarly, the bed that was in the "cooling" mode switches to be the primary bed for adsorption, and the bed that was steaming begins the cooling cycle.

In any regenerative adsorption system, some amount of pollutant remains in the adsorbent after regeneration is complete. This leftover portion is referred to as the *heel*. The magnitude of the heel can be changed by adjusting the temperature, pressure, or length of the regeneration step or by using activated carbon from a different source, but it is economically unfeasible to eliminate it altogether. A balance must be struck between the amount of heel left in the solid and the amount of steam used for the desorption. It is not cost effective or possible under normal time constraints to continue the steaming past an optimum time at which the vast majority of the adsorbate has been removed.

In addition to the amount of adsorbent available, the type of adsorbent also significantly impacts the effectiveness of the system. A wide variety of solids are available for use in adsorption systems. The most common is activated carbon, also referred to as *activated charcoal*. Due to its nonpolar nature, it is the adsorbent best suited for most VOC abatement systems and is the only one discussed in the remainder of this section. Other adsorbents include silica gel; zeolites (or molecular sieves), which have a very precise crystal structure and pore size; and carbon molecular sieves [30], an activated carbon made from polymeric materials that also has uniform pores.

The phenomenon in which higher boiling compounds replace low-boiling compounds has led to the development of "sacrificial" or "suicidal" carbon beds. These are beds that are placed in front of another control device to remove heavier compounds that might otherwise destroy the final control equipment [6]. The sacrificial bed is replaced or regenerated well before breakthrough of the heavier compound occurs. One potential application for such a system is in the manufacture of ammunitions containing nitroglycerine. Organic solvents used in the production of propellents are driven off by evaporation. In propellant formulas that also contain nitroglycerine, the nitroglycerine is also emitted, making it unsafe to collect or control the solvent emissions. It has been suggested that a sacrificial bed could be used to collect the heavier nitroglycerine vapors. This would be followed by a separate carbon bed to recover the other solvents. Such a system would also have to include an acid bath or other means of destroying the nitroglycerine once it accumulates in the carbon, but the sacrificial bed offers an alternative to the current method of exhausting the vapors without control [31].

There are several methods to determine the appropriate size of a carbon bed. One of the simplest methods, given by Vatavuk et al. [25], uses Freundlich isotherms to predict the amount of carbon necessary to accomplish the desired removal. The following equation is used to determine the maximum amount of pollutant that can be adsorbed onto the carbon under a given set of conditions:

$$w_e = k(C_F P)^m \tag{4.2}$$

where

w_e = the maximum amount of pollutant that is adsorbed under the given conditions, expressed as kg (or lb) of adsorbate/kg (or lb) of adsorbent

C_F = conversion factor equal to 1.45×10^{-4} for pressures in Pa (1 for pressures in psia)

P = partial pressure of the pollutant in Pa (or psia)

k, m = pollutant-specific isotherm parameters (Table 4.3 [25, 32]). Values should not be extrapolated outside of the applicable range listed in Table 4.3

The total amount of pollutant to be removed from the gas stream is calculated by the following equation:

$$M = C \cdot A \cdot MW/F \tag{4.3}$$

where

M = Mass of pollutant to be adsorbed, kg/h (lb/h)

C = Concentration of pollutant, ppmv (ppmv)

A = Air flow, Nm^3/h, corrected to 0°C (scfm, the air flow corrected to 70°F)

MW = Molecular weight of pollutant, kg/kg-mol (lb/lb-mol)

F = Conversion factor of 2.24×10^7 (6.42×10^6)

TABLE 4.3 Isotherm Parameters

Adsorbate	Adsorption Temperature			Isotherm Parameters		Applicable Partial Pressure Range	
	K	°C	°F	k	m	Pa	psia
Benzene	298	25	77	0.597	0.176	0.7–340	0.0001–0.05
Chlorobenzene	298	25	77	1.05	0.188	0.7–70	0.0001–0.01
Cyclohexane	311	38	100	0.508	0.210	0.7–340	0.0001–0.05
Dichloroethane	298	25	77	0.976	0.281	0.7–280	0.0001–0.04
Phenol	313	40	104	0.855	0.153	0.7–210	0.0001–0.03
Trichloroethane	298	25	77	1.06	0.161	0.7–280	0.0001–0.04
Vinyl chloride	311	38	100	0.200	0.477	0.7–340	0.0001–0.05
m-Xylene	298	25	77	0.708	0.113	0.7–7	0.0001–0.001
	298	25	77	0.527	0.0703	7–340	0.001–0.05
Acrylonitrile	311	38	100	0.935	0.424	0.7–100	0.0001–0.015
Acetone	311	38	100	0.412	0.389	0.7–340	0.0001–0.05
Toluene	298	25	77	0.551	0.110	0.7–340	0.0001–0.05

From Vatavuk et al. [25].

Once the amount of pollutant to be adsorbed is determined, the required amount of carbon is calculated by dividing the mass of pollutant (M) by the mass loading (w_e). Because the resulting carbon requirement represents the amount needed to adsorb the pollutant under ideal equilibrium conditions, a safety factor of 100% is customary [25]. Therefore, the total amount of carbon in the system is generally twice that which would be required under ideal conditions.

Other design equations and procedures are given by LeVan et al. [33], and Buonicore and Theodore [10, 34]. Although these procedures can be used for initial cost estimates and to estimate the size of the adsorbers, empirical data are essential to properly design a carbon adsorption system [35]. Any potential application should be tested on a laboratory scale not only to determine the feasibility and efficiency that can be achieved but also to evaluate the effectiveness of various carbons for the given application. Two carbons of the same size, even if produced from the same raw material and the same process, but at different plants, will behave differently in the same recovery system [J. Hoskins, personal communication, 1985–1988]. For the best performance, different types of carbons should be tested with the VOC of interest to determine not only the efficiency with which each adsorbs the compound but also the degree of desorption that can be accomplished, the susceptibility of the carbon to abrasion, and the incremental pressure drop through the carbon.

4.4 ECONOMIC ESTIMATES

Capital costs of the adsorption system depend on the gas flow, concentration, and required efficiency. A procedure for estimating the installed capital cost of an adsorption system is given by Vatavuk et al. [25]. Any estimates derived from these calculations should be checked with a vendor for confirmation. The estimates also need to be scaled up to current values. Scaling factors can be found on the EPA CATC website at http://www/epa.gov/ttn/catc. Once the total amount of carbon that is required and the number of adsorbers to be used in the system is determined, the capital cost of each adsorber vessel can be calculated based on the surface area of each vessel according to the equations given below. Note that the equations are based on the amount of carbon in each vessel and the gas flow rate through each actively adsorbing vessel. In SI units, the vessel diameter is calculated from

$$D = 7.93 \times 10^{-3} M_{CA} v_b / Q_A \tag{4.4}$$

In English units, the equation is

$$D = 0.127\ M_{CA} v_b / Q_A \tag{4.5}$$

where

D = diameter of the required adsorber vessel (or ft)
M_{CA} = mass of carbon in each adsorber, kg (or lb)
v_b = superficial velocity through the bed, m/min (ft/min)
Q_A = actual flowrate of gas through each adsorber that is on line, m^3/min (acfm)

The adsorber length is calculated in SI units from the equation

$$L = 126\ Q_A{}^2/(M_{CA} v_b{}^2) \tag{4.6}$$

In English units, the equation is

$$L = 7.87\ Q_A{}^2/(M_{CA} v_b{}^2) \tag{4.7}$$

where

L = length of the adsorber vessel, m (or ft)

Once the diameter and length of the vessel are determined, the surface area is calculated from the following equation:

$$S = \pi D (L + D/2) \tag{4.8}$$

where

S = surface area, m^2 (or ft^2)

Then the vessel cost is estimated from

$$C_v = 9.71\ S^{0.778} \text{ for SI units} \tag{4.9}$$

$$C_v = 271\ S^{0.778} \text{ for English units} \tag{4.10}$$

where C_v is the cost of each vessel in fall, 1989 dollars.

These estimates are valid for horizontal adsorbers made of 304 stainless steel which are larger than approximately 0.7–1 meter (2–3 ft) in diameter, with surface areas between 9 and 196 m^2 (97 and 2110 ft^2). Other materials of construction may significantly alter the cost. See Chapter 3 for factors for other metals. The cost of ductwork, stacks, pumps, fans, condensers, instrumentation, piping, and a decanter are not included in this cost. The ratio of the total system cost (with the exception of instrumentation and site-specific requirements such as long ductwork) to the vessel costs are estimated from the following formulas:

In SI units:

$$R_C = 3.62\ Q^{-0.133} \tag{4.11}$$

In English units:

$$R_C = 5.82\ Q^{-0.133} \tag{4.12}$$

where

R_C = ratio of total equipment cost to adsorber cost
Q = actual flowrate of entire system, m^3/min (or ft^3/min)

The total system cost can then be estimated from

$$CC = R_C(NC_C M_{CA} + NC_V) \tag{4.13}$$

where

CC = capital cost of adsorption equipment
N = number of vessels required
C_C = cost of carbon, approximately \$4.40/kg (\$2/lb) in fall, 1989 dollars

The total capital investment can be estimated from the factors in Table 4.4. Particulate removal devices and site-specific construction requirements are not included. If the adsorption system is to be placed a long distance from the generation site, both the cost of additional ductwork and the cost of additional booster fans should be included in the cost estimate. As in the other chapters in this book, the cost estimates are to indicate a general order of magnitude. They are not specific enough to use as construction costs and should be confirmed with actual prices from vendors.

Annual costs can be approximated with the guidelines in the *OAQPS Cost Control Manual* [25], as given in Table 4.5. Expected pressure drop through a bed of 4 × 10 mesh Calgon "PCB" carbon can be estimated as [25, 32] follows:

In SI units:

$$\Delta P_b = (2.12 \times 10^{-4} v_b + 2.03 \times 10^{-6} v_b{}^2) 0.0333\ M_{AC}/(L\,D) \tag{4.14}$$

In English units:

$$\Delta P_b = (0.03679 v_b + 1.107 \times 10^{-4} v_b{}^2) 0.0333\ M_{AC}/(L\,D) \tag{4.15}$$

where

ΔP_b = pressure drop through the adsorber bed, kPa (or in of water column)

In addition to the pressure drop calculated above, 0.25 kPa (1 in of water column) should be included to account for losses through the ductwork. If long or complex ductwork is used, the miscellaneous pressure drop could be much higher. Electricity requirements can then be estimated from the following formulas:

In SI units:

$$\text{Power} = 1.65 \times 10^{-2} Q\,\Delta P/\varepsilon \tag{4.16}$$

In English units:

$$\text{Power} = 1.17 \times 10^{-4} Q\,\Delta P/\varepsilon \tag{4.17}$$

TABLE 4.4 Factors for Estimating the Total Installed Capital Cost (TCC) of Adsorption Systems

Cost Element	Estimated Cost
Direct Costs	
Purchased equipment cost (PEC)	
Adsorber system and auxiliary equipment	CC, Estimated from equation
Instrumentation	0.10 CC
Taxes	0.03 CC
Freight	0.05 CC
Total PEC	1.18 CC
Installation costs	
Foundations and supports	0.08 PEC
Erection and handling	0.14 PEC
Electrical	0.04 PEC
Piping, insulation, and painting	0.04 PEC
Total installation cost	0.30 PEC
Site preparation (SP)	As required for specific site,
Buildings (BLDG)	As required for specific site,
TOTAL DIRECT COSTS	1.30 PEC + SP + BLDG
Indirect Costs	
Engineering and supervision	0.10 PEC
Construction, field expenses, and fees	0.15 PEC
Start-up and performance test	0.03 PEC
Contingency	0.03 PEC
TOTAL INDIRECT COSTS	0.31 PEC
Total Installed Capital Costs (TCC)	1.61 PEC + SP + BLDG

From Vatavuk et al. [25].

where

ΔP = pressure drop across the entire system, kPa (in of water column)
ε = combined efficiency of motor and fan, usually 35–65%

Operating and maintenance labor requirements are usually minimal, at approximately one half hour per shift each.

4.5 POTENTIAL PROBLEMS

Proper operation of an adsorption system depends on adequate contact between the adsorbent and the adsorbate. Sudden changes in air velocity or poor provisions

TABLE 4.5 Annual Operating Costs for Adsorption Systems

	Unit Cost	Estimated Requirement
	Direct Annual Costs	
Utilities		
Electricity	\$0.05/ kW-h	Calculated above
Steam	\$13/1000 kg \$6/1000 lb	3–4 kg steam/kg of VOC adsorbed (lb steam/lb VOC adsorbed)
Cooling water	\$40–80/1000 m^3 \$0.15–0.30/1000 gallons	3.43 times the steam requirement
Operating labor	\$12.96/h	1/2 hour per shift
Maintenance labor	\$14.26/h	1/2 hour per shift
Maintenance materials	Equal to maintenance labor	
Supervisory labor	15% of operating labor	
Replacement carbon	\$4.40/kg · 1.08 (shipping and taxes) · CRF_5 (see below for calculation, assuming a 5 year life) \$2.00/lb · 1.08 · CRF_5	Carbon requirement calculated above
Replacement carbon labor	CRF_5 · \$0.11/kg CRF_5 · \$0.05/lb	Carbon requirement calculated above
	Indirect Annual Costs	
Overhead	60% of maintenance costs	
Property tax	1% of TCC	
Insurance	1% of TCC	
Administrative costs	2% of TCC	
Capital recovery ($CRF = i[1 + i]^n / [(1 + i)^n - 1]$), where i = pretax marginal rate of return and n = economic life of equipment, usually 10–20 years	CRF · TCC	
TOTAL ANNUAL OPERATING COSTS	Direct + indirect operating costs	

Adapted From Vatavuk et al. [25].

for air distribution in a carbon bed can result in turbulent airflow patterns in which the adsorbent is physically moved and a hollowed spot in the bed results. Air flows more readily through the thin spot in the bed so that a portion of the carbon quickly becomes saturated and the overall capacity of the adsorption system is decreased significantly. Such a problem can be detected if the system is equipped with an analyzer and recorder of the organic vapor concentration in the exhaust from the adsorber.

Normally, such analyzers are used to monitor exhaust concentration from each bed at the very end of the cycle. For suspected problems, however, the analyzer should be capable of continuously measuring the exhaust concentration from any one bed so that problems can be identified and isolated as quickly as possible. The normal exhaust concentration curve will start high when the bed is wet but will quickly decrease as the bed dries out and begins to adsorb. It will reach a minimum where it should remain until just before the adsorber is removed from service for regeneration. At that point, the curve will increase very sharply as breakthrough occurs. By comparing the shape of the concentration curve in a troublesome adsorber with the anticipated shape, conclusions about what might be wrong can be made without having to remove the adsorber from service. For example, if the curve comes down steeply when the adsorption is initiated but starts back up gradually, it may indicate that there is a thin spot in the bed, allowing the polluted gas to bypass the majority of the bed. Once the carbon in the thin spot becomes saturated, the removal efficiency starts to decline. If the curve never reaches the anticipated minimum, it may indicate that there are leaking steam valves so the bed is remaining wet all of the time or that there are leaking isolation valves that allow untreated air to be directly exhausted instead of traveling through the adsorber bed.

The potential for fouling due to lubricant oils or other heavy organic contaminants has already been discussed. Another source of problems is particulate matter in the gas stream. Particulates also block access of the gases to the pores and cause a gradual increase in pressure drop as they accumulate in the bed. In gas streams where solids are likely to be present, a simple particulate filter should be installed before the adsorbers [36].

As noted earlier, the presence of organic vapors, carbon, oxygen, and heat from regeneration raises concerns about the possibility of fires in the adsorption beds. It is advisable to ensure that any time steam is used to regenerate carbon, the bed is adequately cooled before it is exposed to air. This is particularly important if high-pressure steam is used to revitalize the carbon. The temperatures achieved in such a scenario are adequate to cause spontaneous combustion of the carbon if oxygen is available [H.R. Dickson, personal communication, 1985–1988].

The heat and presence of the carbon can cause some compounds such as acrylic acid, phenol, 2-ethylhexanol, and toluene diisocyanate [6] to polymerize. Polymerized compounds are heavier than the original chemical and hence more difficult to desorb from the carbon. As with other contaminants, the polymeric substance occupies available sites on the carbon and decreases overall effectiveness.

TABLE 4.6 Advantages and Disadvantages of Adsorption Systems

Advantages	Disadvantages
Allows for recovery/concentration of the pollutant	Effectiveness is pollutant specific, depends on vapor pressure of compound
Usually no additional pollutants formed	Potential corrosion with some organic solvents (ketones, halogenated compounds)
High efficiency over a wide range of concentrations and pollutants	Treatment of wastewater stream required

4.6 INSTALLATION, OPERATION, AND MAINTENANCE CONSIDERATIONS

Perhaps the most important detail in maintaining the efficiency of an adsorber unit is to eliminate any duct inleakage or unnecessary dilution air. An equilibrium is established in the carbon as the gas makes its way through the bed. The equilibrium is concentration dependent, so, if the pollutant is more dilute, not as much adsorbate will be retained on the carbon. Additional air also means that the gas must travel faster through the bed, and the retention time is decreased. This combination of factors makes the daily or weekly check for duct damage critical. In addition to the typical inspection for obvious holes or creases, old ductwork should also be checked for structural integrity because the duct can be abraded to only a very thin porous shell without noticeable damage [7]. The concentration in the ductwork should be routinely checked to ensure that it never approaches unsafe levels.

Temperature is another important factor in maintaining the equilibrium. The lower the temperature, the more pollutant is adsorbed in a given system [2]. Table 4.6 lists the overall advantages and disadvantages of adsorption systems.

REFERENCES

1. S. Lowell; J.E. Shields, *Powder Surface Area and Porosity*, Chapman and Hill, New York, **1991**.
2. G.T. Joseph; D.S. Beachler, *APTI Course 415 Control of Gaseous Emissions: Student Manual*, US EPA, Research Triangle Park, NC, **1981**, pp. 5-1–5-39.
3. J.C. Mycock; J.D. McKenna; L. Theodore, "Adsorption," in *Handbook of Air Pollution Control Engineering and Technology*, Lewis Publishers, New York, **1995**, Ch. 10, pp. 173–198.
4. J.C. Crittenden; R.D. Cortwright; B. Rick; S. Tang; D. Perram, "Using GAC to Remove VOCs from Air Stripper Off-Gas," *Am. Water Works Assoc. J.*, 80(5), 73–84.

5. J.J. Carberry, *Chemical and Catalytic Reaction Engineering*, McGraw-Hill, New York, **1976**.
6. T.E. Cannon, "Carbon Adsorption Applications," in *Air Pollution Control and Design Handbook* (in Two Parts), P.N. Cheremisinoff; R.A. Young, Eds., Marcel Dekker, New York, **1977**, pp. 421–429.
7. R.Y. Purcell; G.S. Shareef, *Control Technologies for Hazardous Air Pollutants*, US EPA, Research Triangle Park, NC, **1986**, p. 24.
8. T.M. Hellman, "Odor Control by Adsorption," in *Air Pollution Control and Design Handbook* (in Two Parts), Part II, P.N. Cheremisinoff; R.A. Young, Eds., Marcel Dekker, New York, **1977**, pp. 991–1003.
9. J.O. Ledbetter, *Air Pollution in Two Parts: Part B—Prevention and Control*, Marcel Dekker, New York, **1974**, Ch. 17, pp. 123–133.
10. A.J. Buonicore; L. Theodore, *Industrial Control Equipment for Gaseous Pollutants*, Volume I, CRC Press, Cleveland, OH, **1975**, pp. 132–202.
11. R.G. Wynne; L.P. Spencer, "Adsorbers," in *Air Pollution Control Equipment: Selection, Design, Operation, and Maintenance*, L. Theodore; A.J. Buonicore, Eds., ETS, Inc., Roanoke, VA, **1992**, Ch. 4, pp. 73–132.
12. A.J. Buonicore, "Adsorption," in *Air Pollution Engineering Manual*, A.J. Buonicore; W.T. Davis, Eds., Van Nostrand Reinhold, New York, **1992**, Ch. 2, pp. 15, 31–52.
13. C.S. Fang; S. Khor, "Reduction of Volatile Organic Compounds in Aqueous Solutions Through Air Stripping and Gas-Phase Carbon Adsorption," *Environ. Prog.*, 8(4), 270–278.
14. K.E. Noll; V. Gounaris; W.S. Hou, *Adsorption Technology for Air and Water Pollution Control*, Lewis Publishers, Inc., Chelsea, MI, **1992**, pp. 1–45, 76–79, 279–297.
15. J.D. Sherman, "Adsorption, Gas Separation," in *Kirk-Othmer Encyclopedia of Chemical Technology*, 4th Ed., Vol. I, *A to Alkoids*, J.I. Kroschwitz; M. Howe-Grant, Eds. John Wiley & Sons, New York, **1991**, pp. 493–528.
16. M.F. LeDuc, "Adsorption Equipment," in *Air Pollution Engineering Manual*, J.A. Danielson, Ed., EPA, Research Triangle Park, NC, **1973**, pp. 189–198.
17. J.P. Tomany, *Air Pollution: The Emissions, the Regulations, & the Controls*, American Elsevier Publishing Company, New York, **1975**, pp. 272–305.
18. H.E. Hesketh, *Air Pollution Control*, Ann Arbor Science Publishers, Ann Arbor, MI, **1979**, Sect. 4.5, 5.6, 6.6.
19. B.B. Crocker; K.B. Schnelle, Jr, "Adsorption," in *Handbook of Air Pollution Technology*, S. Calvert; H.M. Englund, Eds., John Wiley & Sons, New York, **1984**, Ch. 3, pp. 162–185.
20. N. Mukhopadhyay; E.C. Moretti, *Current and Potential Future Industrial Practices for Reducing and Controlling Volatile Organic Compounds*, Center for Waste Reduction Technologies, American Institute of Chemical Engineers, New York, **1993**, pp. 17–19.
21. R.M. Bethea, *Air Pollution Control Technology: An Engineering Analysis Point of View*, Van Nostrand Reinhold, New York, **1978**, Ch. 13, pp. 366–393.
22. D.E. Painter, *Air Pollution Technology*, Reston Publishing Company, Reston, VA, **1974**, pp. 185–188.
23. T. Godish, *Air Quality*, 2nd Ed., Lewis Publishers, Chelsea, MI, **1991**, pp. 325–326.

24. T.A. Dorling, "Carbon Adsorption," in *Odour Control: A Concise Guide*, F.H.H. Valentin; A.A. North, Eds., Warren Spring Laboratory, Stevenage, Hertfordshire, England, **1980**.
25. W.M. Vatavuk; W.L. Klotz; R.L. Stallings, "Carbon Adsorption," in *OAQPS Control Cost Manual*, 4th Ed., W.M. Vatavuk, Ed., US EPA, Research Triangle Park, NC, **1990** (and subsequent amendments); Chapter 4, available at http://www.epa.gov/ttn/catc.
26. J. Laznow; A. Patkar, "Gaseous Toxic Air Pollutant Control Technologies," in *Toxic Air Pollution Handbook*, D.R. Patrick, Ed., Van Nostrand Reinhold, New York, **1994**, Ch. 21, pp. 373–397.
27. G.E. Keller, "Adsorption: Building Upon a Solid Foundation," *Chem. Eng. Prog.*, Oct. **1995**, pp. 56–68.
28. A. Turk, "Adsorption," in *Air Pollution, Third Edition, Volume IV: Engineering Control of Air Pollution*, A.C. Stern, Ed., Academic Press, New York, **1977**, Ch. 8, pp. 329–363.
29. G.T. Joseph; D.S. Beachler, *APTI Course 415: Control of Gaseous Emissions: Student Workbook*, EPA, Research Triangle Park, NC, **1981**.
30. C.D. Cooper; F.C. Alley, *Air Pollution Control: A Design Approach*, 2nd Ed., Waveland Press, Prospect Heights, IL, **1994**, Ch. 12, pp. 379–409.
31. J. Joseph; D. Beachler; J. Jahnke; M. Leslie; J. Henry, *APTI Course 415 Control of Gaseous Emissions: Instructor's Guide*, US EPA, Research Triangle Park, NC, **1981**.
32. *Adsorption Handbook*, Calgon Corporation, Pittsburgh, PA, **1980**.
33. M.D. LeVan; G. Carta; C.M. Yon, "Adsorption and Ion Exchange," in *Perry's Chemical Engineers' Handbook*, 7th Ed., D.W. Green; J.O. Maloney, Eds., McGraw-Hill, New York, **1997**, Sect. 16, p.16-1–16-66.
34. L. Theodore; A.J. Buonicore, *Air Pollution Control Equipment, Volume II: Gases*, CRC Press, Boca Raton, **1988**, Ch. 10, pp. 42–77.
35. D.M. Ruthven, "Adsorption," in *Kirk-Othmer Encyclopedia of Chemical Technology*, 4th Ed., Vol. I, *A to Alkoids*, J.I. Kroschwitz; M. Howe-Grant, Eds., John Wiley & Sons, New York, **1991**, pp. 493–528.
36. K. Noll; W.J. Franek; L. Stander, *APTI Course 482: Sources and Control of Volatile Organic Pollutants: Instructor's Guide*, US EPA, Research Triangle Park, NC, **1987**.

5

ABSORPTION

5.1 TECHNOLOGY DESCRIPTION

Absorption (also known as *scrubbing*) is a process in which pollutants in a contaminated gas stream are dissolved into a solvent liquid stream by mass transfer [1]. The driving force is the difference in the partial pressure of the pollutant in the gas stream and the vapor pressure (or solubility) of the same component in the liquid. A low vapor pressure of the component in the liquid stream indicates that the component is less likely to come out of the liquid and is therefore soluble [2]. Because this equilibrium establishes the effectiveness and size of the absorption equipment, it is very important that a solvent be chosen in which the organic contaminant is as soluble as possible. The size of absorption systems can also be minimized by using a liquid that results in a reaction with the pollutant. This process, called *chemical absorption*, reduces the concentration of a pollutant in the liquid and drives the absorption toward completion [3]. The process that does not use a chemical reaction but relies on the solubility to remove the pollutant is called *physical absorption*.

Absorption systems are capable of removal efficiencies greater than 98% [3]. Because of their high efficiencies, they are frequently used to remove odorous compounds from gaseous streams. Applications for indoor air treatment have also been described in the literature [4]. Water is commonly used as the scrubbing liquid [5], or *sorbent*. For example, scrubbers are used at rendering plants to remove contaminants in the plant exhaust before it is released into the atmosphere [6]. In some cases, seawater, chlorine, hypochlorite, hydrochloric acid, sodium hydroxide, or sodium bicarbonate is added to enhance the scrubbing efficiency [7, 8]. Absorption systems are most suited to plants that already have an oversized wastewater treatment plant or to pollutants that can be removed by

TABLE 5.1 Typical Applications of Absorption

Odor control in rendering plants or fish processing
Light hydrocarbon absorption at petroleum refineries
Plating operations
Sewage treatment odor removal
Organic binder emissions from foundries

Data adapted from refs. 7, 10, 11.

simple chemical reaction with little effect on the chemical composition of the liquid stream. Simple absorber designs are also useful for streams that contain large amounts of particulate because the particulate can more easily foul a packed bed. It not usually possible to optimize removal of both particulate and gaseous components in the same vessel [9]. Table 5.1 lists some typical applications of absorption systems for air pollution control [7, 10, 11].

There are some significant disadvantages to absorption systems. The primary disadvantage is that absorption usually eliminates a problem with a contaminated gas stream only to create a contaminated liquid stream. The liquid must then be treated either to recover the pollutant and reuse the liquid or to dispose of the spent solvent.

5.2 TYPES OF ABSORPTION SYSTEMS

There are many types of absorber designs, each with unique methods of increasing the contact between the liquid and vapor phases. Some are vertical columns with packing (packed towers) or trays (sieve plate, bubble cap, float valve, or impingement plate columns, all similar to those used for distillation or extraction columns), and others are horizontal chambers with packing (cross-flow), open vessels (spray chambers), or agitated tanks with sparge pipes [12]. Packed columns can be further divided according to the direction of liquid flow: countercurrent or co-current.

A typical flow diagram of a countercurrent packed tower absorber is shown in Figure 5.1. Countercurrent flow columns are similar to distillation columns; the vapor flows upward while the liquid flows down through the column. This provides for the longest contact time and therefore the highest efficiencies. The columns are prone to fouling, however, and have higher pressure drops than columns in which the gas and liquid flow in the same direction [14]. Even with pressure drops approximately twice that of a cross-flow absorber [15], the countercurrent packed tower is the most common type of absorber used in air pollution control applications because it is efficient, versatile, has a broad range of operability with relatively low pressure drop, and can be designed to handle corrosive fluids relatively inexpensively [2, 14, 16]. In this type of system, the vapor and liquid flow in opposite directions so that the treated vapor exiting the absorber contacts the

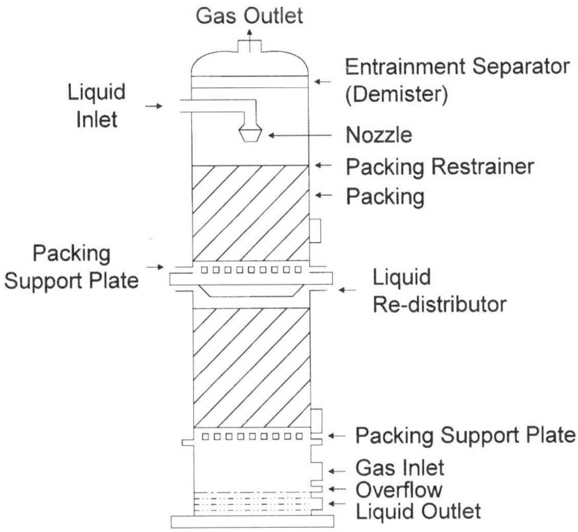

Figure 5.1 Countercurrent packed tower absorber. (Adapted from Theodore and Buonicore [13].)

cleanest liquid [17]. This results in the greatest concentration difference between the vapor and liquid states that can be achieved in an absorber.

The primary component in the system is a vertical cylindrical tower that contains ceramic, plastic, or metal packing. The tower itself may be metal or reinforced plastic. The packing is supported by a plastic or metal plate, which allows for proper distribution of air. A "hold-down" plate or packing restrainer may also be at the top of the packing to keep packing in place during periods of high gas flow [17].

The liquid distribution system above the top of the packing consists of one or more spray nozzles, designed to spread liquid over the entire cross-sectional area of the tower as evenly as possible. Liquid is collected at the bottom of the tower and may be recirculated to the tower with a pump. There may also be a demister at the top of the tower to remove droplets of liquid from the gas stream before it exits the scrubber. Some tall towers contain more than one section of packing, each with its own support plate and liquid distribution system. Multiple sections are used to reduce liquid channeling that can occur if the packing gets too high (above 3 m or 10 ft as a general guide) [18].

In addition to these components, it may be advantageous to have a particulate collector upstream of a packed tower scrubber to eliminate potential problems of poor liquid or gas distribution and elevated pressure drop, which can be caused by buildup of particulate in the column. Packed columns can typically accommodate approximately 11.4 g/m^3 (5 grains/ft^3 where 7000 grains equal 1 lb) of particulate if the particulate dissolves in the scrubbing liquor [19].

Countercurrent columns are designed to ensure that there is plenty of space for the gas stream and liquid stream to pass in the column. As the gas stream velocity increases through the column, it reaches a point where the liquid is actually prevented from flowing down the column in any significant amount. Instead, the liquid builds up in the column. This phenomenon is referred to as *flooding* and must be avoided to ensure safe and effective column operation.

In some countercurrent packed columns, a lightweight packing is used that is partially or completely fluidized by the gas stream. This type of scrubber is referred to as a *mobile packing* scrubber. In each stage, the plastic packing spheres are contained between two grids approximately 0.6–1.5 m (2–5 ft) apart. For difficult separations, as many as five or more stages may be used. The movement of the spheres allows for solids to be flushed through the system and also permits the use of more viscous scrubbing liquids [15]. At air flows of 120–150 m/min (400–500 ft/min), the design has air-flow capacities approximately five to six times that of a conventional packed or plate scrubber [7]. The associated pressure drop is approximately 0.5–0.75 kPa (2–3 in of water) per stage [7].

Countercurrent columns may also have sieve plate or bubble cap trays instead of packing to provide for gas–liquid contact. Scrubbers containing sieve plate or bubble cap trays generally have a more narrow operating range than packed towers [20]. Sieve plates are horizontal disks that fit tightly in the tower. The liquid and gas pass through holes (typically 3 mm or 1/8 in diameter, placed about 10 mm or 3/8 in apart [21]) in the plates. The column may also have downcomers to allow the liquid to flow from plate to plate more easily. The plates are typically placed about 0.3 m (12 in) apart if there are no downcomers and 0.38 m (15 in) apart if there are downcomers [18]. Sieve plates can be placed closer together in the column than bubble cap trays because sieve plates normally only have about one third as much entrainment [22].

A bubble cap tray has a more complex structure that collects the gas coming up through a tray and forces it back down into the liquid to establish good gas–liquid contact. Bubble cap trays are usually set about 0.46 m (18 in) apart [22]. It is important to maintain an even liquid level on both sieve plates and bubble cap trays so that there is equal resistance to gas flow across the entire cross-sectional area of the absorber. Therefore, the plates should be level, and the bubble cap columns should have an adequate number of downcomers to ensure that the liquid head across each plate is even [21]. Typical pressure drops through these columns are approximately 0.5–1 kPa (2–4 in of water column) per plate or tray [16]. These absorbers are better suited to processes in which heat is evolved during solution of the pollutant in the liquid (e.g., ammonia in water) because they can be designed with cooling coils between the plates. They are also better suited to gas streams with fouling tendencies or particulate matter because they can be cleaned easier [23]. They are less suitable than packed columns, however, for foaming liquids [12]. Additional information comparing high-performance trays with structured packings is available [24].

In co-current columns, the gas and liquid flow in the same direction, usually downward through a packed column [14]. The amount of contaminant that can be removed from the vapor stream with this method is limited because there is no supply of fresh liquid toward the end of the contact time to drive the equilibrium further toward removal of the contaminant from the gas stream. These vessels, however, can handle more gas flow than countercurrent columns because increased gas flow facilitates liquid flow through the column, and there is no possibility of flooding [25].

The gas in cross-flow vessels flows horizontally while the liquid flows downward through the packing, as shown in Figure 5.2 [23]. It has both a low pressure drop and improved solids handling capability. Cross-flow vessels are currently used for pollutants with moderate- to high-solubility [14]. This type of scrubber effectively removes particulates 5 μm and larger from the gas stream [25].

Baffle towers are not common, but are a type of scrubber with simple contact mechanisms between the gas and liquid. The liquid cascades down through the column, flowing from one plate to the other as shown in Figure 5.3 [19]. The gas must pass through the liquid cascade as it makes its way up the column. Very little information is available on the design of these types of columns [19].

Some equipment that has historically been used to remove particulate from gas streams is also occasionally used to remove very soluble organic compounds. Spray chambers may be vertical vessels (as shown in Fig. 5.4) or horizontal vessels that contain multiple levels of spray nozzles to provide a liquid mist or fog through which the gas passes. Figure 5.5 shows a venturi scrubber, which operates by injecting liquid into the gas stream. The liquid may be injected at a venturi, or it may be injected upstream of the venturi. The difference in velocity between the gas stream and the liquid droplets provides for contact between the phases. Although venturi scrubbers are well suited for gaseous streams that contain a significant amount of particulate, they do not provide for long contact times and are therefore only suitable for use with very soluble compounds [26]. The venturi scrubber must be followed by a gas–liquid separation device; most installations have a cyclonic separator in which liquid droplets are removed by inertia.

Agitated tanks are also sometimes used as absorption vessels when the gas stream contains significant amounts of particulate [12]. The contact between the gas and liquid is enhanced by an agitator and baffle plates in the tank. Centrifugal

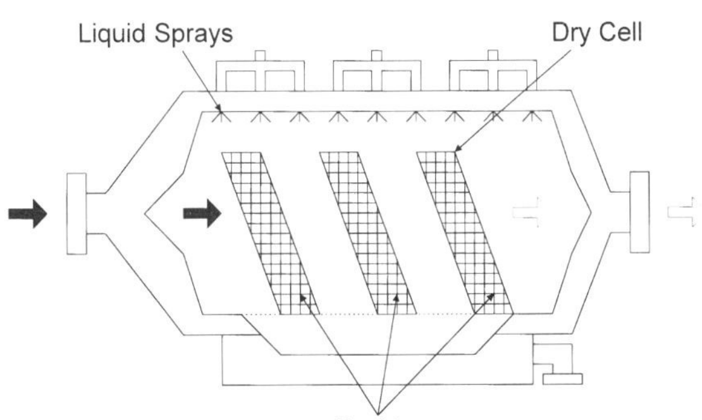

Figure 5.2 Cross-flow absorber [23].

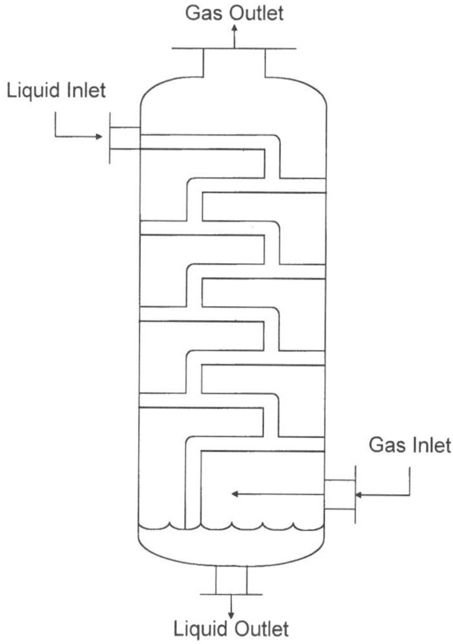

Figure 5.3 Baffle tower [19].

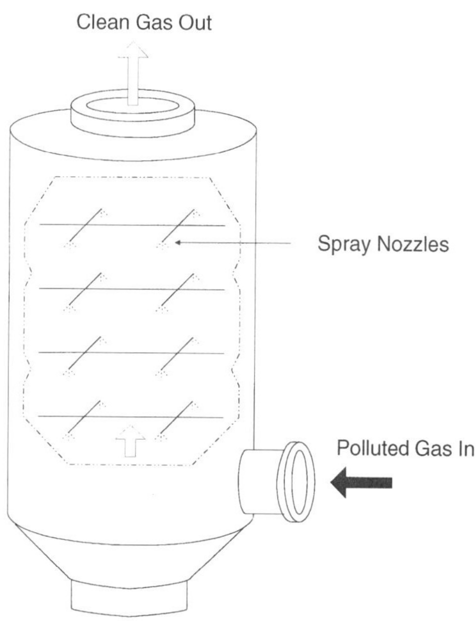

Figure 5.4 Vertical spray chamber. (Adapted from Joseph and Beachler [23].)

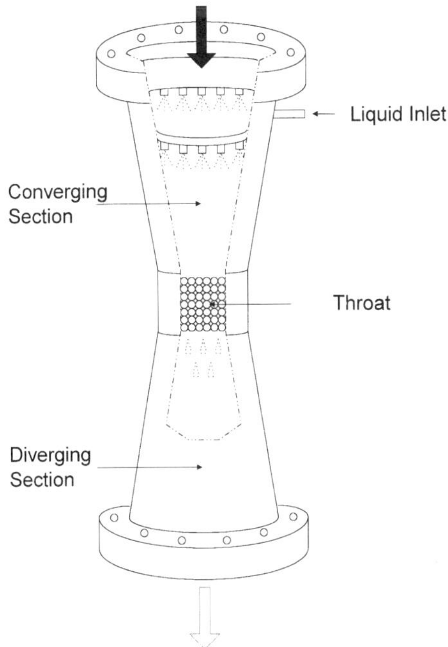

Figure 5.5 Venturi scrubber. (Redrawn from Joseph and Beachler [23].)

absorbers, such as the one shown in Figure 5.6, have been used in Europe to remove contaminants from gas streams. These units contain a disk in the middle that spins at high velocity. The rotation of the disk forces liquid to spray into the gas stream, resulting in contact between the vapor and liquid gas streams [19].

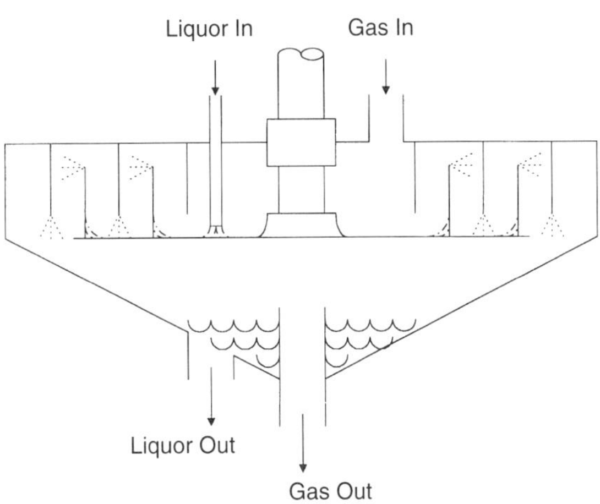

Figure 5.6 Centrifugal absorber. (Redrawn from Crocker and Schnelle [19].)

5.3 ABSORPTION SYSTEM DESIGN

The design methods given here are for packed bed absorbers because most scrubbers are packed beds. They are included as general guides to facilitate understanding of absorption systems, to enable quick cost estimates, and to assist in identifying problems with existing absorption systems. Detailed design considerations are given in *Perry's Chemical Engineers' Handbook* [1, 27], *Indoor Air Quality and Control* [4], *Air Pollution Control Equipment*, Volume II [13], *Air Pollution Control and Design Handbook*, Part II [8], *Industrial Control Equipment for Gaseous Pollutants*, Volume I [22], *Odor and VOC Control Handbook* [28–30], *Kirk-Othmer Encyclopedia of Chemical Technology* [31], and the *OAQPS Control Cost Manual* [17], available on the Internet at http://www.epa.gov/ttn/catc/products.html. Additional information useful in estimating tray efficiencies can be found in *Bubble Tray Design Manual* [32] and *Equilibrium-Stage Separation Operations in Chemical Engineering* [33]. Additional information on designing cross-flow units can be found in "The Crossflow Scrubber—A Digital Model for Absorption" [34]. Formulas for calculating removal efficiencies of particulate in spray towers are given by Hesketh [35]. It is also advantageous to consult manufacturers in the design of any absorption system because they can frequently offer specific experience with similar applications.

The efficiency of a given absorber is usually predicted using Henry's law, which relates the concentration of a compound in the liquid stream to the concentration (or partial pressure) of the compound in the gas stream at equilibrium [23]:

$$p_A = Hx_A \tag{5.1}$$

where

- A = pollutant to be absorbed
- p_A = partial pressure of A in the vapor stream (molar concentration, mol/L or M)
- X_A = concentration of A in the liquid (molar concentration, mol/L or M)
- H = Henry's law constant (molar concentration in gas/molar concentration in liquid, M/M)

Several Henry's law coefficients are given in Table 5.2 [36]. Additional information is given by Schwartz and White [37] and Martin [38]. A more comprehensive list of values is contained in *Sources and Control of Air Pollution* [39] and on the Internet at http://chemfinder.camsoft.com. Various equations for Henry's law are used. When using Henry's law coefficients from other sources, care must be exercised to identify the units [21]. Some texts define a Henry's law coefficient that is the inverse of the one given here. Henry's law applies when the partial pressure of the pollutant does not exceed one atmosphere of pressure [40] and when compounds do not react during absorption. The coefficients are temperature dependent; they may be adjusted by van't Hoff's

TABLE 5.2 Henry's Law Coefficients

Pollutant	Solvent	H, M (gas)/M (liquid)	Temp. (K)
Acetylene (C_2H_2)	Water	0.823	288
Ammonia (NH_3)	Water	4.71×10^{-4}	288
Ethane (C_2H_6)	Water	17.2	288
Ethylene (C_2H_4)	Water	6.82	288
Hydrogen chloride (HCl)	Water	2.82×10^{-9}	288
Hydrogen sulfide (H_2S)	Water	0.322	288
Methane	Water	25.7	288
Ozone (O_3)	Water	2.19	298
Propylene (C_3H_6)	Water	3.86	288
Sulfur dioxide (SO_2)	Water	7.85×10^{-3}	288

From Peters [36].

equation [41, 42] which simplifies to

$$\ln \frac{H(T_2)}{H(T_1)} = \frac{\Delta H}{R}\left(\frac{1}{T_1} - \frac{1}{T_2}\right) \tag{5.2}$$

where

ΔH = heat of solution of the gaseous pollutant, kJ/kmol
R = gas constant, 8.314 kJ/kmol K
$H(T_1)$ and $H(T_2)$ = Henry's law coefficient at two different temperatures,
T_1 and T_2 = absolute temperatures

This approximation is only valid when ΔH is relatively constant, that is, over relatively small temperature ranges.

Lower Henry's law coefficients indicate improved solubility because the pollutant is less likely to evaporate into the gas stream. The absorption efficiency is enhanced if the pollutant reacts with the liquid, thereby decreasing the concentration of pollutant in the liquid and driving the absorption reaction toward completion. For this reason, weak acid solutions, caustic solutions, and ozone are sometimes used in scrubbers to improve the removal of organic compounds from a gas stream.

The selection of solvent is generally the first step in absorption design [13, 27]. Other important considerations in determining an appropriate solvent are [7, 13, 19, 26, 43] as follows:

1. Solubility. The pollutant should be as soluble as possible in the liquid at the expected operating temperature to enhance the mass transfer rate [44].
2. Low volatility. The solvent should generally have a volatility less than approximately 0.1 mm Hg at the outlet temperature to avoid excessive evaporation of the solvent from the column [19]. Evaporation of the solvent

can cause unnecessary operating expense, in addition to creating a different air pollution problem. The prominent exception to this is water, which is used in most absorption systems [5] because evaporation of water is not generally a significant cost or pollution issue, and other design considerations such as solubility and disposal outweigh the concerns.

3. Corrosivity. The solvent selected should have a low corrosivity if possible so that less expensive materials of construction can be utilized. In addition, the corrosivity of the pollutant and any reaction products between the pollutant and the solvent should be considered when determining appropriate construction materials.
4. Cost. Cost is an obvious concern for systems in which the liquid is only used once. In some absorption systems, the solvent is continuously regenerated and reused by distilling the pollutant, by reacting it to form a removable solid, or by liquid–liquid extraction. Even though the liquid is recycled, there are inevitably losses through equipment leaks, cleaning, equipment blowdown, and degradation.
5. Availability. The liquid should be easy to obtain in a relatively short time period so that the equipment is not out of service if a column is accidently released to a sewer or in case of a malfunction. For solvents that may have a long delivery time, it may be feasible to use an additional storage tank to ensure that there is always an adequate supply available.
6. Safety. To simplify the operating procedures and to protect workers, the liquid should have as low a toxicity and flammability as possible.
7. Low viscosity. A low viscosity reduces electrical demands for pumping the liquid, reduces the column pressure drop (which results in lower electrical demands for moving the gas stream), decreases the likelihood of exceeding the capacity of the column during periods of high gas flow (flooding), and increases the efficiency of the column by improving liquid distribution.
8. Stability. The liquid should not degrade with use or expected temperature.
9. Low melting point. A low melting point is desirable to keep the column from freezing in cold weather, particularly if the column is taken out of service.

Mineral oil, heavy hydrocarbon oils, and aqueous solutions are also used as sorbents in absorption systems [3, 45].

Before the overall size of the absorber can be determined, the packing material must be identified. The packing material will affect the degree of contact that can be achieved in the scrubber, the weight of the scrubber, and the materials of construction. Packing may be "dumped" or structured. Despite the name, "dumped" packing is not randomly thrown into the absorber. It must be installed carefully to avoid breaking the packing and to ensure that it is spread uniformly. Structured packing is generally stacked in the vessel. In addition to sophisticated dumped or structured packing material, simple materials such as broken rock, coke, broken bottles, aggregate, marbles, and wood slats have also been used

[8, 9, 15, 21]. The selected packing should provide for maximum amount of wetted area per volume of packing, reasonable weight, chemical resistance to the gas stream and liquid stream constituents, good distribution of liquid and gas phases, minimal amount of liquid retained on the packing (holdup), low gas pressure drop through the packing, structural strength, design flexibility (to allow changes in operation without substantial effects on the removal efficiency), and low cost [21, 46]. The chosen packing should have a diameter less than about 1/30th of the column diameter [2], and it should not force the gas stream to contract or expand because that would increase the pressure drop through the column unnecessarily [8].

The packing supports must have the physical strength to support the weight of the packing and liquid, they must be able to withstand the expected temperature and corrosivity of the gas and liquid streams, and they must provide adequate passage of the gas and liquid so that there is not a significant pressure drop. The support plates may be as simple as a grid or expanded metal plate, or they may be designed to separate the gas and liquid flows as in "gas-injection" support plates (Fig. 5.7) [8].

An estimate of the absorber size can be obtained with U.S. Stoneware's correlation [9, 22, 47] once the liquid and gas mass flow rates have been determined. It is assumed that the gas flow rate, initial concentration, and required removal efficiencies are specified by the application. The concentration of pollutant in the liquid stream and exiting the column can then be determined from the Henry's law correlation. It is usually assumed that the concentration of pollutant in the entering liquid stream is zero, although this can also be estimated from Henry's law if the liquid is reused after recovery of the pollutant. Furthermore, the required

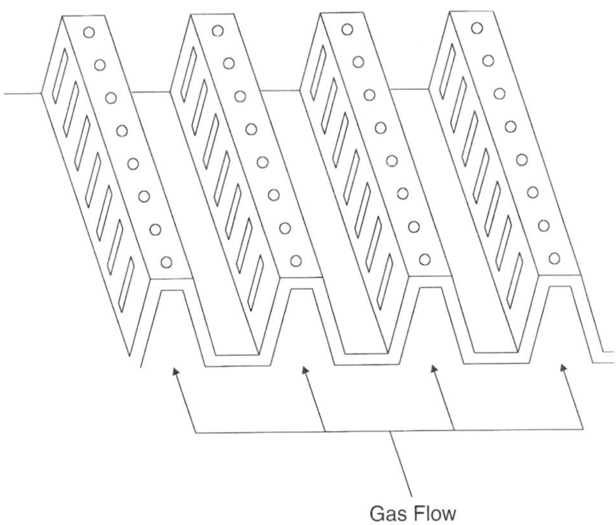

Figure 5.7 "Gas-injection" support plate. (Adapted from Bhatia [8].)

liquid flow rate can be calculated by an overall mass balance from the overall amount of pollutant that must be removed and the final concentration of that pollutant in the gas stream. Assuming that the pollutant is the only constituent of the gas stream that is soluble in the liquid and that the solvent is nonvolatile, the minimum liquid to gas ratio can be determined from the following equations [17]. If the concentration of pollutant is more than a few percent of the volume, these assumptions are no longer valid.

$$Y_o = Y_i(1 - \eta/100) \tag{5.3}$$

where

Y_o = pollutant concentration in gas stream leaving absorber, kg-mol pollutant/kg-mol pollutant-free gas (lb-mol/lb-mol)
Y_i = pollutant concentration in gas stream entering absorber, kg-mol pollutant/kg-mol pollutant-free gas (lb-mol/lb-mol)
η = removal efficiency, %
X_o^e = maximum concentration of pollutant in the liquid stream leaving the absorber, assuming that the liquid is in equilibrium with the gas stream entering the absorber, kg-mol pollutant/kg-mol pollutant-free solvent (lb-mol/lb-mol)
X_i = concentration of pollutant in liquid stream entering the absorber kg-mol pollutant/kg-mol pollutant-free solvent (lb-mol/lb-mol). Henry's law is generally valid for determining this value for small concentrations of pollutant.

The minimum liquid to gas ratio is then

$$\left(\frac{L}{G}\right)_{min} = \frac{Y_i - Y_o}{X_o^e - X_i} \tag{5.4}$$

and the actual liquid to gas ratio can be calculated from

$$\left(\frac{L}{G}\right)_{act} = \left(\frac{L}{G}\right)_{min} AF \tag{5.5}$$

where

L = molar flow rate of pollutant-free solvent, kg-mol/h (lb-mol/h)
G = molar flow rate of pollutant-free gas, kg-mol/h (lb-mol/h)
AF = adjustment factor to compensate for the fact that true equilibrium is not realistic, typically between 1.2 and 1.5 [48]

Once the pollutant-free liquid and gas flows are known, the total molar flow-rates of the liquid and gas streams are calculated by the following equations:

$$G_{mol,T} = G(1 + Y_i) \tag{5.6}$$

$$L_{mol,T} = L(1 + X_i) \tag{5.7}$$

Then the cross-sectional area that is required for the expected gas and liquid flow rates is determined. The value of the following expression is calculated and located on the x-axis of the chart given in Figure 5.8.

$$\frac{L_m}{G_m}\left(\frac{\rho_G}{\rho_L}\right)^{0.5} \tag{5.8}$$

where

L_m = liquid flow rate, mass/h units
G_m = gaseous flow rate, mass/h units; must correspond with those for the liquid flow rate [9]
ρ_G = density of gas stream, 1.17 kg/m^3 (0.073 lb/ft^3) for air at 303 K (30°C, 86°F) [23]
ρ_L = density of liquid stream, 1000 kg/m^3 (62.4 lb/ft^3) for water at 303 K (30°C, 86°F); units must correspond to those of ρ_G

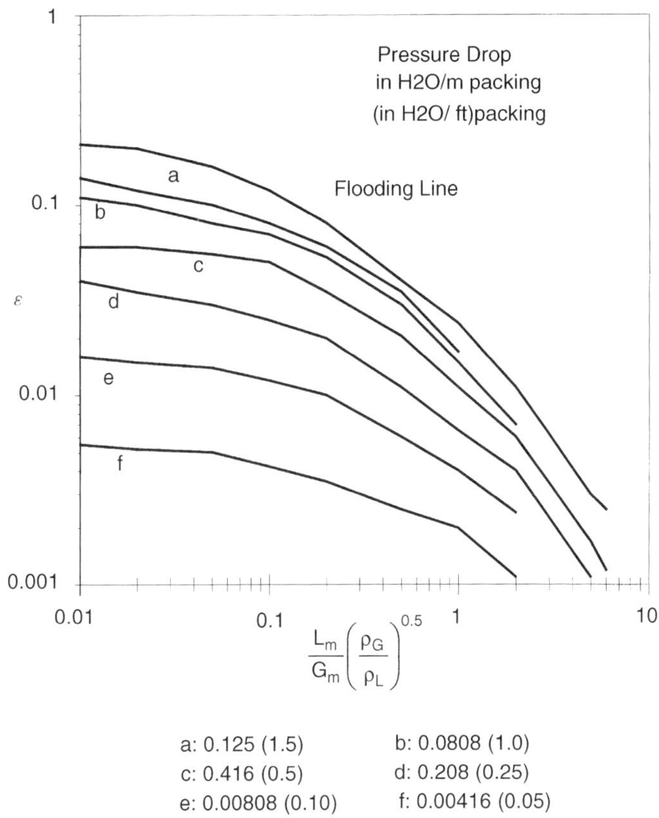

Figure 5.8 U.S. Stoneware's pressure drop correlation (Redrawn from Joseph and Beachler [23].)

The point on the flooding line in Figure 5.8 is then located that corresponds to the abscissa value calculated above. The value on the y-axis that results is ε. With this value and the packing factors listed in Table 5.3, the following equation can be solved for G'_{flood} [23]:

With a packing factor in SI units:

$$G'_{flood} = \frac{1.81\left[\varepsilon \rho_G \rho_L g_c\right]^{0.5}}{\left(F\phi\mu_L^{0.2}\right)^{0.5}} \tag{5.9}$$

With a packing factor in English units:

$$G'_{flood} = \frac{\left[\varepsilon \rho_G \rho_L g_c\right]^{0.5}}{\left(F\phi\mu_L^{0.2}\right)^{0.5}} \tag{5.10}$$

where

G'_{flood} = the mass flow rate of gas per unit of column cross-sectional area that will just cause the column to flood, g s^{-1} m^{-2} (lb $s^{-1}ft^{-2}$)
ε = the y-axis value from Figure 5.8
g_c = the gravitational constant, 9.82 m/s^2 (32.2 ft/s^2)
F = packing factor, m^2/m^3 (ft^2/ft^3)
ϕ = ratio of liquid specific gravity to that of water
μ_L = liquid viscosity, cP, 0.8 cP for water

It is usually assumed that the column will operate at a gas flow between 50 and 75% of the flooding velocity [13, 22], although well-designed columns can operate at flows up to 90% of flooding [19]. The gas flow per unit area through the column under normal operating conditions can be estimated from the following equation [9]:

$$G'_{oper} = fG'_{flood} \tag{5.11}$$

where

G'_{oper} = mass flow of gas stream per unit area of column under normal operation, g $s^{-1}m^{-2}$ (lb $s^{-1}ft^{-2}$)
f = acceptable fraction of flooding for normal operation, typically 0.6 [2, 22]

The required column cross-sectional area can be obtained by [9]:

$$A = G/G'_{oper} \tag{5.12}$$

where

A = Column cross-sectional area, m^2 (ft^2)

TABLE 5.3 Packing Data

Packing	Size (in)	Weight		Surface Area		Void Fraction	Packing Factor, F		Price (1990 $)	
		lb/ft^3	kg/m^3	ft^2/ft^3	m^2/m^3	(%)	ft^2/ft^3	m^2/m^3	$/ft^3	$/m^3
Raschig rings, ceramic or porcelain	$\frac{1}{2}$	52	833	114	374	65	580	1903		
	1	44	705	58	190	70	155	509	12.8	452
	$1\frac{1}{2}$	42	673	36	118	72	95	312	10.3	364
	2	38	609	28	92	75	65	213	9.4	332
	3	34	545	19	62	77	37	121	7.8	275
Raschig rings, carbon steel	$\frac{1}{2}$ × 1/32	77	1233	128	420	84	300	984		
	1 × 1/32	40	641	63	207	92	115	377	36.5	1289
	2 × 1/16	38	609	31	102	92	57	187	20.5	724
Raschig rings, stainless steel	1								155	5474
	1.5								117	4132
	2								87.8	3101
Raschig rings, carbon	1								52.0	1836
	1.5								46.2	1631
	2								33.9	1197
	3								31.0	1095

(continued)

TABLE 5.3 continued

Packing	Size (in)	Weight		Surface Area		Void Fraction	Packing Factor, F		Price (1990 $)	
		lb/ft^3	kg/m^3	ft^2/ft^3	m^2/m^3	(%)	ft^2/ft^3	m^2/m^3	$/ft^3	$/m^3
Berl saddles, Ceramic or porcelain	$\frac{1}{4}$	55	881	274	899	63	900	2953		
Prices given for porcelain	$\frac{1}{2}$	54	865	155	509	64	240	787		
	1	48	769	79	259	68	110	361	33.5	1183
	1.5	40	641	46	151	71	65	213	21.5	759
	2	38	609	32	105	75	45	148	15.6	551
Intalox saddles, Ceramic	$\frac{1}{4}$	54	865	300	984	75	725	2379		
	$\frac{1}{2}$	45	721	190	623	78	200	656		
	1	44	705	78	256	77	98	322	17.6	622
	1.5	39	625	59	194	76	52	171	13.0	459
	2	42	673	36	118	79	40	131	11.8	417
	3	36	577	28	92	79	22	72	10.7	378
Intalox saddles, plastic	1	6.00	96	63	207	91	30	98	21.2	749
Prices given for polypropylene	2	3.75	60	33	108	93	20	66	13.1	463
	3	3.25	52	27	89	94	15	49	7.0	247

Pall rings (plastic)	5/8	7.0	112	104	341	87	97	318		
Prices given for polypropylene	1	5.5	88	63	207	90	52	171	21.2	749
	1.5	4.8	77	39	128	91	40	131	14.4	509
	2	4.5	72	31	102	92	25	82	13.1	463
Pall rings steel	5/8 × 0.018 thick	38	609	104	341	93	73	239		
	1	30	481	63	207	94	48	157		
	$1\frac{1}{2}$ × 0.03 thick	24	384	39	128	95	28	92		
	2	22	352	31	102	96	20	66		
Tellerettes[a]	1	7.5	120	55	180	87	40	131		
	2	3.9	62	38	125	93	20	66	11.5	406
	3	5	80	30	98	92	15	49		

[a]Tellerette pricing given is an average of prices from \$100 to 700/m^3 (\$3 to 20/ft^3).
Data adapted from refs. 17, 23, 46.

Because absorption columns are usually round, the diameter is calculated from the formula for the cross-sectional area of a circle [49]:

$$D = \sqrt{\frac{4A}{\pi}} \tag{5.13}$$

where

D = column diameter, m (ft)

The calculated column diameter should be rounded up to a readily available diameter. Increments of 0.15 m (or $\frac{1}{2}$ foot) are common.

Typical liquid flow rates are approximately 0.04 to 2 m^3/min per square meter (1–50 gal/min/ft^2) of column area. Superficial gas velocity through a packed tower is generally about 30–150 m/min (100–500 ft/min) [15].

For extremely soluble pollutants or pollutants that react with the scrubbing liquid, the number of theoretical stages (N_{OG}) can be estimated from [9, 17]:

$$N_{OG} = \ln(Y_1/Y_2) \tag{5.14}$$

where

Y_1 = mole fraction of pollutant in the entering gas
Y_2 = mole fraction of pollutant in the exiting gas

The height of a theoretical transfer unit (H_{OG}) should be obtained from vendor information for the specific application, but it usually ranges between 0.3 and 1.2 m (1 and 4 ft). A rough estimate of 0.6 m (2 ft) can be used if necessary [9]. The total packing height is calculated from

$$H_{pack} = N_{OG} \cdot H_{OG} \tag{5.15}$$

It may be necessary to install the packing in sections, each with a separate packing support plate and liquid distribution system in tall towers. Generally, packing is installed in sections 3 m (10 ft) high. For columns packed with Raschig rings, the maximum depth of each section should be no more than approximately three column diameters, to a maximum of 3–5 m (10–15 ft). For saddle packings, the depth of each section should be no more than approximately 5–10 column diameters, to a maximum of 4–6 m (12–20 ft) [13].

Other important factors in designing an absorption system are to ensure that the liquid and gases are distributed evenly throughout the column. Otherwise, gas channeling can result, causing areas of low gas or liquid flow, which cause losses in removal efficiency.

Two types of demisters are shown in Figure 5.9. While chevrons are frequently used, particularly in dirty gas streams, they are usually considered less efficient in removing small particles than are mesh pads. The very small droplets (less than 5 μm in diameter) are not captured as effectively as they are in a mesh pad, increasing the likelihood that water droplets will escape the collector. Demisters

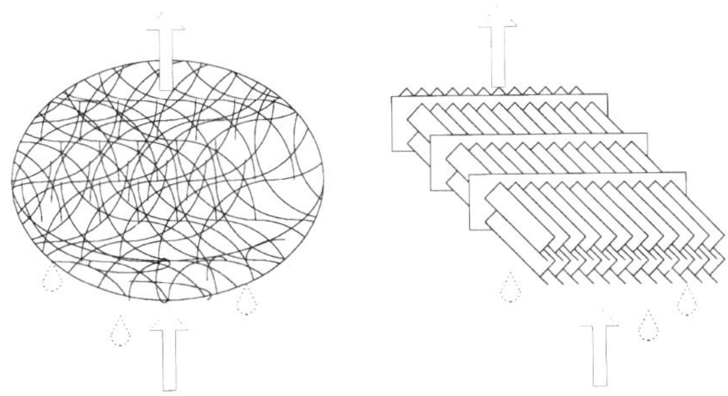

Figure 5.9 Two types of demisters. (Adapted from Joseph and Beachler [23].)

may be installed in horizontal or vertical piping; horizontal is preferred [21]. Figure 5.10 [8] gives a graphical representation of the range of droplet sizes and velocities that result in effective separation in a mesh pad.

New developments in absorption techniques include cryogenic absorption [45], ultraviolet enhancement, and combinations of scrubbing columns and catalytic beds designed to destroy odorous compounds. In the Odorgard process from ICI Katalco, compounds are absorbed into a dilute hypochlorite solution, and then the liquid is passed over a catalyst to oxidize the pollutants. The process is designed to reduce the amount of free chlorine emitted from scrubbers and to achieve lower outlet concentrations of the pollutants [50]. Caternary grid scrubbers use suspended webs of wire mesh that are suspended from the column sides and fall in a hyperbolic shape. One design includes an ultraviolet lamp and hydrogen peroxide to destroy methanol in the wastewater from paper production [51]. Table 5.4 lists instrumentation that should be considered for absorption systems [21].

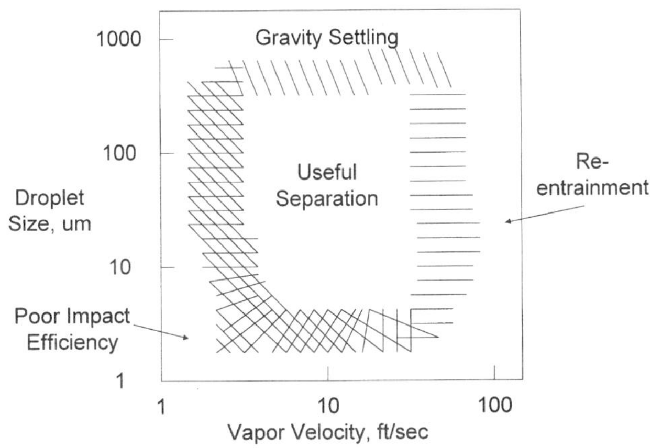

Figure 5.10 Range of droplet sizes and velocities for effective separation [8].

TABLE 5.4 Recommended Instrumentation

Gas stream temperature and pressure (inlet and outlet)
Liquid flow rate and pressure
Inlet gas stream relative humidity
Fan speed
pH of liquid effluent
Mist eliminator wash rate and frequency

Data adapted form Hesketh [21].

5.4 ECONOMIC ESTIMATES

One method of estimating the cost of an absorption system is given in the *OAQPS Control Cost Manual* and is considered to be accurate to within 30% [17]. The total column surface area is first calculated:

In SI Units:

$$S = \pi D(1.4H_{pack} + 1.52D + 0.856)$$

In English Units:

$$S = \pi D(1.4H_{pack} + 1.52D + 2.81) \qquad (5.16)$$

where

S = surface area of column, m^2 (or ft^2), valid for diameters between 0.6 and 3.7 m (2 and 12 ft) and packing heights between 1.2 and 3.7 m (4 and 12 ft)

Then the cost of the components of a fiberglass-reinforced plastic column (in third-quarter 1991 dollars) can be estimated by multiplying the surface area in m^2 (ft^2) by a factor of \$1240/m^2 (\$115/ft^2) [17]. This factor includes the cost of the column, inlet and outlet ports for gas and liquid, sump space, two liquid distributors, two packing supports, a mist eliminator, internal piping, and ladders and access platforms, but it does not include packing, instrumentation, fans, or pumps. This cost estimate is valid for columns with surface areas from 6.4 to 140 m^2 (69 to 1507 ft^2). For other materials of construction, the cost can be scaled by a factor of 0.8–1.1 for a polypropylene column, a factor of 0.5–0.9 for a polyvinyl chloride (PVC) column, and a factor of 1.1–1.75 for a stainless steel column (304 SS) [17]. This, as well as any other costs, should be scaled up to current value using procedures given in the EPA's "Escalation Indexes for Air Pollution Control Costs," which can be found on the Internet at http://www.epa.gov/ttn/catc [17].

In addition to the items included in the cost of the column, the cost of the packing, instrumentation, fans, and pumps must also be included in the total equipment cost. Approximate packing costs are given in Table 5.3. The cost of instrumentation is estimated at 10% of the cost of the column, packing, and auxiliary equipment, following the method in the *OAQPS Control Cost*

Manual. As noted there, however, the cost of instrumentation and controls varies depending on the complexity of the system [17].

General information on other types of scrubbers indicates that a bubble cap column costs about two thirds as much as a packed tower with ceramic rings for the same application using the same materials of construction. A sieve plate tower is about 45% of the cost of the packed tower [21]. The total installed cost of the system can be estimated from the factors in Table 5.5.

Annual costs may be approximated with the method given in the *OAQPS Cost Control Manual* [17], as shown in Table 5.6. Total electricity requirements can be estimated from the formulas for the fan and pump requirements. A method for calculating the electricity requirement for the fan is given in Chapter 4. The electricity required for the pumps can be estimated from the following equation:

$$E_{pump} = \frac{0.746 \cdot 0.252 L_i P_L}{eff_{pump}} \tag{5.17}$$

where

E_{pump} = electricity requirements for the liquid pump, kW
L_I = inlet liquid flow, gal/min
P_L = liquid pressure, ft of water
eff_{pump} = efficiency of the pump and motor combined, approximately 0.4–0.7

TABLE 5.5 Factors for Estimating the Total Installed Capital Cost of Absorption Systems

Direct Costs	
Purchased Equipment Cost (PEC)	
Absorber, packing, and auxiliary equipment	Equipment cost (EC), estimated from equation
Instrumentation and controls	0.1 · EC
Taxes	0.03 · EC
Freight	0.05 · EC
Total PEC	1.18 · EC
Installation costs	
Foundations and supports	0.12 PEC
Erection and handling	0.40 PEC
Electrical	0.01 PEC
Piping, insulation, and painting	0.32 PEC
Total installation cost	0.85 PEC
Site preparation (SP)	As required for specific site
Buildings (BLDG)	As required for specific site
TOTAL DIRECT COSTS	1.85 PEC + SP + BLDG
Indirect Costs	
Engineering and supervision	0.10 PEC
Construction, field expenses, and fees	0.20 PEC

(continued)

TABLE 5.5 continued

Indirect Costs continued	
Start-up and performance test	0.02 PEC
Contingency	0.03 PEC
TOTAL INDIRECT COSTS	0.35 PEC
Total Installed Capital Costs, TCC	2.2 PEC + SP + BLDG

Adapted from Barbour et al. [17].

TABLE 5.6 Annual Operating Costs for Absorber Systems

	Unit Cost	Estimated Requirement
	Direct Annual Costs	
Utilities		
Electricity	\$0.05/kW	Calculated Chapter 4 and Eqn. 5.17
Solvent	\$40–80/1000 m^3 \$0.15–0.30/1000 gallons	Application specific
Wastewater treatment or solvent disposal	\$1000/1000 m^3 \$3.80/1000 gal	Application specific
Chemical use for regenerating solvent	Application specific	Application specific
Operating labor	\$12.96/h	$\frac{1}{2}$ h per shift
Maintenance labor	\$14.26/h	$\frac{1}{2}$ h per shift
Maintenance materials	100% of Maintenance labor	
Supervisory labor	15% of Operating labor	
	Indirect Annual Costs	
Overhead	60% of total operating and maintenance costs	
Property tax	1% of TCC[a]	
Insurance	1% of TCC	
Administrative costs	2% of TCC	
Capital recovery [52, 53]	CRF · TCC where CRF = $[i(1+i)^m]/[(1+i)^m - 1]$ i = interest rate m = control system economic life CRF = 0.1315 for 15 year equipment life and 10% interest rate	
Total Annual Operating Costs	Direct + Indirect Operating Costs	

Adapted from Barbour et al. [17].
[a] Total installed capital costs.

Solvent requirements can be estimated from the liquid flowrates determined earlier. If the solvent is recycled, an estimate should be made of the amount of solvent that is evaporated during absorption or lost during the remaining processing steps. Operating and maintenance labor requirements vary depending on the complexity and dependability of the system, but can be estimated as one half hour per shift, following the suggestion in the *OAQPS Control Cost Manual*.

5.4 POTENTIAL PROBLEMS

The capital cost of the absorber depends on the volume of gas to be treated. Similarly, the efficiency of the absorption process depends on the concentration of the pollutant in the gas stream (the higher the concentration in the gas stream, the better the removal is expected). Therefore, it is critical that air infiltration be kept to a minimum by ensuring that seals in the ductwork remain airtight, eliminating unnecessary pickup points, maintaining valves and valve seats in excellent condition, and reducing the amount of air going into the capture system as much as possible. Care must be taken to ensure that the concentration in the ductwork does not exceed safe levels.

As with other gaseous control systems, corrosion can become a problem if the selection of construction materials is not deliberate. Because the absorption system saturates the gas stream with liquid, condensation in the remaining ductwork and fan can also lead to corrosion [18]. If a lining or plastic material is chosen to reduce corrosion, care must be taken to ensure that the temperature of the gas stream will not exceed the maximum allowable temperature of the material [54].

Even if corrosion is not anticipated as a problem, the gas stream characteristics may require other considerations for the materials of construction. For example, if the gas or liquid stream contains high levels of particulate, organic linings or construction materials may be particularly susceptible to erosion [55].

For the column to be as effective as possible, the contact between the gas phase and the liquid phase must be excellent. Clogged liquid distribution systems, plugged air holes in plates, and bypassing must be avoided. Proper instrumentation can assist in diagnosing such problems; a relatively small amount of money spent for additional pressure or flow measurements when the system is installed can save a significant amount of time in troubleshooting later.

The column should also be designed with adequate turndown ratios that correspond to typical production levels in the process so that the column operates as efficiently as possible under various sets of operating conditions. Adequate excess capacity should, however, be included in the initial design to ensure that the capacity of the column can be increased if necessary without increasing the physical size [7]. This can be done by either providing a flanged connection on the shell to allow for extra shell sections to be added or designing the shell can with extra height to allow for additional packing to be added [18]. If the shell is designed with extra height, the design should ensure that the liquid distribution system is never so high above the packing level that it makes the liquid distribution inefficient by allowing all of the liquid to reach the tower walls. Similarly, overpacking the column can result in poor distribution [56].

5.5 INSTALLATION, OPERATION, AND MAINTENANCE CONSIDERATIONS

During installation, it is critical that the tower be carefully leveled to ensure that the liquid and gas flow are not altered by a slight tilt. The liquid and gas lines should be traced to ensure that connections are correct and are leak tight, and rotation of the pumps should be checked. In lines that may have significant concentrations of solids, provisions may be needed for recirculation to prevent settling in the lines. Packing should be installed by careful stacking if it is structured packing. In the case of dumped packing, liquid should be added to the column to protect the packing from damage [13, 21]. Any damage to the packing could result in liquid channeling or in increased pressure drop on the gas stream. A week or more may be required for polypropylene packings to become "conditioned"; because of the molding process, the surface may not be immediately wettable [56].

Approximately 2 weeks after the scrubber is put into operation, it should be inspected for pluggage and to determine if the packing has settled [18] or shifted. If the packing has settled, packing should be added to the design level, taking care to ensure that the packing is evenly distributed over the bed [21].

The warm, humid environment in wet scrubbers can be a preferred location for bacterial growth. Excessive bacteria can result in a soap bubble-like appearance and clogging of the column internals or mist eliminators. It is frequently necessary to add bacteriostats to ensure that this problem does not occur. Before bacteriostats, are added, however, the Material Safety Data Sheet should be reviewed to make sure that the ingredients in the bacteriostat are compatible with the construction materials of the column.

Proper operation of mist eliminators ensures that liquid droplets do not escape from the scrubber. If the demisters become clogged, the gas flow through the remaining demister area may speed up and may carry over additional liquid with it. This possibility can be reduced in scrubbers that have significant particulate loading in the gas stream by installing nozzles above the demisters to wash them in place. Washing is typically done while the absorber is off line to ensure that the washing liquid is not entrained in the exhaust gas [18].

Effective pollutant removal depends on establishing intimate contact between the gas phase and the liquid phase. If the scrubber is operating perfectly, the gas stream that leaves the scrubber is completely saturated with solvent. For scrubbers that use water as the solvent, the gas stream leaving the scrubber contains the maximum amount of moisture that it can hold under the existing temperature conditions. Depending on gas stream temperatures, the gas stream exiting the scrubber may begin to cool as it travels through the stack and into atmosphere.

The gas cannot continue to hold all of the moisture at lower temperatures, so the vapor begins to condense.

Condensing liquid particles agglomerate into droplets. Therefore, on days when the outdoor temperature is cooler than the operating temperature of the scrubber, it is typical to have droplets of water in the gas stream exiting the stack. Droplets in the gas stream on days when the weather is warmer than the scrubber operating temperature indicate that there is carryover through the demisting system. Additional discussion of this phenomenon and potential methods of addressing it are given in the *Electric Utility Engineer's FGD Manual* [57]. While the referenced article specifically addresses flue gas desulfurization systems, most of the concepts apply to other types of wet scrubbers as well, with the exception that flue gas desulfurization systems usually use a lime or limestone slurry as the scrubbing liquid. This concept is also documented in the *Handbook of Air Pollution Control Engineering and Technology* [9], which states that "conceivably, during wintertime operation some moisture due to condensation may be observed and considered a problem with the entrainment separator. However, it should be recognized that this is not uncommon, since the air off an absorber [scrubber] is saturated with water vapor, and any difference or lowering of the temperature will cause condensation to occur." Other methods of reducing condensation in the stack exhaust is to cool the gas stream to below the ambient temperature or to heat it to approximately 27 K (27°C, 50°F) above saturation [15].

Carryover of droplets through the demisting system can be caused by several conditions:

1. Most packed bed scrubbers are designed for vertical velocities of approximately 1–2 m/s (3–6 ft/s) [58]. If the velocity of the gas stream approaches 2 m/s (6 ft/s) at the design flow rate, any clogging can cause the air to bypass the portions of the demisters that are clogging, increasing the linear velocity through the remainder of the demister. This can cause liquid to be entrained in the gas stream as it leaves the demisting zone.
2. If mesh pad demisters are used, they are typically installed on a slight angle to allow captured droplets to drain properly [23].
3. Velocities below approximately 1 m/s (3 ft/s) [7], or about 30% of the design velocity [21], can result in inefficient demister performance for droplets less than approximately 100 μm in diameter because the water droplets do not have sufficient inertial energy to impact on the scrubber demister material. Instead, they follow the streamlines of the gas flow and exit the scrubber with the gas.
4. Torn or misplaced mesh pads allow bypassing of the gas stream, resulting in entrained liquid exiting with the gas stream. It is possible for demisters to be chemically degraded by the constituents of the liquid or gaseous streams. In the case of polypropylene mesh pads, the polypropylene can be attacked by strong oxidizing acids, peroxide, gasoline, benzene, carbon tetrachloride, acetone, alcohols, or wet chlorine [48] streams. Chemical damage to plastic material can be verified by examining one of the old mesh pads to determine

the condition of the fiber. If either chemical or heat damage has occurred, the fiber should be brittle, maleable, or deformed.

5. Poorly functioning nozzles or liquid distribution systems can also result in carryover of liquid through the demister. The nozzles should be examined to identify small pieces of plastic or solids that can clog the distributors.

Increases and decreases in pressure drop through a column can indicate improper operation. Increases in pressure may be due to increased liquid flow, plugged or decomposing packing, plugging demister, blockages in the packing support plate or air distribution systems, or increased air flow through the scrubber [18]. Liquid and gas flow rates should be monitored during scrubber operation to detect any gradual changes in flow. To check for plugging, it is advantageous to take pressure drop readings without any liquid flow immediately after the packing and demisters are installed, cleaned, or replaced. Such data will help determine whether future episodes of high pressure drop are associated with increased pressure drop in the vapor stream (such as in the demisters or packing) or are the result of increased liquid flow. The dry pressure drop during periods of questionable operation can be compared with the dry pressure drop when the demisters and packing are known to be clean. If there is a significant difference between the two, it is an indication that the increase in overall pressure drop is due to clogging of the demisters or column [9]. If there is not a significant difference, then other factors should be examined to identify the reason for the excessive pressure drop. Other factors that affect the demister operation, pressure drop, and the overall operation of the scrubber are excessive vapor or liquid flow rates. Care should be taken to ensure that the gas flow is neither excessive nor so low that it results in an excessive liquid flow rate.

Decreases in pressure drop can indicate low liquid flows, low air flows, channeling of liquid through the scrubber that may be resulting from poor liquid distribution, collapsed packing support plates, or changes in the density of the gas [18, 21]. Changes in the pressure or flow of liquid may be attributable to plugged filters or pipes in the liquid system, plugged nozzles, low sump level (which may cause pump cavitation), a worn pump impeller, changes in valve position, internal piping leaks or breaks, or eroded spray nozzles [18]. Changes in gas flow may result from plugging in the packing or demister, changes in liquid flow, poor fan operation due to loose belts or worn impeller, partially plugged ductwork, changes in damper settings, leaks in the ductwork or scrubber, or damage to the packing or packing support plates [18]. If the efficiency of an absorption system decreases, it may be attributable to low liquid flow, insufficient additive feed, pH probe malfunction, or liquid channeling due to pluggage in the scrubber or liquid distribution system [18].

It is also recommended that copies of the fan curves for the systems be retained so that the gas flow rate can be determined as the pressure drop through the scrubber changes. The curves should be prepared at approximately the same temperature of the gas as it goes through the fan.

TABLE 5.7 Advantages and Disadvantages of Absorption

Advantages	Disadvantages
Allows for recovery/concentration of the pollutant	Effectiveness is pollutant specific, depends on solubility of compound
High efficiencies possible	Transfers air pollution problem to liquid waste problem
No additional pollutants formed	Potential corrosion of ductwork due to additional moisture in gas stream

Particulate that builds up in the tower, demister, or distribution systems increases the pressure drop and hinders efficient operation, so any particulate collection devices before the scrubber should be maintained in good operating condition. Because it is possible that media replacement will be required, the system should be designed and constructed with adequate space and access for demisters and packing or trays to be removed as easily as possible.

Once the column is in place and operational, there are still changes that can be made to facilitate improved operation. If increased gas flows must be accommodated, it is possible to convert a countercurrent column into a co-current one, where the gas and liquid both travel down through the column. Co-current columns do not experience the same difficulties with flooding, but efficiency would be sacrificed [7]. Changes in packing material can also affect the pressure drop and efficiency achieved in the column for a cost significantly below that required to replace the column or to install an additional unit.

If liquid is channeling through the column but the liquid distributors are working well, the problem may be due to poor liquid flow characteristics in the packing. It may help to replace a portion of the top packing layers with smaller size packing to improve liquid distribution [8]. Newer packing designs have been developed to reduce lateral flow [56]; these may also be advantageous in such a situation. These packings are also less likely to redistribute liquid flow, so proper liquid distribution is even more critical than with older style packings. An overall summary of the advantages and disadvantages of absorption systems is given in Table 5.7.

REFERENCES

1. J.R. Fair; D.E. Steinmeyer; W.R. Penney; B.B. Crocker, "Gas Absorption and Gas–Liquid System Design," in *Perry's Chemical Engineers' Handbook*, 7th Ed., D.W. Green; J.O. Maloney, Eds., McGraw-Hill, New York, **1997**, Sect. 14.
2. T.T. Shen, "For VOC Control Seminar—Absorption," in *Control of Gaseous Emisions—VOC's Handbook*, ETS Professional Training Institute, ETS International, Inc., Roanoke, VA, **1995**.

3. R.W. Purcell; G.S. Shareef, *Control Technologies for Hazardous Air Pollutants*, US EPA, Research Triangle Park, NC, **1986**.
4. "Absorption Applications," in *Indoor Air Quality and Control*, A.L. Hines; T.K. Ghosh; S.K. Loyalka; R.C. Warder, Jr., PTR Prentice-Hall, Inc., Englewood Cliffs, NJ, **1993**, pp. 205–234.
5. T. Godish, *Air Quality*, 2nd Ed., Lewis Publishers, Chelsea, MI, **1991**, pp. 327–328.
6. K. Noll; W.J. Franek; L. Stander, *APTI Course 482: Sources and Control of Volatile Organic Air Pollutants*, US EPA Research Triangle Park, NC, **1987**, Lesson 24, draft.
7. R.M. Bethea, *Air Pollution Control Technology: An Engineering Analysis Point of View*, Van Nostrand Reinhold, New York, **1978**, pp. 331–365.
8. M.V. Bhatia, "Packed Tower and Absorption Design," in *Air Pollution Control and Design Handbook* (in Two Parts), Part II, P.N. Cheremisinoff; R.A. Young, Eds., Marcel Dekker, New York, **1977**, pp. 609–697.
9. "Absorption," in *Handbook of Air Pollution Control Engineering and Technology*, J.C. Mycock; J.D. McKenna; L. Theodore, Lewis Publishers, Boca Raton, **1995**, Ch. 9, pp. 149–171.
10. R.W. Boubel; D.L. Fox; D.B. Turner; A.C. Stern, "Control Devices and Systems" in *Fundamentals of Air Pollution*, 3rd Ed., Academic Press, San Diego, **1994**, p. 478.
11. G. Gschwandtner; S. Fairchild, *Controlling Odorous Emissions from Iron Foundries*, US EPA, Research Triangle Park, NC, **1992**.
12. "Gas Absorption," in *Air Pollution Technology*, D.E. Painter, Reston Publishing Company, Reston, VA, **1974**, pp. 179–185.
13. L. Theodore; A.J. Buonicore, "Absorption," in *Air Pollution Control Equipment, Volume II: Gases*, CRC Press, Boca Raton, **1988**, pp. 79–113.
14. J.W. Macdonald, "Absorbers," in *Air Pollution Control Equipment: Selection, Design, Operation, and Maintenance*, L. Theodore; A.J. Buonicore, Eds., ETS, Inc., Roanoke, VA, **1992**, pp. 45–71.
15. J.P. Tomany, *Air Pollution: The Emissions, the Regulations, & the Controls*, American Elsevier Publishing Company, New York, **1975**, pp. 302–344.
16. S. Calvert, "Scrubbing" in *Air Pollution, Third Edition, Volume IV: Engineering Control of Air Pollution*, Academic Press, New York, **1977**, pp. 257-291.
17. W. Barbour; R. Oommen; G.S. Shareef; W.M. Vatavuk, "Gas Absorbers," in *OAQPS Control Cost Manual*, 5th Edition, US EPA, Research Triangle Park, NC, **1996**, Ch. 9.
18. A.J. Buonicore, "Absorption," in *Air Pollution Engineering Manual*, Van Nostrand Reinhold, New York, **1992**, pp. 15–31.
19. B.B. Crocker; K.B. Schnelle, Jr., "Control of Gases by Absorption, Adsorption, and Condensation" in *Handbook of Air Pollution Technology*, S. Calvert; H.M. Englund, Eds., John Wiley & Sons, New York, **1984**, pp. 135–161.
20. Unit Operations Laboratory, Chemical Engineering Department, Clemson University, Clemson, SC, **1984**.
21. H.E. Hesketh; *Air Pollution Control*, Ann Arbor Science, Ann Arbor, MI, **1979**.
22. A.J. Buonicore; L. Theodore, *Industrial Control Equipment for Gaseous Pollutants, Vol. I*, CRC Press, Cleveland, OH, **1975**, pp. 59–138.
23. G.T. Joseph; D.S. Beachler, *APTI Course 415: Control of Gaseous Emissions: Student Workbook*, US EPA, Research Triangle Park, NC, **1981**, Chs. 5, 6, and 7.

24. J.L. Bravo, "Select Structured Packings or Trays?" *Chem. Eng. Prog.*, July **1997**, pp. 36–41.
25. J.W. MacDonald, "Packed Wet Scrubbers," in Air Pollution Control and Design Handbook (in Two Parts), Part II, P.N. Cheremisinoff; R.A. Young, Eds., Marcel Dekker, New York, **1977**, pp. 699–720.
26. J.A. Danielson, *Air Pollution Engineering Manual*, EPA, Research Triangle Park, NC, **1973**.
27. L. Theodore; A.J. Buonicore; J.D. McKenna; I.J. Kugelman; J.S. Jeris; J.J. Santoleri; T.F. McGowan, "Waste Management," in *Perry's Chemical Engineers' Handbook*, 7th Ed., D.W. Green; J.O. Maloney, Eds., McGraw-Hill, New York, **1997**, Sect. 25.
28. C.M. McGinley, "Packed Columns," in *Odor and VOC Control Handbook*, H.J. Rafson, Ed., McGraw-Hill, New York, **1998**, pp. 8.102–8.110.
29. H.J. Rafson, "Mist Scrubbing" in *Odor and VOC Control Handbook*, H.J. Rafson, Ed., McGraw-Hill, New York, **1998**, pp. 8.111–8.130.
30. R.P.G. Bowker, "Other Designs" in *Odor and VOC Control Handbook*, H.J. Rafson, Ed., McGraw-Hill, New York, **1998**, pp. 8.131–8.138.
31. U. Von Stockar; C.R. Wilke, "Absorption," in *Kirk-Othmer Encyclopedia of Chemical Technology, Fourth Edition, Volume 1: A to Alkaloids*, J.I. Kroschwitz; M. Howe-Grant, Eds., John Wiley & Sons, New York, **1991**, pp. 38–93.
32. Research Committee, *Bubble Tray Design Manual*, American Institute of Chemical Engineers, New York, **1958**.
33. E.J. Henley; J.D. Seader, *Equilibrium-Stage Separation Operations in Chemical Engineering*, John Wiley & Sons, New York, **1981**.
34. S.V. Cabibbo; A.J. Teller, "The Crossflow Scrubber — A Digital Model for Absorption", Paper number 69-186, APCA 62nd Annual Meeting, New York, **1969**.
35. H.E. Hesketh, "Predict Particle Collection in Spray Towers," *Chem. Eng. Prog.*, Oct. **1995**, pp. 98–100.
36. L. Peters, "Topics in Air Pollution Control," class notes, Virginia Tech, Blacksburg, VA, **1995**.
37. S.E. Schwartz; W.H. White, "Solubility Equilibria of the Nitrogen Oxides and Oxy-Acids in Dilute Aqueous Solution," *Adv. Environ. Sci. Technol.*, 4, 1–45.
38. L.R. Martin, "Kinetic Studies of Sulfite Oxidation in Aqueous Solution," in *SO_2, NO, NO_2 Oxidation Mechanisms: Atmospheric Considerations*, J.G. Calvert, Ed., Butterworth, Boston, **1984**, pp. 63–100.
39. R. J. Heinsohn; R.L. Kabel, *Sources and Control of Air Pollution*, Prentice Hall, Upper Saddle River, NJ, **1999**.
40. R.L. Pigford; A.P. Colburn, "Gas Absorption," in *Chemical Engineers' Handbook*, 3rd Ed., McGraw-Hill, New York, **1950**, pp. 673–676.
41. J.H. Seinfeld, *Atmospheric Chemistry and Physics of Air Pollution*, John Wiley & Sons, New York, **1986**, pp. 195–246.
42. K.G. Denbigh, *The Principles of Chemical Equilibrium*, 3rd Ed., Cambridge University Press, **1971**.
43. J.O. Ledbetter, *Air Pollution: In Two Parts; Part B — Prevention and Control*, Marcel Dekker, New York, **1974**, pp. 99–121.
44. K. Wark; C.F. Warner; W.T. Davis, "Absorption," in *Air Pollution: Its Origin and Control*, Addison Wesley Longman, Menlo Park, CA, **1998**, pp. 314–340.

45. N. Mukhopadhyay; E.C. Moretti, *Current and Potential Future Industrial Practices for Reducing and Controlling Volatile Organic Compounds*, Center for Waste Reduction Technologies, American Institute of Chemical Engineers, New York, **1993**. pp. 19–20, 50.

46. "Gas Absorption," in *Air Pollution Control: A Design Approach*, 2nd Ed., C.D. Cooper; F.C. Alley, Waveland Press, Prospect Heights, IL, **1994**, pp. 411–447.

47. "Generalized Pressure Drop Correlation, Chart Number GR-109, Revision 4," U.S. Stoneware Co., Akron, Ohio, **1963**.

48. R.H. Perry; D. Green, Eds., *Perry's Chemical Engineers' Handbook*, 6th Ed., McGraw Hill, New York, **1984**.

49. R.N. Robinson, *Chemical Engineering Reference Manual for the PE Exam*, 5th Ed., Professional Publications, Belmont, CA, **1996**.

50. "Process Destroys Odors in Gas Streams," *Chem. Eng. Prog.*, June **1995**, pp. 19, 21.

51. M. Valenti, "Scrubbing away pollution," *Mech. Eng.*, 118, (**1996**), 12, pp. 72–74.

52. W.M. Vatavuk, *Estimating Costs of Air Pollution Control*, Lewis Publishers, Boca Raton, **1990**, pp. 30–31.

53. W.M. Vatavuk, "Cost Estimating Methodology," in *OAQPS Control Cost Manual*, 5th Ed., US EPA, Research Triangle Park, NC, **1996**, Ch. 2.

54. H.E. Hesketh, "Wet Scrubbers," in *Air Pollution Control*, Ann Arbor Science, Ann Arbor, MI, **1979**, p. 217–235.

55. M. Borenstein, "Special Construction Materials for Scrubbers," in *Air Pollution Control and Design Handbook* (in Two Parts), Part II, P.N. Cheremisinoff, R.A. Young; Eds., Marcel Dekker, New York, **1977**, pp. 949–961.

56. A.D. Nelson; R.J. Schmitt; D. Dickeson, "Upgrading the Performance of Groundwater VOC Air Strippers," *Environ. Prog.*, **1997**, 16, 1, 43–46.

57. *Electric Utility Engineer's FGD Manual*, US Dept of Energy, Pittsburgh, Energy Technology Center, prepared by Radian International, LLC, pp. I.4.5-1 through I.4.6-19.

58. A.J. Buonicore, "Wet Scrubbers" in *Air Pollution Control Equipment: Selection, Design, Operation and Maintenance*, L. Theodore; A.J. Buonicore, Eds., ETS, Inc., Roanoke, VA, **1992**, pp. 233–281.

6

THERMAL INCINERATION

6.1 TECHNOLOGY DESCRIPTION

Incinerators are relatively simple devices in which organic material in a gas stream is destroyed by combustion. Gaseous incinerators are also sometimes referred to as '*afterburners*', particularly if they are installed downstream of a primary fuel combustion unit [1], like a medical waste incinerator. This name is indicative of the equipment's function: to provide adequate time and temperature for compounds to finish burning. If designed and operated properly, the exhaust from gaseous incinerators consists primarily of carbon dioxide and water, along with inert gases from the atmosphere like nitrogen. Because of their capability for high destruction efficiency and their ability to destroy a wide variety of compounds, incinerators are very common control devices [2]. For situations that are particularly sensitive to emissions of nitrous oxides, other control devices should be considered, because there is inevitably some formation of nitrous oxides at combustion temperatures, although methods and equipment are available to minimize their formation.

Care must be taken to properly identify the constituents of a gas stream before incineration is selected as the control method. Sulfur compounds in the gas stream (including mercaptans, which are added to natural gas and propane to make sure that leaks are detectable [3]) react to form sulfur dioxide, which may need to be removed before the gas is vented to the atmosphere. Likewise, halogenated compounds in the contaminated gas stream react to form acids that can severely corrode ductwork and downstream equipment if not removed immediately after the incineration process. The gas stream must also be relatively clean, that is, there should not be material present that would clog nozzles if the gas stream is used as primary combustion air. Metals present in the gas stream can also affect the combustion process, especially in units where combustion is promoted by the presence of a catalyst. Vaporized species from elements such as iron and silicon can bind to the catalyst and render it inactive. In addition, materials that might coat the surface of the catalyst without reacting must be avoided in catalytic incinerators.

There are two primary types of incinerators: *thermal incinerators* (or *thermal oxidizers*) and *catalytic incinerators* [4]. Thermal incinerators rely on a high operating temperature and usually a direct flame to ignite the pollutants. They generally operate in the range of 973–1073 K (700°–800°C, 1300°–1500°F) [5], but field testing has shown that most incinerators should maintain approximately 1143 K (870°C, 1600°F) for at least 0.75 s to achieve 98% reduction of nonhalogenated compounds [6–9]. Halogenated compounds are more difficult to destroy [10]. Incineration temperatures as high as 1477 K (1204°C, 2200°F) are required for effective destruction of some chlorinated compounds [11]. Catalytic incinerators, as described in Chapter 8, contain a surface of noble metal or other catalyst that brings the reactants together in close proximity or otherwise facilitates the oxidation reaction so that it occurs at a lower temperature. Catalytic units generally operate in the range of 620–870 K (340°–590°C, 650°–1100°F) [12]. Flares, which basically function as thermal incinerators with no combustion chamber, are discussed in Chapter 7.

In thermal incinerators, the organic removal efficiency that can be achieved is governed by the "three Ts" [13]: time, temperature, and turbulence. In other words, the combustion reaction depends on maintaining the proper temperature long enough for the combustion process to occur. To ensure that the temperature and reactants are distributed evenly, the incinerator must provide adequate velocity and turbulence to mix the combusted gases with the pollutants. The proper conditions are already present in many boiler or process heater fireboxes; these, too, can be used for incineration of gases if the gaseous stream introduction is properly positioned [14]. It is well documented that there is flexibility between the temperature and residence time requirements, that is, the required time can be shortened by elevating the temperature, and a lower temperature can be used if the mixture is allowed a longer time to react. Figure 6.1 [15] gives a typical correlation of the relationship between temperature and retention time. Well-designed and maintained incineration units are capable of achieving efficiencies of 99% at pollutant concentrations greater than approximately 100 ppm [16].

Incineration is most appropriate for processes in which there is no physical or economic benefit to recovering the organic material. Examples of typical uses for thermal incineration are given in Table 6.1 [17–21].

6.2 TYPES OF THERMAL INCINERATION SYSTEMS

Incinerators are further characterized by the method used to transfer heat from the gaseous exhaust to the incoming waste gas stream. Older thermal incinerator units and catalytic incinerators commonly use *recuperative* heat exchangers (as shown

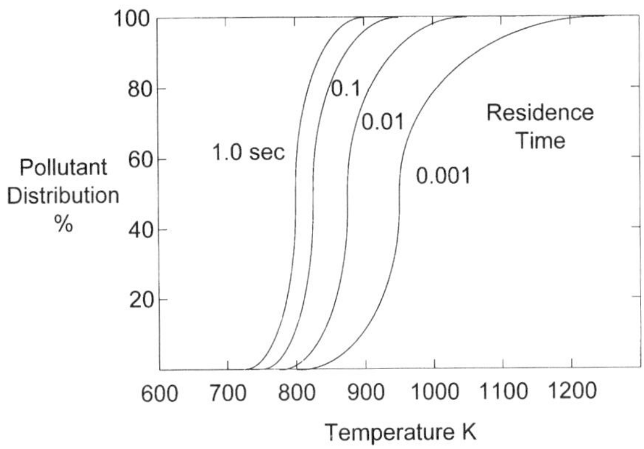

Figure 6.1 Effect of temperature and residence time on destruction efficiency. (Adapted from Rolke [15].)

TABLE 6.1 Typical Applications of Thermal Incineration for VOC Removal

Landfill offgases
Printing operations
Medical waste incinerator exhaust
Bakeries
Paint bake ovens
Foundry core ovens
Paper operations
Coffee roasting
Sewage sludge incineration offgas
Rendering plants
Smokehouse operations
Textile drying
Asphalt blowing and asphalt paper manufacture

Data adapted from refs. 17–21.

in Figs. 6.2 [22] and 6.3 [23]) in which the exhaust gases from the incinerator pass through a shell-and-tube or plate-type heat exchanger to recover the heat of combustion and increase the fuel efficiency of the VOC destruction process. The incoming gases pass through the heat exchanger, too, but without directly contacting the exhaust gas. Recuperative units are generally only able to recover about 70% of the available heat [24], although at least one manufacturer reports primary heat recovery in a recuperative unit of up to 82% [25]. Some units are designed with a secondary heat recovery unit to preheat air or water or heat transfer oil for other plant processes such as drying ovens [26].

More recently, *regenerative* systems have been introduced (see Fig. 6.4 [27]) that use beds of rock or packed ceramic material as a heat sink. Regenerative

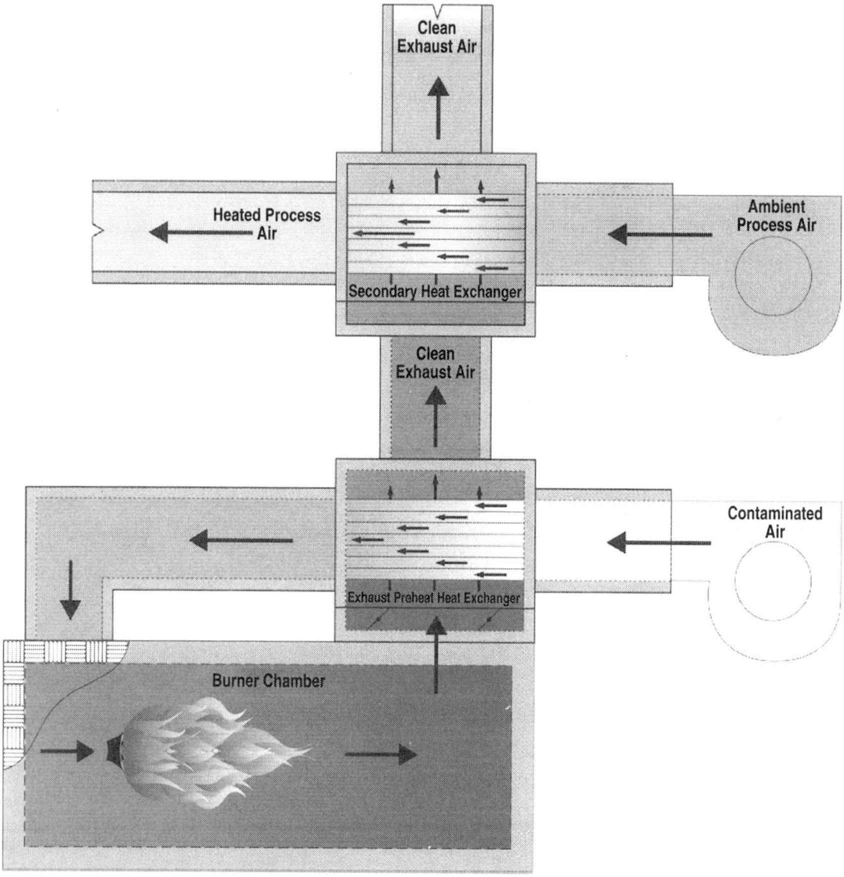

Figure 6.2 Glenro, Inc., recuperative oxidizer with secondary heat recovery [22].

incinerators allow for better heat recovery, meaning that dilute vapor streams can be incinerated at lower costs [28]. The postcombustion gases pass through one bed and heat up the ceramic material. Then the flow shifts, and the postcombustion gas stream is used to heat up a second bed of ceramic material while precombustion air flows through the first hot bed, gaining heat as it travels to the burner. Flow is shifted back and forth between the beds as they heat up and cool down. Heat transfer efficiencies of approximately 95% can be achieved with regenerative units [6]. The average number of cycles is between 15 and 60/h [24]. The alternating sequence decreases the removal efficiency that can be achieved in such a system; control efficiencies are usually below 99% [29, 30]. At least one manufacturer reports destruction efficiencies of 95–99% [31]. Depending on the required removal efficiencies, more than two beds may be needed. In some cases a third bed is used as a "purge" bed to capture the untreated gases that would

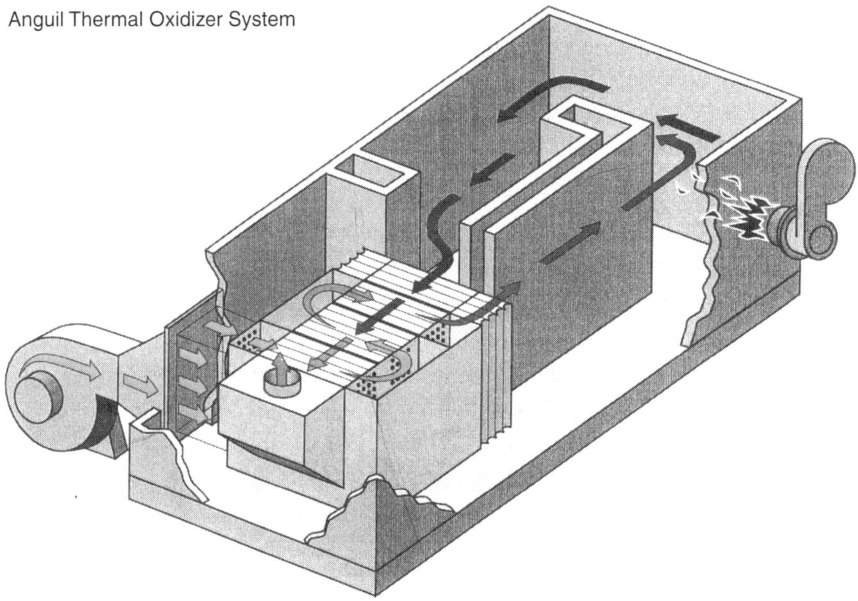

Figure 6.3 Anguil recuperative thermal oxidizer [23].

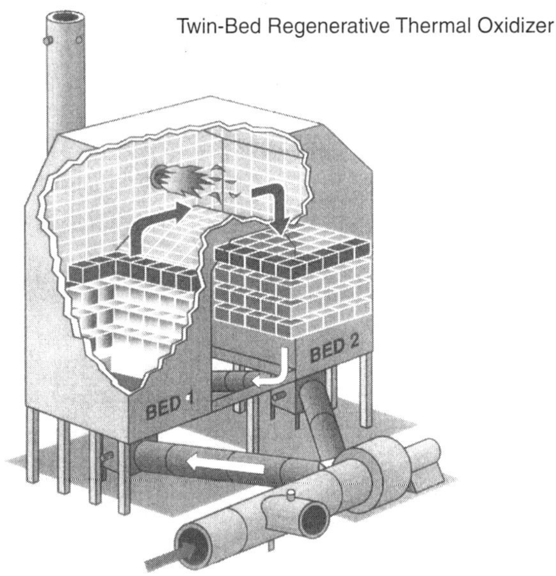

Figure 6.4 Anguil Environmental Systems, Inc., regenerative oxidizer [27].

normally be exhausted from the ducting when the valves change position [24]. For systems demanding removal efficiencies above 98%, the valving mechanism becomes more complex, as double valves are used to reduce the opportunity for any leakage. The space between the valves is slightly pressurized with fresh air to ensure that no untreated gas is released to the stack [24].

Horizontal flow regenerative incinerators have been introduced recently that result in lower space requirements and make the technology more suitable for retrofit situations [29]. Figure 6.5 shows an example of the flow through a horizontal regenerative incinerator [29]. In addition, some manufacturers have introduced a system that combines an adsorbing wheel with thermal or catalytic incineration so that the inlet gas to the incinerator is more concentrated. This results in better thermal efficiency, because the amount of air that must be heated to the combustion temperature is significantly decreased. An example is shown in Figure 6.6 [23].

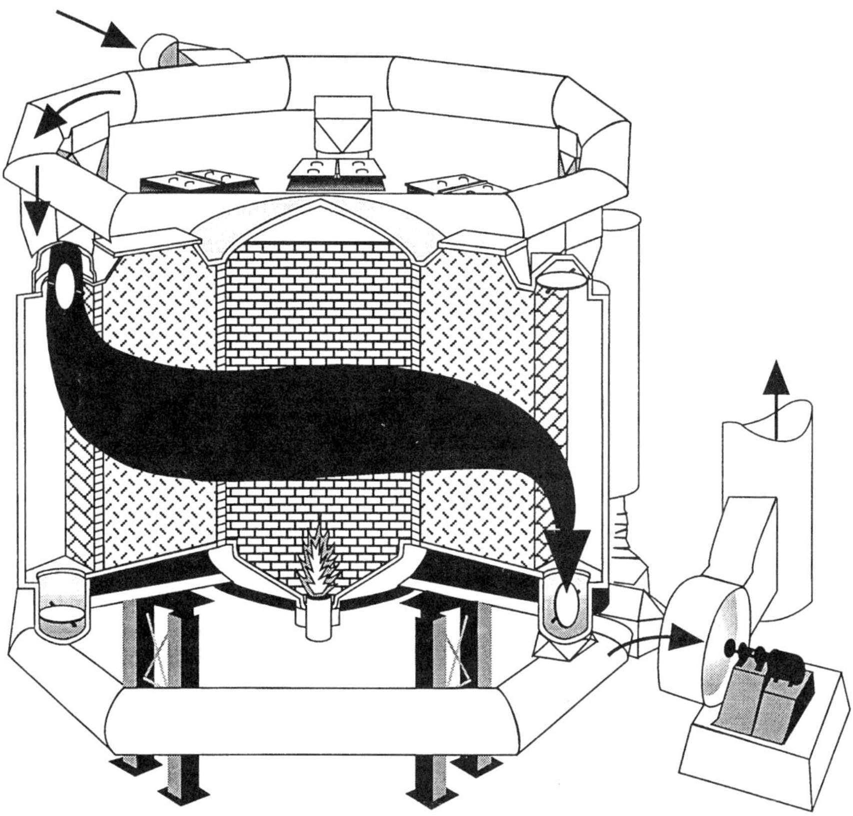

Figure 6.5 REECO horizontal flow regenerative incinerator [29].

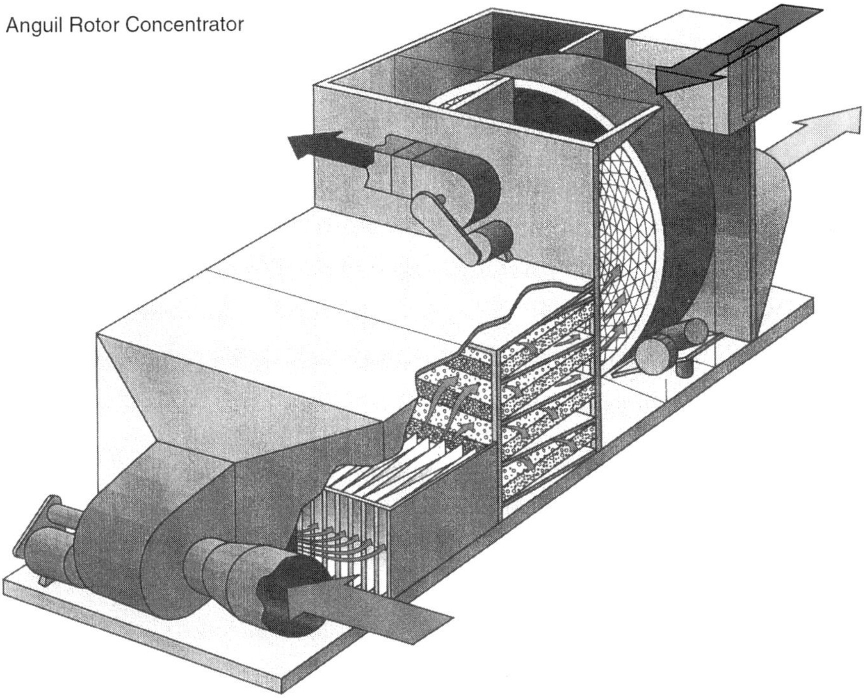

Figure 6.6 Anguil Environmental Systems, Inc., roto-concentrator [23].

6.3 THERMAL INCINERATOR SYSTEM DESIGN

It is a generally accepted practice to model the complex free radical reactions that occur during combustion as a single first-order reaction for many compounds [32–35], so that the final concentration (or the overall destruction efficiency) can be estimated if the temperature, T, and residence time, τ, are known:

$$\ln\left(\frac{C_B}{C_{B0}}\right) = -k\tau \tag{6.1}$$

where

$$k = Ae^{-E/RT} \tag{6.2}$$

A = compound-specific constant (s^{-1})
E = apparent activation energy, kJ/mol (or Btu/lb-mol if English units are used throughout the calculation)
R = gas constant, 8.314 kJ/mol K (or 1.987 Btu/lb-mol °R)
τ = residence time, s
T = absolute temperature, K (or °R)

Values of the apparent activation energy and compound-specific constants are given in Table 6.2. The average residence time of an incinerator can be estimated

TABLE 6.2 Pre-Exponential Factors and Apparent Activation Energies

Pollutant	Pre-Exponential Factor, A (s^{-1})	Apparent Activation Energy, E (kJ/g-mol)
Acrolein	3.30×10^{10}	150
Acrylonitrile	2.13×10^{12}	218
Allyl alcohol	1.75×10^{6}	89.5
Allyl chloride	3.89×10^{7}	122
Benzene	7.43×10^{21}	401
Butene-1	3.74×10^{14}	244
Chlorobenzene	1.34×10^{17}	320
Cyclohexane	5.13×10^{12}	199
1,2-Dichloroethane	4.82×10^{11}	191
Ethane	5.65×10^{14}	266
Ethanol	5.37×10^{11}	201
Ethyl acrylate	2.19×10^{12}	192
Ethylene	1.37×10^{12}	213
Ethyl formate	4.39×10^{11}	187
Ethyl mercaptan	5.20×10^{5}	61.5
Hexane	6.02×10^{8}	143
Methane	1.68×10^{11}	218
Methyl chloride	7.34×10^{8}	171
Methyl ethyl ketone	1.45×10^{14}	244
Natural gas	1.65×10^{12}	206
Propane	5.25×10^{19}	356
Propylene	4.63×10^{8}	143
Toluene	2.28×10^{13}	236
Triethylamine	8.10×10^{11}	181
Vinyl acetate	2.54×10^{9}	150
Vinyl chloride	3.57×10^{14}	265

Adapted from Buonicore [32].

from knowledge of the outlet volumetric flow rate, Q, and the volume of the incinerator, V:

$$\tau = \left(\frac{V}{Q}\right) \tag{6.3}$$

where

V = internal volume of the incinerator chamber, m^3 (or ft^3)
Q = flow rate of gases out of the incinerator, m^3/s (or ft^3/s)

As with other VOC control devices, the amount of gas to be treated should be minimized to reduce the capital and operating costs of the incinerator. Care must be exercised, however, to ensure that the concentration of pollutant in the gas does not exceed a safe level in the ducting, even during process upsets or start-up conditions. This is particularly important, because the incinerator itself provides a source of heat and ignition. Most insurance companies consider 25% of the *lower explosive limit* (LEL) to be the maximum acceptable concentration of VOC in the ducting as a general guide. Although this value can be exceeded in some situations, it typically requires additional instrumentation to monitor the level of VOCs at all times and to instantaneously shut down the process or vent the contaminated gas to the atmosphere at a given threshold. Information regarding the LEL should be available from Material Safety Data Sheets or the chemical manufacturer. Additional information about the LEL is also available from the U. S. Bureau of Mines [36, 37]. The heating value of hydrocarbons at the LEL has been determined to be approximately 2 mJ/Nm^3 (50 Btu/scf) [6].

To ensure an effective incinerator design, the operating temperature should be approximately 90°–150°C (200°–300°F) above the highest autoignition temperature of any of the components [38]. The autoignition temperature is the temperature at which a mixture of organics and air will ignite without a spark or flame [39]. Modulating controls should be used to regulate the flow of fuel to tne burner (Ray Goetz, Department of Environmental Quality, Roanoke, VA, personal communication). In the past, units have been designed with on/off controls for the pilot ignition source. This type of control is ineffective because it allows short intervals of time during which no flame is available for the combustion process. Although the intervals are short, they can substantially affect the overall performance of the incinerator. It is essential that the proper temperature and retention time be maintained so that contaminants are completely destroyed. Incomplete combustion can lead to compounds such as aldehydes, dioxins, and furans with more severe health effects than the original hydrocarbon gas components [3, 40].

Most of the gas streams treated with incinerators have fairly low loadings of VOCs. For more concentrated gas streams, a regenerative system may not be advantageous. Depending on the heat content of the vapor, heat recovery above 90% may not be necessary and may actually be detrimental, because too high a temperature in the combustion air can reduce the efficiency of the incinerator [41].

Halogenated or sulfur compounds in the gas stream form acidic gases upon combustion, and the incinerator system must be designed to handle the corrosion that may result. Chlorinated compounds may also form dioxins when they are treated by thermal incineration [2, 42]. A scrubber following the incinerator is necessary under most circumstances to remove the acid gases before the gas stream is released to the atmosphere. The scrubber should be considered as an integral part of the incinerator system and should not be overlooked even in preparing rough cost estimates because the cost of the scrubber can be a significant portion of the overall system cost.

Some heavy metals such as mercury and cadmium may become vapors during high temperature incineration [3]. Special considerations are required for process streams containing metals that may be released into the atmosphere during incineration; additional heat recovery and particulate control may be warranted in these circumstances.

Procedures for calculating the total flue gas flow (that is, the flow rate of gases leaving the incinerator) are given in the *OAQPS Control Cost Manual* [6] and by Vatavuk [43]:

$$Q = Q_w + Q_f \tag{6.4}$$

where

Q = Total flue gas flow rate, normal m^3/ min abbreviated Nm^3/ min (in English units, standard ft^3/ min abbreviated scfm)
Q_w = Waste gas flow rate at inlet to incinerator, Nm^3/ min (scfm)
Q_f = flow rate of auxiliary fuel, Nm^3/ min (scfm)

The flow rate of auxiliary fuel is calculated from the equation

$$Q_f = Q_W X/Y \tag{6.5}$$

where

$$X = 1.1\, C_{po}\, (T_c - T_r) - C_{pi}\, (T_i - T_r) - h_w \tag{6.6}$$

$$Y = h_f - 1.1\, C_{po}\, (T_c - T_r) \tag{6.7}$$

C_{po} = mean heat capacity of exhaust stream from the incinerator, kJ min/Nm^3 °C (Btu scf^{-1} $°F^{-1}$). The heat capacity of air is typically used because most of the gas stream is air. At 1144 K (871°C, 1600°F), the heat capacity of air is 1.50 kJ/Nm^3 °C (0.0194 Btu/scf °F)
T_c = combustion temperature, °C (or °F)
T_r = reference temperature; 21°C (70°F) is typical
C_{pi} = mean heat capacity of inlet streams to the incinerator, kJ min/Nm^3 °C (Btu scf^{-1} $°F^{-1}$). Again, the heat capacity of air is typically used because most of the incoming stream is air. At 311 K (38°C, 100°F), the heat capacity of air is 1.4 kJ/Nm^3 °C (0.018 Btu/scf °F)
T_i = inlet gas stream temperature, after preheating, °C (°F). T_i may be calculated from the equation $T_i = T_W + \text{Eff}\,(T_c - T_W)$
T_w = temperature of waste gas stream before preheating, °C (°F)
Eff = anticipated fractional thermal efficiency of the incinerator (usually between 0.5 and 0.95)
h_w = heat content of waste gas stream, kJ/Nm^3 (Btu/scf)
h_f = heat content of fuel, kJ/Nm^3 (Btu/scf). For natural gas, h_f is approximately 37,000 kJ/Nm^3 (900 Btu/scf)

h_f may be calculated in SI units from the following equation if the heat content is known in kJ/kg:

$$h_f = \text{heat content in kJ/kg} \times \text{fuel molecular weight in kg/kg-mol}/22.4\ Nm^3 \text{ per kg-mol} \tag{6.8}$$

In English units, h_f can be calculated from the known heat content in Btu/lb:

$$h_f = \text{heat content in Btu/lb} \times \text{fuel molecular weight in lb/lb-mol}/386.5\ \text{ft}^3\ \text{per lb-mol} \quad (6.9)$$

These equations are considered valid for gas streams with an oxygen content above 16% that contain VOCs at a concentration less than 25% of the LEL. They assume a 10% loss of energy from the incinerator [43]. Note that no allowance is included for volume changes due to changes in the total number of moles in the system that may occur during reaction. Additional correction may be needed for gas streams that contain large concentrations of organic compounds, particularly if the organic compounds contain significant amounts of oxygen.

Additional capital cost can sometimes be avoided if an existing boiler or furnace can be used to control VOCs in addition to its original function [17]. Using a VOC-containing gas stream as combustion air can benefit the overall process the organic compounds serve as fuel when the gas stream is combusted. The following criteria must be satisfied before a boiler or process heater is used as an incinerator for waste gases [17, 44, 45]:

1. There should be very few particles present in the waste gas stream or formed during combustion because any particulate may build up on heat exchanger surfaces, decreasing the overall fuel efficiency that can be achieved. If this occurs, the additional cleaning required to return the boiler or furnace to peak efficiency may be extremely expensive and may require significant equipment outage time. In addition, solids that make it beyond the heat exchanger can contribute to the overall particulate loading in the exhaust stream and may make it more difficult for the combustion device to meet any air pollution control requirements.
2. The gas stream should be only a small part of the total amount of air required for combustion. If the gas stream represents a significant amount of the combustion air for the boiler or process furnace, additional design is required to ensure that the oxygen balance is maintained, that mixing is adequate, and that there is an alternative source of combustion air if the VOC process is not operating.
3. The contaminated gas stream should have an oxygen content similar to that of air. If the oxygen content is insufficient, incomplete combustion could occur. The products of incomplete combustion can be more toxic than the original VOC contaminant. Products of incomplete combustion may also be sticky material that fouls heat exchanger surfaces.
4. Using a boiler or process furnace to control VOC emissions may inadvertently tie the VOC emitting process and the boiler operation together because the boiler needs to be operating whenever the emissions are required to be controlled. This may have significant impacts on the VOC-emitting process when a boiler is taken off line for annual or biannual maintenance.

5. There should be no compounds in the waste gas stream that will damage the boiler. Halogenated compounds in the gas stream should be avoided, because they may contribute to corrosion in the boiler or flue gas equipment.
6. Additional design changes may be needed to the boiler to ensure adequate mixing. For example, baffling may be necessary to ensure that the waste gas does not bypass the flame. Changes to the flame patterns and radiation should be avoided.
7. It is preferable to use a base-loaded boiler to combust the waste gas. A base-loaded boiler is one that operates at a consistent rate for long periods of time. Base-loaded boilers are normally accompanied by swing-load boilers, which alter the fuel input to respond to changing steam or energy demands.
8. The concentration of the organic compound should be less than 25% of the LEL. For more concentrated streams, the waste gas stream may need to be treated as an auxilliary fuel, particularly if the concentration is above the LEL.

Any type of thermal combustion device generally results in the formation of some nitrous oxides (NO_X). In some cases, the formation of NO_X compounds is more of a concern than the presence of VOCs. Therefore, careful examination of the NO_X emissions and applicable requirements should be conducted before incineration is selected as the control method. A packed bed flameless thermal oxidizer is reportedly being developed that has a relatively low combustion temperature, resulting in minimal NO_X formation [2, 46]. The concentration of NO_X that results from the combustion process is highly dependent on the peak temperature of combustion because the NO_X is formed relatively quickly at high temperatures but does not readily degrade at lower temperatures. Therefore, the NO_X concentration in the exhaust gases from combustion devices are generally the same magnitude as equilibrium concentrations at the peak combustion temperature [47]. The formation of NO_X from the nitrogen and oxygen in the air is very fast when the combustion temperature is above approximately 1920 K (1650°C, 3000°F) [48]. Most thermal incinerators operate at temperatures lower than the combustion chambers in boilers, and NO_X formation is minimal (A.B. Nunn, personal communication, 1999).

Instrumentation that should be considered is listed in Table 6.3. While the amount of instrumentation may seem excessive during the design phase, it is often helpful for monitoring any deterioration in performance and for troubleshooting.

TABLE 6.3 Recommended Instrumentation

Combustion temperature
Inlet/outlet concentration
Fan speed
Pressure
Auxilliary fuel rate
Pressure drop across any particulate collection devices

6.4 ECONOMIC ESTIMATES

Estimates of the capital cost of a package recuperative thermal incineration system are given by Vatavuk [43]. In SI units, the equation is

$$CC = a\ (Q/0.026)^b \tag{6.10}$$

In English units, the equation is

$$CC = a\ Q^b \tag{6.11}$$

where

CC = purchased price, in 1988 dollars
Q = flow rate of gases after combustion, Nm^3/min (scfm)

a and b are parameters dependent on the amount of heat recovery required. For 0% heat recovery, a = 3,120 and b = 0.360. For 50% heat recovery, a = 4,920 and b = 0.389. At 70% heat recovery, a = 5,690 and b = 0.408 [49].

These estimates are valid for incinerators between 130 and 1300 Nm^3/min (5,000 and 50,000 scfm) operating at approximately 1090 K (820°C, 1500°F). In addition to the combustion chamber and burner, the price includes a heat exchanger, fan, motor, instrumentation and controls, and a 10 ft stack. Ductwork, particulate removal devices, and site-specific construction requirements are not included. If the incinerator is to be placed a long distance from the generation site, the costs of both ductwork and additional booster fans should be considered. As with the other chapters in this book, the cost estimates are to indicate a general order of magnitude. They are not specific enough to use as construction costs and should be confirmed with actual prices from vendors.

For comparison, Ruddy and Carroll [50] give estimates of capital costs for recuperative incinerators as \$35.3–7060 per cubic meter per minute (\$10–200 /cfm). The capital cost for regenerative systems is estimated as \$1060–16,000/(m^3/min) (\$30–450 /cfm). It is not specified whether these costs are for actual or standard flow rates.

For a regenerative incinerator with 95% thermal efficiency, the *installed* cost may be estimated based on the inlet gas flow according to the following equations [43]. In SI units:

$$C = 383{,}000 + 589\ Q_W \tag{6.12}$$

In English units:

$$C = 383{,}000 + 15.3\ Q_w \tag{6.13}$$

This equation is valid for waste gas stream flows between 130 and 1300 Nm^3/min (5,000–50,000 scfm). The costs of foundations, freight, taxes, and ductwork are not included, but the labor to install the incinerator is. Approximately 20% additional cost is required for the foundations, freight, taxes, and ductwork [51]. To estimate the total installed capital expenditure, the factors in Table 6.4 may be used, which follow the methodols in the EPA's *OAQPS Cost Control Manual*.

Annual costs may be approximated using the guidelines in the *OAQPS Cost Control Manual* [6], as given in Table 6.5. Electricity requirements can be estimated from the formulas given in Chapter 4, assuming a pressure drop of 1 kPa (4 in of water column) for the combustion chamber itself. The heat exchanger adds another 1–3.7 kPa (4–15 in of water column), corresponding to heat recoveries of 35–70% [6]. Operating and maintenance labor requirements are usually minimal with thermal incinerators, at approximately one half hour per shift.

Rough estimates of annual operating costs range from \$530–3200 per m^3/min (\$15–90/cfm) for recuperative units to \$710–5300 per m^3/min (\$20–150/cfm) for regenerative systems according to Ruddy and Carroll [50].

TABLE 6.4 Factors for Estimating the Total Installed Capital Cost of Thermal Incineration Systems

Cost Element	Estimated Cost
Direct Costs	
Purchased equipment cost (PEC)	
Incinerator and auxiliary equipment	CC, Estimated from equation
Taxes	0.03 CC
Freight	0.05 CC
Total PEC	1.08 CC
Installation costs	
Foundations and supports	0.08 PEC
Erection and handling	0.14 PEC
Electrical	0.04 PEC
Piping, insulation, and painting	0.04 PEC
Total installation cost	0.30 PEC
Site preparation (SP)	As required for specific site
Buildings (BLDG)	As required for specific site
TOTAL DIRECT COSTS	1.30 PEC + SP + BLDG
Indirect Costs	
Engineering and supervision	0.10 PEC
Construction, field expenses, and fees	0.15 PEC
Start-up and performance test	0.03 PEC
Contingency	0.03 PEC
TOTAL INDIRECT COSTS	0.31 PEC
Total Installed Capital Costs, TCC	1.61 PEC + SP + BLDG

Adapted from van der Vaart et al. [6].

TABLE 6.5 Annual Operating Costs for Incinerator Systems

	Unit Cost	Estimated Requirement
	Direct Annual Costs	
Utilities		
Electricity	\$0.05/kW-h	See Chapter 4
Natural gas	\$3.30/1000 ft^3	Calculated above
Operating labor	\$12.96/h	1/2 h per shift
Maintenance labor	\$14.26/h	1/2 h per shift
Maintenance materials	Equal to maintenance labor	
Supervisory labor	15% of operating labor	
	Indirect Annual Costs	
Overhead	60% of maintenance costs	
Property tax	1% of Total Installed Capital Costs (TCC)	
Insurance	1% of TCC	
Administrative costs	2% of TCC	
Capital recovery (see Chapter 4 for formula to calculate capital recovery factor)	CRF · TCC	
Total Annual Operating Costs	Direct + Indirect costs	

Adapted from van der Vaart et al. [6].

6.5 POTENTIAL PROBLEMS

Like any equipment, incinerators are susceptible to damage caused by severe temperature changes. Although they are designed to be operated at high temperature, large fluctuations in temperature can cause thermal stress and lead to premature failure of the refractory lining. Therefore, incinerators are best suited to operations that maintain a fairly constant emission rate [29] and to those that operate continuously instead of those that operate for only a few hours each day. The manufacturer's start-up procedure normally specifies a warm-up sequence in which the incinerator is brought to operating temperature gradually. Following this procedure lengthens the life of the refractory, but also uses additional fuel and may take longer than is feasible for the process. It is not unusual for maintenance or operating staff to develop a shorter start-up procedure. Using such a shortcut lowers the overall gas usage but is likely to result in considerably higher maintenance requirements. For cyclical processes, a regenerative incinerator should be considered, which can be maintained at operating temperature with minimal natural gas consumption. Some of these units can even be put in an "idle" mode in which little or no gas is pulled from the process, but the incinerator recycles exhaust air to the inlet to conserve heat (Tellkamp Industries, personal communication with T. Rowe 1996, Sante Fe Springs, CA).

If a dilute waste gas stream is mixed with the required amount of fuel to achieve the flame temperature, it is possible that the resulting mixture will not have adequate energy concentration to propogate a flame. This can result in flame quenching in the incinerator, which produces additional contaminants due to the partial oxidation of the fuel and pollutant gas. Therefore, usually about one half of the waste gas stream is mixed with the fuel, and the remaining waste gas stream is introduced after the initial flame and mixed with the products of combustion [52].

6.6 INSTALLATION, OPERATION, AND MAINTENANCE CONSIDERATIONS

The main problem with recuperative incinerators tends to be ensuring that the heat exchange surface remains clean and that the heat exchanger does not develop cracks [53]. In addition to difficulties with the refractory linings in the incinerator, heat stress can take a toll on the heat recovery equipment. At best, heat stress or buildup of material may keep butterfly valves on regenerative units from seating properly, a common problem that allows leakage of the contaminated gas stream into the outlet gas stream, decreasing efficiency. At worst, heat stress can crack metal joints in heat exchangers so that gas is allowed to escape from the clean side to the dirty side or vice versa, depending on the position of the fan. Such cracks can be difficult to locate because the metal contracts as it cools to room temperature. The incinerator system should be designed considering the potential leakage of material. An additional fan is usually required to move the air from the process into the incinerator. This fan may be installed either before or after

TABLE 6.6 Advantages and Disadvantages of Thermal Incineration

Advantages	Disadvantages
Simple, straightforward operation	Some NO_x formation, although it can be minimized
Heat recovery for other processes is feasible	Explosion potential
High efficiencies possible with proper design	Fuel consumption can be significant
Technology is applicable for a wide variety of gas concentrations and constituents.	No ability to reuse pollutants
Useful for multi-component streams	Approximately 3 times higher fuel usage than in catalytic incinerators

Data adapted from refs. 54, 55.

the heat exchangers and incinerator. By careful analysis of the system and fan placement, it can be ensured that any leakage occurs in an acceptable direction. Any inorganic particulate that may block nozzles, interfere with the combustion flame, or foul heat transfer surfaces should be removed before entering the incineration system. Advantages and disadvantages of thermal incineration are listed in Table 6.6 for comparison with other technologies during a technical and economic evaluation [54, 55].

REFERENCES

1. "Combustion," in *Air Quality*, 2nd Ed., T. Godish, Lewis Publishers, Chelsea, MI, **1991**, pp. 321–325.
2. R.L. Berglund, "Exhaust Control, Industrial," in *Kirk-Othmer Encyclopedia of Chemical Technology, 4th Ed., Vol. 9: Elastomers, Polyisoprene to Expert Systems*, J.I. Kroschwitz; M. Howe-Grant, Eds., John Wiley & Sons, New York, **1994**, pp. 1022–1060.
3. "Control by Combustion," in *Air Pollution Control Engineering*, N. De Nevers, McGraw-Hill, New York, **1995**, pp. 300–315.
4. "Incineration," in *Standard Handbook of Environmental Engineering*, R.A. Corbitt, McGraw-Hill, New York, **1990**, pp. 4.69–4.72.
5. G.T. Joseph; D.S. Beachler, "Combustion Equipment Used for Control of Gaseous Emissions," in *APTI Course 415: Control of Gaseous Emissions, Student Manual*, United States Environmental Protection Agency, Air Pollution Training Institute, Research Triangle Park, NC, **1981**, pp. 3-15–3-31.
6. D.R. van der Vaart; J.J. Spivey; W.M. Vatavuk; A.H. Wehe, "Thermal and Catalytic Incinerators," in *OAQPS Control Cost Manual*, 5th Ed., United States Environmental Protection Agency, Office of Air Quality Planning and Standards, Research Triangle Park, NC, **1995**, Chapter 3; available at http://www.epa.gov/ttn/catc/products.html.
7. D.C. Mascone, "Thermal Incinerator Performance for NSPS," memo to J.R. Farmer, dated June 11, **1980**, as described in *OAQPS Control Cost Manual*, 5th Ed., United States Environmental Protection Agency, Office of Air Quality Planning and Standards, Research Triangle Park, NC, Chapter 3.
8. D.C. Mascone, "Thermal Incinerator Performance for NSPS Addendum," memo to J.R. Farmer, dated July 22, **1980**, as described in *OAQPS Control Cost Manual*, 5th Ed., United States Environmental Protection Agency, Office of Air Quality Planning and Standards, Research Triangle Park, NC, Chapter 3.
9. D.C. Mascone, "Thermal Incinerators and Flares," memo to J.R. Farmer, dated August 22, **1980**, as described in *OAQPS Control Cost Manual*, 5th Ed., United States Environmental Protection Agency, Office of Air Quality Planning and Standards, Research Triangle Park, NC, Chapter 3.
10. R.Y. Purcell; G.S. Shareef, "Thermal Incineration," in *Control Technologies for Hazardous Air Pollutants*, Air and Energy Engineering Research Laboratory, Office of Research and Development, U.S. Environmental Protection Agency, Research Triangle Park, NC, **1986**, pp. 47–59.

11. "Thermal Oxidation," in *Assessment and Control of VOC Emissions from Waste Treatment and Disposal Facilities*, T.T. Shen; C.E. Schmidt; T.R. Card, Van Nostrand Reinhold, New York, **1993**, pp. 164–166.
12. S.M. Weiss, "Catalytic Afterburners," in *Air Pollution Engineering Manual*, 2nd Ed., J.A. Danielson, Ed., United States Environmental Protection Agency, Research Triangle Park, NC, **1973**, pp. 179–181.
13. "Equipment Characteristics — Thermal Incineration," in *Air Pollution: The Emissions, The Regulations, and The Controls*, J.P. Tomany, American Elsevier Publishing Company, New York, **1975**, pp. 231–251.
14. W.L. Polglase, "Boilers Used as Afterburners," in *Air Pollution Engineering Manual*, 2nd Ed., J.A. Danielson, Ed., United States Environmental Protection Agency, Research Triangle Park, NC, **1973**, pp. 183–189.
15. Rolke, 1972 as cited by G.T. Joseph; D.S. Beachler, "Combustion Equipment Used for Control of Gaseous Emissions," in *APTI Course 415: Control of Gaseous Emissions, Student Manual*, United States Environmental Protection Agency, Air Pollution Training Institute, Research Triangle Park, NC, **1981**, pp. 3-15–3-31.
16. R.Y. Purcell; G.S. Shareef, "Vapor Emissions Control," in *Control Technologies for Hazardous Air Pollutants*, Air and Energy Engineering Research Laboratory, Office of Research and Development, U.S. Environmental Protection Agency, Research Triangle Park, NC, **1986**, pp. 23–33.
17. "Air Pollution Control Methods," in *Kirk-Othmer Encyclopedia of Chemical Technology, 4th Edition, Volume 1: A to Alkaloids*, J.I. Kroschwitz; M. Howe-Grant, Eds., John Wiley & Sons, New York, **1991**, pp. 749–825.
18. J.M. Tyler, "Combustion Systems," in *Handbook of Air Pollution Control Engineering and Technology*, J.C. Mycock; J.D. McKenna; L. Theodore, Eds., Lewis Publishers, Boca Raton, **1995**, pp. 199–221.
19. B.J. Glowiak; K. Grefen; H. Jüntgen; K. Kirchner; E. Pilarczyk; F. Sporenberg; E. Weber, "Thermal Combustion," in *Ambient Air Pollutants from Industrial Sources: A Reference Handbook*, M.J. Suess; K. Grefen; D.W. Reinish, Eds., Published for World Health Organization Regional Office for Europe, Elsevier, Amsterdam, **1985**, pp. 714–722.
20. "Afterburner Systems Study," Report EPA-R2-062, United States Environmental Protection Agency, Research Triangle Park, NC, **1972**.
21. "Technological Fact Sheet: Biothermica International, Inc.; Regenerative Thermal Oxidizer (BIOTOX®) Treatment of Air and Gas," available at http://www.enviroaccess.ca/fiches_2/F2-01-95a.html, **1998**.
22. "Fume Oxidizers for Destruction of VOCs, Odors, and Smoke," Glenro Inc. sales literature, Glenro Inc., Paterson, NJ, **1998** (http://www.glenro.thomasregister.com).
23. "Anguil Thermal Oxidizer System," sales literature from Anguil Environmental Systems, Inc., Milwaukee, WI, **1998**.
24. J.M. Klobucar, "Choose the Best Heat Recovery Method for Thermal Oxidizers," *Chem. Eng. Prog.*, April **1995**, pp. 57–63.
25. "Cor-Pak Recuperative Oxidizer Value Added Advantages," in ABB Air Preheater, Inc., Wellsville, NY, sales literature, number TS-25.
26. W. Elnicki, "Thermal Incineration Economics," in *Air Pollution Control and Design Handbook* (in Two Parts), P.N. Cheremisinoff; R.A. Young, Eds., Marcel Dekker, New York, **1977**, pp. 469–482.

27. "Twin-Bed Regenerative Oxidizer System," sales literature from Anguil Environmental Systems, Inc., Milwaukee, WI, **1998**.
28. Chemical Section Engineers, New Source Review Division of Texas Natural Resource Conservation Commission, *Air Permit Technical Guidance for Chemical Sources: Flares and Vapor Oxidizers*, Texas Natural Resources Conservation Commission, **1998**, available at http://www.tnrcc.state.tx.us/air/nsr_permits/guidedoc.html.
29. R.A. Zerbonia; J.J. Spivey; S.K. Agarwal; A.S. Damle; C.W. Sanford, "Incineration," in *Survey of Control Technologies for Low Concentration Organic Vapor Gas Streams*, Control Technology Center, U.S. Environmental Protection Agency, Research Triangle Park, NC, **1995**, Section 2.
30. Martin et al., **1993**, as cited in *Survey of Control Technologies for Low Concentration Organic Vapor Gas Streams*.
31. N. Mukhopadyay; E.C. Moretti, "Thermal Oxidation (Thermal Incineration)," in *Current and Potential Future Industrial Practices for Reducing and Controlling Volatile Organic Compounds*, Center for Waste Reduction Technologies, American Institute of Chemical Engineers, New York, **1993**, pp. 48–49.
32. A.J. Buonicore, "Incineration," in *Air Pollution Engineering Manual*, A.J. Buonicore; W.T. Davis, Eds., Van Nostrand Reinhold, New York, **1992**, pp. 58–62.
33. "Incineration," in *Air Pollution Control Equipment, Volume II: Gases*, L. Theodore; A.J. Buonicore, CRC Press, Boca Raton, **1988**, pp. 115–136.
34. J. Wei; C.D. Prater, "A New Approach to First Order Chemical Reaction Systems," *Am. Inst. Chem. Eng. J.*, 9, 77, **1963**.
35. J. Wei; C.D. Prater, *Advances in Catalysis*, Vol. 13, Academic Press, New York, **1962**.
36. U.S. Bureau of Mines, "Limits of Flammability of Gases and Vapors," U.S. Bureau of Mines, Washington, DC, Bulletin Number 503 [no date].
37. R.D. Ross, "Thermal Incineration," in *Air Pollution Control and Design Handbook* (in Two Parts), P.N. Cheremisinoff; R.A. Young, Eds., Marcel Dekker, New York, **1977**, pp. 449–468.
38. "Concepts of Combustion as an Emission Control Method," in *APTI Course 415: Control of Gaseous Emissions Course*, United States Environmental Protection Agency, Richmond, Virginia, **1991**.
39. C.D. Cooper; F.C. Alley, "VOC Incinerators," in *Air Pollution Control: A Design Approach*, 2nd Ed., Waveland Press, Prospect Heights, IL, **1994**, pp. 337–378.
40. "Incinerators for Vapors and Gases," in *Kirk-Othmer Encyclopedia of Chemical Technology, 4th Ed., Vol. 14, Imaging Technology to Lanthanides*, J.I. Kroschwitz; M. Howe-Grant, Eds., John Wiley & Sons, New York, **1995**, pp. 118–120.
41. S.M. Weiss, "Preheating of Afterburner Inlet Gases," in *Air Pollution Engineering Manual*, 2nd Ed., J.A. Danielson, Ed., United States Environmental Protection Agency, Research Triangle Park, NC, **1973**, pp. 182–183.
42. *Catalytic Control of VOC Emissions, A Cost Effective Viable Techonology for Industrial, Commercial, and Waste Processing Facilities*, Manufacturers of Emission Controls Association, Washington, DC, **1992**.
43. "Incinerators," in *Estimating Costs of Air Pollution Control*, W.M. Vatavuk, Lewis Publishers, Boca Raton, **1990**, pp. 141–156.
44. K. Noll; W.J. Franek; L. Stander, "Lesson 21: Methods of Emission Control—Incineration," in *APTI Course 482: Sources and Control of Volatile Organic Air*

Pollutants, United States Environmental Protection Agency, Manpower and Technical Information Branch of the Office of Air Quality Planning and Standards, Research Triangle Park, NC, **1987**, pp. 21-1–21-7.

45. EPA, *Controlling Pollution from the Manufacturing and Coating of Metal Products*, US Environmental Protection Agency, Research Triangle Park, NC, **1977**.
46. R.J. Martin; R.E. Smyth; J.T. Schofield, "Elimination of Petroleum Industry Air Toxic Emissions with a Flameless Thermal Oxidizer," Petro-Safe '93, Houston Texas, January 29, **1993**.
47. R.J. Heinsohn; R.L. Kabel, "Formation of Nitrogen Oxides," in *Sources and Control of Air Pollution*, Prentice Hall, Upper Saddle River, NJ, **1999**, pp. 300–308.
48. "Combustion By-Products," in *Steam: Its Generation and Use*, Babcock & Wilcox, New York, **1978**, Chapter 18, pp. 18-1–18-6.
49. V.S. Katan; W.M. Vatavuk; A.H. Wehe, "Incineration Techniques for Control of Volatile Organic Compound Emissions, Part II: Capital and Annual Operating Costs," *J Air Poll. Cont. Assoc.*, **1987**, 37, (2), pp. 198–201.
50. E.N. Ruddy; L.A. Carroll, "Select the Best VOC Control Strategy," *Chem. Eng. Prog.*, July **1993**, pp. 28–35.
51. W.M. Vatavuk, *Estimating Costs of Air Pollution Control*, Lewis Publishers, Boca Raton, **1990**, pp.141–156.
52. C.F. Gottschlich, "Combustion," in *Air Pollution, 3rd Ed., Vol. IV: Engineering Control of Air Pollution*, A.C. Stern, Ed., Academic Press, New York, **1977**, pp. 365–376.
53. T.T. Shen; C.E. Schmidt; T.R. Card, "Thermal Oxidation," in *Assessment and Control of VOC Emissions from Waste Treatment and Disposal Facilities*, Van Nostrand Reinhold, New York, **1993**, pp. 164–166.
54. L. Theodore; A.J. Buonicore, "Thermal and Catalytic Combustion Reactors and Flares," in *Industrial Control Equipment for Gaseous Pollutants*, Vol. II, CRC Press, Cleveland, OH, **1975**, pp. 75–87.
55. L. Theodore; A.J. Buonicore; J.D. McKenna; I.J. Kugelman; J.S. Jeris; J.J. Santoleri; T.F. McGowan, "Waste Management," in *Perry's Chemical Engineers' Handbook*, 7th Ed., D.W. Green; J.O. Maloney, Eds., McGraw-Hill, New York, **1997**, Ch. 25, pp. 25-1–25-111.

7

FLARING

7.1 TECHNOLOGY DESCRIPTION

Flaring is a type of combustion process, referred to as *complete combustion*, in which the combustion reaction occurs instantaneously at the burner tip [1], with the air diffusing across the combustion boundary toward the core of fuel gas [2]. The operating temperature is generally between 1370 and 1920 K (1090° and 1650°C, 2000° and 3000°F) [3, 4]. Flares are frequently used to control emergency venting of pollutants during process upsets [5]. Although the mechanisms are similar to those of thermal oxidation, flares have historically been treated separately from thermal oxidizers because it is often extremely difficult to sample the emissions from a flare. Instead, emissions are frequently calculated from emission factors developed in the early 1980s [6]. Mixing and temperature are the critical parameters of properly designed flares, because there is no residence chamber to continue the combustion reaction [7]. Although flares constitute relatively simple control devices, there are many concerns about their operation, including noise, smoke, energy consumption, and safety [1]. Safety concerns encompass both the large amounts of heat released and the potential for releasing combustion by-products [8]. A typical flow diagram of a flare is given in Fig. 7.1 [9].

Flares are prevalent in the petrochemical industry [10, 11] for both control of emissions at steady state and destroying pollutants during unstable operating conditions such as start-ups and shutdowns [12]. They are particularly well suited to processes with variable gas flow because they can be designed to operate over a wide range of flow (Table 7.1). Flares can handle a gas stream with a maximum flow rate 1000 times as high as the minimum flow rate [7]. In addition, they can handle gas flows at least as high as 2 million pounds per hour [12]. If properly designed, flares are capable of efficiencies rivaling those of incinerators and

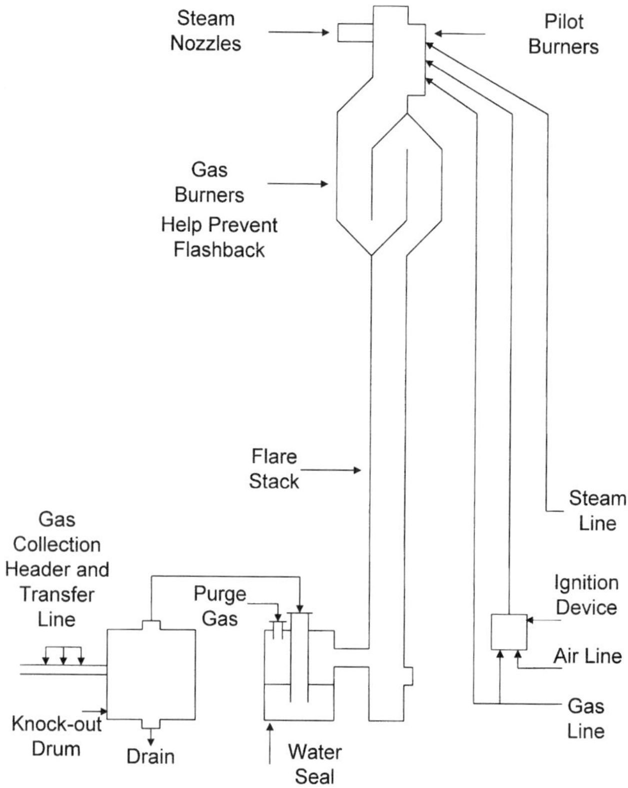

Figure 7.1 Typical flare diagram [9].

TABLE 7.1 Typical Applications of Flaring

Acrylonitrile product loading
Refineries
Dangerous gases such as hydrogen cyanide and ammonia
Landfills
Sewage treatment plants
Dimethylamine odors from soap manufacturing

Data adapted from refs. 7, 15–18.

absorbers. On gas streams with heat contents of 12.2 MJ/Nm^3 (300 Btu/scf) or higher, 98% destruction is feasible. In fact, this is required by EPA regulations for some flares [2, 13].

The color of the flare flame indicates the mechanisms by which combustion is occurring. A blue flame indicates that the fuel and air mixture is being heated gradually to form aldehydes and ketones, which are then reacted to form carbon dioxide and water. In contrast, a yellow-orange flame indicates that cracking of the fuel is occurring and that the hydrocarbons have not had adequate opportunity to react with available oxygen so that elemental carbon and hydrogen are formed. The carbon particles cause the flame to glow yellow. If the flame is not adequately mixed or of sufficient temperature, the carbon particles cool below the carbon ignition temperature and form black smoke or soot [2, 7].

As with other combustion techniques, the application of flaring is limited to situations in which the pollutants are not valuable enough to make recovery economically viable. Furthermore, it is generally also limited to gas streams with relatively concentrated organic constituents, although less concentrated streams can be flared if the additional cost of auxiliary fuel can be justified due to the toxicity of the pollutant or to the sensitivity of local citizens.

In general, flaring is not a suitable control technology for halogenated compounds because these compounds require a longer residence time to achieve complete combustion and accelerate corrosion of the nozzle [14]. In addition, it is impossible to treat the acid by-products that result from burning halogenated compounds. If thermal incineration is employed instead, the off-gases can be treated by a scrubber before they are released to the atmosphere.

7.2 TYPES OF FLARES

There are four types of flares, classified according to the method used to provide for adequate contact at the burner tip. The most common is the steam-assisted flare, which is generally used in chemical processes and refinery operations [2, 19, 20]. Steam injected into the flame provides mixing between the air and the gas for better combustion and helps to pull air into the combustion zone [2, 13]. The steam also assists in the combustion efficiency by reacting with intermediate combustion products to form oxygenated compounds that combust at lower temperatures [2, 7]. It also helps maintain a lower temperature in the flame to reduce thermal cracking [2]. The type and concentration of pollutants determine the amount of steam needed for good combustion; the mass ratio of steam to air is usually in the range of 0.01–0.6 [2, 7, 10]. Steam-assisted or smokeless flares are used to control emissions of organic compounds that have a lower ratio of hydrogen to carbon [18].

Other types of flares are air assisted, in which air is blown into the combustion zone to improve mixing and to provide oxygen for combustion; nonassisted, in which a gas stream with low heat content and easily combustible gases that are resistant to thermal cracking is burned without any mixing enhancement; and pressure assisted, in which the gas stream itself is pressurized to enhance the mixing. Air- and pressure-assisted flares are advantageous in that they can be used in areas without available steam. [2]

7.3 FLARE DESIGN

A flaring system generally includes a collection header for the waste gas, followed by a knock-out pot (if necessary) to remove liquid droplets. A water seal is used to prevent the flame from propagating back down the stack and into the process if the velocity of the gas is reduced. As the gas travels through the flare, pilot burners ignite the gas stream to destroy the VOCs [21]. The diameter of the flare stack and nozzle must be sized appropriately so that the flame does not propagate back into the stack because of too low a stack gas velocity or get extinguished because of too high a flow rate. The EPA has developed regulations to ensure that flares are designed to achieve high destruction efficiencies [22]. The EPA guidelines for steam-assisted flares include the following [2]:

1. For gas streams with heat contents greater than 41 MJ/Nm^3 (1000 Btu/scf), the maximum exit velocity is 122 m/s (400 ft/s). For other gas streams, the maximum exit velocity is determined from the net heating value (Bv) by the following equation:

   In SI units:

   $$V_{max} = 10^{(Bv/0.0404+1214)/852} \tag{7.1}$$

   In English units:

   $$V_{max} = 10^{(Bv+1,214)/852} \tag{7.2}$$

   where

   V_{max} = maximum exit velocity, m/s (or ft/s for English units)
   Bv = net heating value, MJ/Nm^3 (or Btu/scf)

2. The net heating value of the gas must be at least 12.2 MJ/Nm^3 (300 Btu/scf).
3. A flame must be present at any time emissions may be vented.
4. No visible emissions (smoke) is allowed from the flare.

The minimum nozzle diameter (D_{min}) can be calculated from the volumetric flow rate (Q) and the maximum velocity. The diameter is normally sized so that the maximum volumetric flow rate is conveyed through the nozzle at approximately 80% of the allowable velocity, V_{max} [2]:

In SI units:

$$D_{min} = 162.5(Q_{min}/V_{max})^{1/2} \tag{7.3}$$

In English units:

$$D_{min} = 1.95\left(\frac{Q}{V_{max}}\right)^{1/2} \tag{7.4}$$

where

D_{min} = minimum nozzle diameter, (mm or in)

Q = outlet maximum volumetric flow rate (actual m^3/ min or actual ft^3/ min)

Safety concerns associated with flares are obvious: Because flares are usually installed on streams in which the gas exceeds the lower explosive limit (LEL), a flame will propagate very quickly through the ducting if adequate suppression systems are not installed. Explosive limits for several compounds are listed in Table 7.2. The flare must also be installed at a safe elevation. A conservative calculation of the required distance from exposed objects can be made from the following equations [2]:

In SI units:

$$L = 0.02078\left(\frac{mB_m}{K}\right)^{1/2} \tag{7.5}$$

In English units:

$$L = \left(\frac{0.2mB_m}{4K}\right)^{1/2} \tag{7.6}$$

TABLE 7.2 Flammability Limits in Air

Pollutant	Lower Explosive Limit (Vol %)	Upper Explosive Limit (Vol %)
Hydrogen	4.00	74.20
Carbon monoxide	12.50	74.20
Methane	5.00	15.00
Ethane	3.00	12.50
Propane	2.12	9.35
n-Butane	1.86	8.41
Isobutane	1.80	8.44
n-Hexane	1.18	7.4
Ethylene	2.75	28.60
Propylene	2.00	11.10
n-Butene	1.75	9.70
Benzene	1.40	7.10
Toluene	1.27	6.75
Xylene	1.00	6.00
Methyl alcohol	6.72	36.50
Ethyl alcohol	3.28	18.95
Ammonia	15.50	27.00
Hydrogen sulfide	4.3	45.50
Gasoline	1.40	7.60

Data from Tyler [7].

TABLE 7.3 Instrumentation to Consider for Flare Systems

Gas stream pressure and flow (particularly if flow is variable)
Concentration or % LEL
Fan speed
Auxilliary fuel flow

where

L = the required distance, m (or ft)
m = mass flow rate, kg/h (or lb/h)
B_m = heat content of the gas stream, kJ/kg (or Btu/lb)
K = allowable radiation, assumed to be 49 kJ h^{-1} m^{-2} (500 Btu h^{-1} ft^{-2})

Specifications in the state of Texas for steam-assisted flares include the guideline that at least 0.4 kg (or lb) of steam be used per kilogram (or pound) of waste gas when the H/C ratio is less than 0.3. In addition, the guideline recommends that an H/C ratio higher than 1 be avoided because the VOC destruction efficiency may be decreased [6]. Additional design information is given by Swithenbank [23]. Instrumentation that should be considered when installing a flare system is listed in Table 7.3.

7.4 ECONOMIC ESTIMATES

As shown in the following equations, the type of support required for the flare stack can substantially affect the cost of the system. Three types of flare structures are included: self-supported flares, derrick-supported flares, and flares supported by guy wires. Figures 7.2 [2], 7.3 [2], and 7.4 [2] show examples of support structures that can be used for flares. Operating costs are governed primarily by the amount of auxiliary fuel, steam, and electricity needed.

Estimates of the capital cost of the flare can be obtained from the following equations [2]. The figures given are in 1990 dollars and should be scaled up to the current year using escalation factors available on the Clean Air Technology Center Web site (http://www.epa.gov/ttn/catc).

In SI units:

$$C_F = a + bD/25.4 + cL/3.2808 \tag{7.7}$$

In English units:

$$C_F = (a + bD + cL)^2 \tag{7.8}$$

where

D = tip diameter, mm (or in)
L = Height of flare stack, m, 9.14 m at a minimum (or ft, 30 feet at a mimimum).

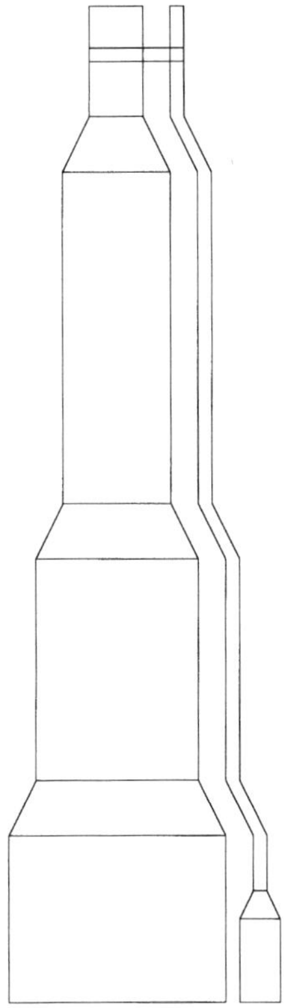

Figure 7.2 Self-supported flare [2].

Factors a, b, and c are specific to the type of support structure as given in Table 7.4.

These cost estimates include a carbon steel flare stack with the upper 1.2 m (4 ft) made of 310 stainless steel, with 310 stainless steel burner tip, pilot lights, liquid and gas seals, access ladders and platforms as required. They also include utility piping from the base of the stack, metering, and control. The equations are generally valid for tip diameters from 25 to 1520 mm (1–60 inches) and for stacks from 9.1 to 152 m (30–500 ft).

TABLE 7.4 Factors for Capital Cost Equation

Type of Support	a	b	c
Self-supported	78.0	9.14	0.749
Guy wires	103	8.68	0.470
Derrick supported	76.4	2.72	1.64

Data from Stone, et al. [2].

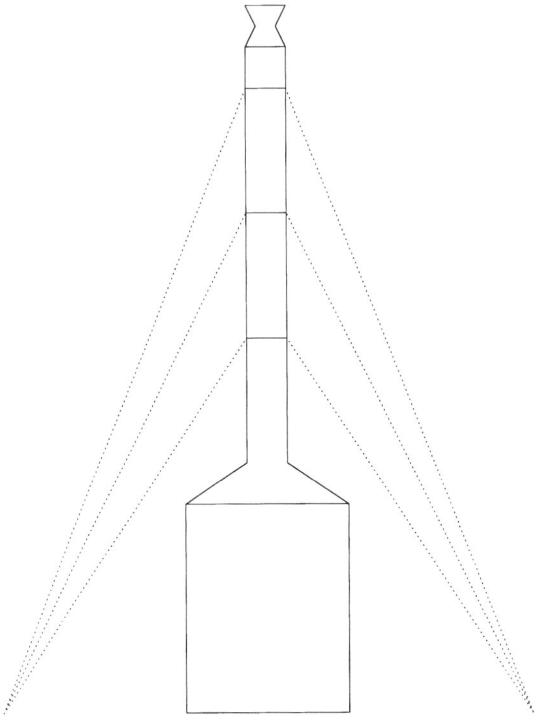

Figure 7.3 Guyed Wire–supported flare [2].

In addition to the cost of the flare itself, costs for vent piping and a knockout pot must also be included in the overall estimate of equipment cost. The approximate size of a vertical knockout pot on a relatively constant flow vent stream can be estimated from the following equations:
In SI units:

$$A = \frac{Q_a}{197G\left(\dfrac{\rho_1 - \rho_v}{\rho_v}\right)^{1/2}} \tag{7.9}$$

In English units:

$$A = \frac{Q_a}{60G\left(\dfrac{\rho_1 - \rho_v}{\rho_v}\right)^{1/2}} \tag{7.10}$$

where

A = cross-sectional area of the knockout vessel, m^2 (or ft^2)
Q_a = actual flow rate of vent stream, m^3/min (or ft^3/min)

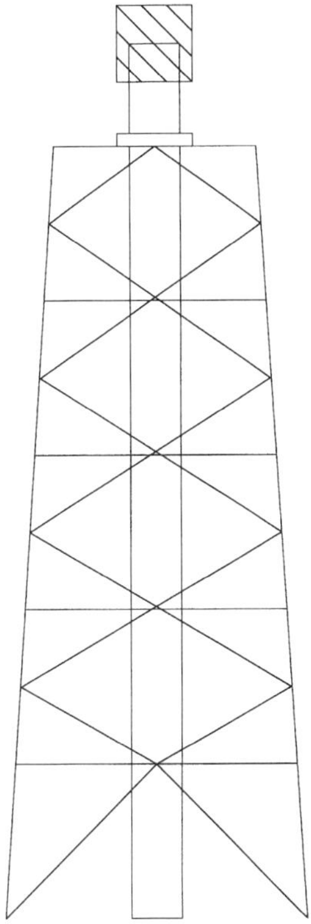

Figure 7.4 Derrick-supported flare [2].

G = vapor velocity factor, usually 0.15–0.25
ρ_l = density of the liquid to be removed in the knock out pot, kg/m^3 (or lb/ft^3)
ρ_g = density of the gas stream, kg/m^3 (or lb/ft^3)

Once the cross-sectional area of the vessel is known, the diameter can be calculated from the following equations [2]:
In SI units:

$$d_{min} = \sqrt{\frac{4}{\pi} A} \tag{7.11}$$

In English units:

$$d_{min} = 13.5\sqrt{A} \tag{7.12}$$

where

d_{min} = minimum diameter of the vessel, m (or ft)

For an actual diameter, d_K, the minimum should be rounded to the next highest standard size. Vessels are usually available in 0.152 m (0.5 ft) increments. The thickness of the vessel, t, is determined from Table 7.5.

The height of the knockout drum, h (m or ft), is estimated as three times the actual diameter [2]. With the diameter and height of the knockout drum, the cost of the knockout drum, C_K, is estimated as [2, 24, 25]:

In SI units:

$$C_K = 2.1\,[d_K\,t(h + 0.812\,d_K)]^{0.737} \tag{7.13}$$

In English units:

$$C_K = 0.364\,[d_K t(h + 0.812\,d_K)]^{0.737} \tag{7.14}$$

The cost of piping the vent stream to the flare stack, C_P, is calculated separately from the following equation [2], assuming that 31 m (100 ft) of piping is required that has the same diameter as the tip of the flare [26]:

In SI units:

$$C_P = 2.53\,D^{1.21} \text{ for diameters between 25 and 610 mm} \tag{7.15}$$

$$C_P = 4.36\,D^{1.07} \text{ for diameters between 760 and 1520 mm} \tag{7.16}$$

In English units:

$$C_P = 127\,D^{1.21} \text{ for diameters between 1 and 24 in} \tag{7.17}$$

$$C_P = 139\,D^{1.07} \text{ for diameters between 30 and 60 in} \tag{7.18}$$

Piping costs are based on Schedule 40 carbon steel pipe.

The total purchased equipment cost (PEC) includes the cost of the flare, the knockout pot, and the piping, plus instrumentation, sales taxes, and freight. Additional fans or blowers are not included in these calculations, but may be required

TABLE 7.5 Determination of Knockout Pot Wall Thickness

Diameter		Thickness, t	
m	ft	mm	in
Less than 0.914	Less than 3	6.4	0.25
At least 0.914 and less than 1.83	At least 3 and less than 6	9.4	0.37
At least 1.83 and less than 2.74	At least 6 and less than 9	12.7	0.5
At least 2.74 and less than 3.66	At least 9 and less than 12	19.1	0.75
3.66 or over	12 or over	25.4	1

Data adapted from Stone et al. [2].

due to the significant pressure drop in some systems. Instrumentation is estimated as 10% of the equipment cost, sales taxes as 3%, and freight as 5%, for a total of [2]

$$PEC = 1.18(C_F + C_K + C_P) \tag{7.19}$$

As shown in Table 7.6, the total capital investment is estimated as 192% of the purchased equipment price [18].

Annual costs may be approximated as given in Table 7.7. Electricity requirements can be estimated from the formulas given in Chapter 4. The pressure drop for flare systems can be as high as 60 in of water [27], depending on the configuration of the equipment and the length of the gas piping. Steam costs can be estimated as 0.6 lb/lb of waste gas [27]

TABLE 7.6 Factors for Estimating the Total Installed Capital Cost of Flares

Cost Element	Estimated Cost
Direct Costs	
Purchased equipment cost	
Flare and auxiliary equipment (EC)	$EC = C_F + C_K + C_P$
Instrumentation	0.10 EC
Taxes	0.03 EC
Freight	0.05 EC
Total purchased equipment cost, PEC	1.18 EC
Installation costs	
Foundations and supports	0.12 PEC
Erection and handling	0.40 PEC
Electrical	0.01 PEC
Piping, insulation, and painting	0.04 PEC
Total installation cost	0.57 PEC
Site preparation (SP)	As required for specific site
Buildings (BLDG)	As required for specific site
TOTAL DIRECT COSTS	1.30 PEC + SP + BLDG
Indirect Costs	
Engineering and supervision	0.10 PEC
Construction, field expenses, and fees	0.20 PEC
Start-up performance test	0.02 PEC
Contingency	0.03 PEC
TOTAL INDIRECT COSTS	0.35 PEC
Total Installed Capital Costs, TCC	1.92 PEC + SP + BLDG

Data adapted from Stone et al. [2].

TABLE 7.7 Annual Operating Costs for Flare Systems

	Unit Cost	Estimated Requirement
	Direct Annual Costs	
Utilities		
Electricity	\$0.05/kW-h	See Chapter 4
Steam	\$10.25/1000 kg	0.6 kg/kg
	\$4.65/1000 lb	0.6 lb/lb
Natural gas (equation is for elevated flares)	\$127/1000 Nm^3	SI units:
	\$3.30/1000 scf	$2.03 \times 10^{-6} Q_A^{0.934}$ Nm^3/h English units: $0.00218\ Q_a^{0.934}$ scf/h [18, 28]
Operating labor	\$12.96/h	1/2 h per shift
Maintenance labor	\$14.26/h	1/2 h per shift
Maintenance materials	100% of Main. labor	
Supervisory labor	15% of Operating labor	
	Indirect Annual Costs	
Overhead	60% of Maintenance costs	
Property tax	1% of TCC	
Insurance	1% of TCC	
Administrative costs	2% of TCC	
Capital recovery (see Chapter 4 for the formula to calculate capital recovery)	CRF · TCC	
Total Annual Operating Costs	Direct + Indirect costs	

Data adapted from refs 2, 18.

7.5 POTENTIAL PROBLEMS

It is helpful to enclose the sides of the flare with a fibrous insulating material held in place by a lightweight shroud. Such protection reduces the impact of wind turbulence, which can make the flare less effective. The enclosure also reduces heat losses [29, 30]. An open-ended refractory chamber can also be used to increase the residence time of the flare to improve destruction efficiency [31].

The pilot flames need to be designed in such a way that they can be relit, even in severe weather such as high wind and hard rain [31]. Precautions must also be taken to ensure that gas streams that are normally above the explosive limit do not become explosive due to increased temperature, pressure, or increased air infiltration.

TABLE 7.8 **Advantages and Disadvantages of Flares**

Advantages	Disadvantages
Relatively simple equipment and operation	No recovery of contaminant
Capable of high efficiency	Potential corrosion of ductwork due to moisture in gas stream

Liquid droplets must be carefully removed from the gas stream to eliminate the possibility that water will cool the flame or that organic droplets will pass through the flare nozzle and fall out in the atmosphere as flaming particles [21]. Solid particulate may also serve to erode the nozzle and should be removed from the gas stream. If inert material is present, it may be necessary to preheat the gas stream by installing a ring burner in addition to the pilot burners [30].

7.6 INSTALLATION, OPERATION, AND MAINTENANCE CONSIDERATIONS

Corrosion is a potential problem with any control device. In flares, it is usually a problem with high concentrations of sulfur compounds or high concentrations of halogenated compounds. These compounds are better suited to incineration followed by scrubbing to remove the corrosive compounds [28].

Texas requires that all proposed flare systems include a continuous pilot ignition system so that process and emergency flares are relit immediately. In addition, it is recommended that a remote infrared monitor be used to monitor the flame [6]. A summary of the advantages and disadvantages of using flaring systems to control VOCs is given in Table 7.8.

REFERENCES

1. G.T. Joseph; D.S. Beachler, "Combustion Equipment Used for Control of Gaseous Emissions," in *APTI Course 415, Control of Gaseous Emissions, Student Manual*, G.T. Joseph; D.S. Beachler, United States Environmental Protection Agency, Office of Air, Noise and Radiation and Office of Air Quality Planning and Standards, Research Triangle Park, NC, **1981**, pp. 3-15–3-37.
2. D.K. Stone; S.K. Lynch; R.F. Pandullo; L.B. Evans; W.M. Vatavuk, "Flares," in *OAQPS Control Cost Manual*, 5th Ed., US EPA, Research Triangle Park, NC, **1995**, Ch. 7.
3. R.M. Bethea, "Combustion," in *Air Pollution Control Technology: An Engineering Analysis Point of View*, Van Nostrand Reinhold, New York, **1978**, pp. 395–431.
4. A.J. Buonicore; L. Theodore, "Thermal and Catalytic Combustion Reactors and Flares," in *Industrial Control Equipment for Gaseous Pollutants*, Vol. II, CRC Press, Cleveland, OH, **1975**, pp. 75–87.

5. N. De Nevers, "Control by Combustion," in *Air Pollution Control Engineering*, McGraw-Hill, New York, **1995**, pp. 300–315.

6. Chemical Section Engineers, *Air Permit Technical Guidance for Chemical Sources: Flares and Vapor Oxidizers*, New Source Review Permits Division, Texas Natural Resource Conservation Commission, Austin, TX, **1998**, available at www.tnrcc.state.tx.us/air/nsr_permits/guidedoc.htm.

7. J.M. Tyler, "Combustion Systems," in *Air Pollution Control Engineering and Technology*, J.C. Mycock; J.D. McKenna; L. Theodore, Eds., Lewis Publishers, New York, **1995**, pp. 199–221.

8. "Waste Management," in *Perry's Chemical Engineers' Handbook*, 7th Ed., D.W. Green; J.O. Maloney, Eds., McGraw-Hill, New York, **1997**, pp. 25-37–25-41.

9. "Flares," in *Survey of Control Technologies for Low Concentration Organic Vapor Gas Streams*, R.A. Zerbonia; J.J. Spivey; S.K. Agarwal; A.S. Damle; C.W. Sanford; R.J. Blaszczak, Eds., Control Technology Center, U.S. Environmental Protection Agency, Research Triangle Park, NC, **1995**, Sect. 2.4.

10. A.J. Buonicore, "Incineration," in *Air Pollution Engineering Manual*, A.J. Buonicore; W.T. Davis, Eds., Van Nostrand Reinhold, New York, **1992**, pp. 58–70.

11. T.T. Shen, "For VOC Control Seminar—Combustion," in *Control of Gaseous Emissions—VOC's Handbook*, ETS Professional Training Institute, Roanoke, VA, **1995**.

12. "Control Device Selection," in *Control Technologies for Hazardous Air Pollutants*, R.Y. Purcell; G.S. Shareef, US EPA, Research Triangle Park, NC, **1986**, pp. 25–26.

13. "HAP Control Techniques," in *Control Technologies for Hazardous Air Pollutants*, R.Y. Purcell; G.S. Shareef, US EPA, Research Triangle Park, NC, **1986**, pp. 60–63.

14. J. Joseph; D. Beachler; J. Jahnke; M. Leslie; J. Henry, *APTI Course 415, Control of Gaseous Emissions, Instructor's Guide*, United States Environmental Protection Agency, Manpower and Technical Information Branch, Office of Air Quality Planning and Standards, Research Triangle Park, NC, **1981**, Lesson 5.

15. R.D. Bell; A.V. Mechtenberg, "Acrylonitrile," in *Air Pollution Engineering Manual*, A.J. Buonicore; W.T. Davis, Eds., Van Nostrand Reinhold, New York, **1992**, pp. 404–409.

16. J.E. Rucker; R.P. Streiter, "The Petroleum Industry," in *Air Pollution Engineering Manual*, A.J. Buonicore; W.T. Davis, Eds., Van Nostrand Reinhold, New York, **1992**, pp. 811–834.

17. J.P. Bilotti; J.K. Sutherland, "Incinerators," in *Air Pollution Control Equipment: Selection, Design, Operation and Maintenance*, L. Theodore; A.J. Buonicore, Eds., ETS, Inc., Roanoke, VA, **1992**, pp. 133–151.

18. W.M. Vatavuk, "Flares," in *Estimating Costs of Air Pollution Control*, Lewis Publishers, Chelsea, MI, **1990**, pp. 179–186.

19. *Guide for Pressure-Relieving and Depressurizing Systems*, 2nd Ed., Refining Department, API Recommended Practice 521, **1982**.

20. V. Kalcevic, "Control Device Evaluation Flares and the Use of Emissions as Fuels," in *Organic Chemical Manufacturing Vol. 4: Combustion Control Devices*, US EPA, Research Triangle Park, NC, **1980**, Report 4.

21. N. Mukhopadhyay; E.C. Moretti, "Flares," in *Current and Potential Future Industrial Practices for Reducing and Controlling Volatile Organic Compounds*, Center for Waste Reduction Techologies, American Institute of Chemical Engineers, New York, **1993**, pp. 21–22.

22. "General Control Device Requirements," 60 CFR §60.18, **1993**.
23. J. Swithenbank, AIChE J., 18, 553, **1972**.
24. A. Mulet, "Estimate Costs of Pressure Vessels Via Correlations," *Chem. Eng.*, Oct. 5, **1981**, pp. 145–150.
25. *Process Plant Construction Estimating Standards*, Richardson Engineering Services, Inc., Vol. 4, **1988**.
26. P. Tkatschenko, letter to D. Stone, May 2, **1990**.
27. W.M. Vatavuk; R.B. Neveril, "Estimating Costs of Air-Pollution Control Systems, Part XV: Estimating Costs of Flares," *Chem. Eng.*, Feb. 21, **1983**, pp. 89–90.
28. D.K. Stone; S.K. Lynch; R.F. Pandullo; L.B. Evans; W.M. Vatavuk, "Flares," in *OAQPS Control Cost Manual*, 4th Ed., United States Environmental Protection Agency, Research Triangle Park, NC, **1990**, Ch. 7.
29. B.B. Crocker, "Air Pollution Control Methods," *Kirk-Othmer Encyclopedia of Chemical Technology, Vol. 1, A to Alkoids*, J.I. Kroschwitz; M. Howe-Grant, Eds., John Wiley & Sons, New York, **1991**, pp. 749–825.
30. C.F. Gottschlich, "Combustion," in *Air Pollution, 3rd Ed., Vol. IV: Engineering Control of Air Pollution*, A.C. Stern, Ed., Academic Press, New York, **1977**, pp. 365–376.
31. "Incinerators," in *Kirk-Othmer Encyclopedia of Chemical Technology, Vol. 14, Imaging Technology to Lanthanides*, J.I. Kroschwitz; M. Howe-Grant, Eds., John Wiley & Sons, New York, **1995**, pp. 749–825.

8

CATALYTIC INCINERATION

8.1 TECHNOLOGY DESCRIPTION

Like thermal incinerators, *catalytic incinerators* destroy organic pollutants by combustion to produce a mixture of carbon dioxide and water. Catalytic incinerators utilize a noble metal or metal oxide solid catalyst material. The catalyst itself does not react, but it enables the combustion reaction to occur at a lower temperature than normal, generally without the presence of a flame. This allows for virtually complete combustion of the pollutants with very little NO_x formation. Although the mechanism of the catalyzed reaction for most organics is not well understood [1], the presence of the catalyst makes a new reaction pathway possible that has a lower activation energy than the uncatalyzed reaction. It is believed that the catalyst promotes the reaction by forming reactive intermediates and bringing them into close proximity on its surface [2, 3].

The catalyst is usually impregnated over a high surface area washcoat like alumina on a porous matrix as shown in Figure 8.1 [4]. As the gas flows through the matrix, the pollutants and oxygen diffuse to the surface of the catalyst, where probably both species (oxygen and the pollutant) adsorb onto the catalyst surface. The type of catalyst and the type of pollutant determine how the species adsorb [1]. At higher temperatures, the combustion reaction can also propagate into the gas phase, initiating homogeneous chain reactions [5]. Catalytic incinerators operate at a lower temperature, usually 620 to 870 K (340°–590°C, 650°–1100°F) [6], than thermal incinerators but, if properly applied, are capable of destruction efficiencies similar to those for thermal incinerators. The lower operating temperature results in less NO_x formation [4] (generally less than 15 ppm [7]) and significantly smaller fuel consumption (reported to be as low as one third) [8–10] than most thermal incinerators.

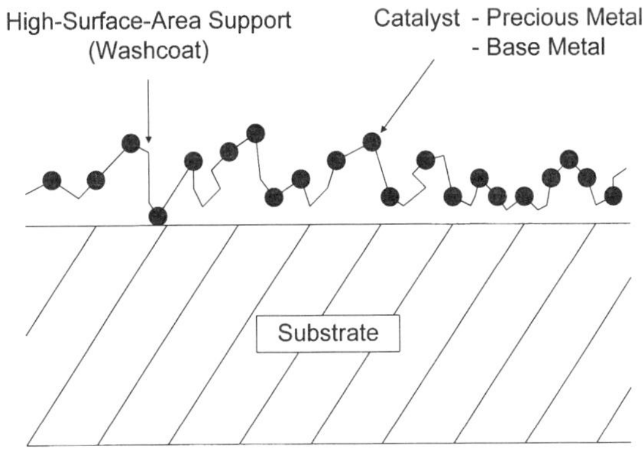

Figure 8.1 Cross-sectional diagram of catalyst composition [4].

A typical flow diagram for a catalytic incinerator is shown in Figure 8.2 [11, 12]. The pollutant-laden gas flows through a heat exchanger to preheat the incoming gas with the heat of the exhaust gases. The incinerator may or may not have a supplemental burner to further preheat the pollutant-laden gas stream. As the preheated gases react going through the catalyst bed, the temperature of the gas stream increases. The reaction generates the same amount of heat as the heat of combustion that is generated in a thermal incinerator [13]. The reacted gas stream then passes back through the heat exchanger and is released to the atmosphere.

Care must be taken to properly identify the constituents of a gas stream before either thermal or catalytic incineration is selected as the control method. Some compounds interfere with the action of the catalyst either by chemically binding to the surface or by adsorbing on to the surface and deactivating sites for future reaction. Compounds that bind chemically to the catalyst are referred to as *poisons* and may be classified as either fast or slow acting. Fast-acting poisons include phosphorus, bismuth, arsenic, antimony, chlorine, fluorine, bromine, and mercury. Zinc, lead, and tin are slow-acting poisons [14, 15]. The portion of the surface

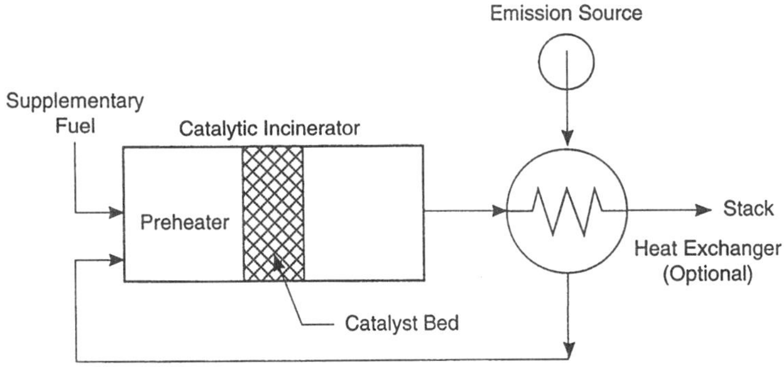

Figure 8.2 Catalytic oxidizer flow diagram [11, 12].

that is bound to the poison is not available for reaction, so the overall reaction rate decreases due to a decrease in the amount of available surface area. Other compounds, such as heavy oils, dust, and silica [16], may coat the surface of the catalyst. If the operating temperature is not high enough to promote combustion of the oily compounds, the surface area will be reduced. Sulfur compounds in the gas stream (including mercaptans, which are added to natural gas and propane to make sure that leaks are detectable [17]) may bind to components in the washcoat to form sulfates (e.g., aluminum sulfate) [4], rendering a portion of the surface unusable. At a minimum, sulfur reacts to form sulfur dioxide, which may need to be removed before venting to the atmosphere. Halogenated compounds in the contaminated gas stream may also interfere with the catalyst action either by binding to the catalyst or by reacting to form acids that can severely corrode the catalyst supports, ductwork, and downstream equipment. Special attention must be given to the catalyst selection to ensure that halogenated compounds do not react to form molecular halogen species like Cl_2 [18]. Catalysts have been developed recently that are resistant to poisoning by halogen compounds [19, 20]. Particulates, oils, sulfur, and halogens are all compounds that can poison the catalyst but that are reversible inhibitors. Once the compound is removed, the activity of the catalyst is restored. Generally, a superheated steam is adequate to remove the compounds from the catalyst surface [21].

Destruction efficiencies of 98% or greater are now common in catalytic incinerators [4]. Destruction efficiencies of 99% are achievable, but the increased temperature or large catalyst volume required may make the technology economically infeasible [22]. The removal efficiency is determined by the type of catalyst, the surface area of the catalyst, the pollutant species, the pollutant concentration, the residence time, the extent of mixing, the uniformity of gas flow through the catalyst, the oxygen concentration, and the temperature [2, 23, 24]. Table 8.1 lists examples of potential uses for catalytic incineration [2, 15, 24–31].

TABLE 8.1 Potential Applications of Catalytic Incineration for VOC Removal

Odor control
Printing operations
Polymer processing
Bakeries
Paint bake ovens
Foundry core ovens
Tire manufacturing
Metal furniture coating
Wire enameling
Textile drying
Gasoline marketing

Data adapted from refs. 2, 15, 24–31.

8.2 TYPES OF CATALYTIC INCINERATION SYSTEMS

Catalytic incinerators can be classified in several different ways. They may be classified according to the type of heat recovery used, the type of catalyst (noble metal or metal oxide catalyst) used, or the shape of the catalyst substrate. Because most gas streams are dilute, heat recovery is normally used to help reduce the amount of supplemental fuel required. Catalytic incinerators may use either recuperative or regenerative heat exchangers (see the discussion in Chapter 6). Most systems use *recuperative* heat exchangers (as shown in Fig. 8.3 [32]) in which the exhaust gases from the incinerator pass through a noncontact heat exchanger to recover the heat of combustion and increase the fuel efficiency of the VOC destruction process. A few *regenerative* systems have been introduced (see Fig. 8.4 [33]) that use beds of inert ceramic material as a heat sink [34, 35]. Regenerative incinerators allow for better heat recovery, meaning that dilute vapor streams can be incinerated at lower costs [36]. The type of heat recovery system must be chosen carefully both to avoid higher than necessary energy costs and to ensure that efficient energy recovery does not result in too high a catalyst temperature.

Catalytic systems may also be classified according to the form of the catalyst material. The primary objective in design of a catalyst is to provide as much surface area as possible to promote the reaction with minimal pressure drop. Both fixed beds and fluid beds are used to accomplish this. A typical catalyst consists

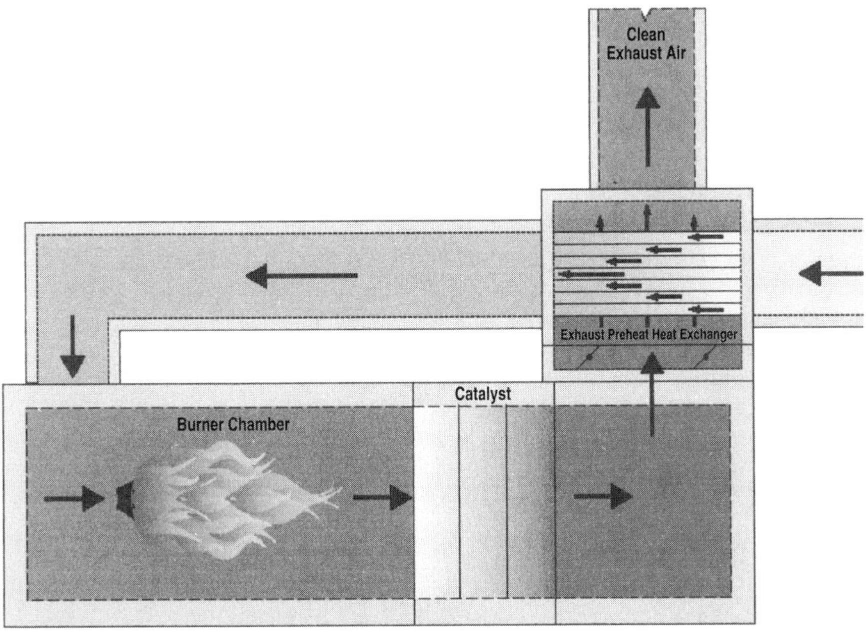

Figure 8.3 Catalytic system with recuperative heat exchange, Glenro, Inc. [32].

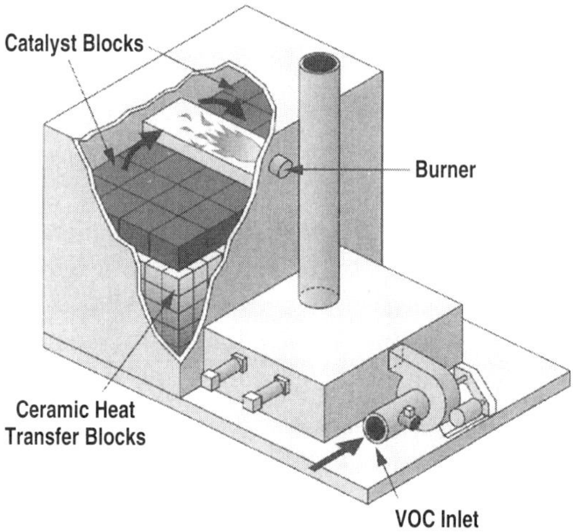

Figure 8.4 Anguil regenerative catalytic oxidizer [33].

of a ceramic or metal substrate coated with an alumina washcoat to provide surface area and then a noble metal or metal oxide on the surfaces [35]. Precious metals such as platinum, palladium, and rhodium are frequently used; they usually allow for more reactivity, which results in a lower operating temperature and shorter residence time [37, 38]. Substrates for fixed beds may be in the shape of honeycombs, rods, or metal ribbon [2]. Fixed beds may also contain catalysts in the form of beads or pellets. Fluid beds generally consist of a tray or chamber filled with spherical or pellet-shaped catalysts. Although pellet-shaped catalysts are more susceptible to attrition, this can be a benefit for gas streams that contain more particulate [4]. The attritive action serves to clean off the surfaces of the catalyst beads.

Hybrid systems have also been introduced that combine adsorption with catalytic combustion to increase the concentration of pollutants for the combustion process. The VOCs from the gas stream are adsorbed onto zeolite; then they are desorbed with a smaller gas stream. This results in lower energy usage because less air must be heated to the combustion temperature. Although the system is more complex, it does provide an alternative that uses less auxiliary fuel [39]. Chapter 6 contains a diagram of one such system [40].

8.3 CATALYTIC INCINERATOR SYSTEM DESIGN

It is generally not feasible to empirically predict the action of a given catalyst and organic compound with the accuracy required for designing process equipment [13, 41]. Therefore, it is highly encouraged that a catalytic system vendor be contacted to determine an appropriate type and volume of catalyst for the specific VOC mixture in the gas stream. It may also be helpful to use a test unit, particularly if poisons or deactivators may be present in the gas stream [4].

The amount of gas to be treated should be minimized to reduce the capital and operating costs of the incinerator. Usually, the gas streams treated with catalytic incineration are so dilute that no additional combustion air is required. If the concentration of VOC is significant, care must be exercised to ensure that the catalyst does not overheat and that the concentration of pollutant in the gas does not exceed a safe level in the ducting, especially during process upsets or startup conditions. See Chapter 7 and the U.S. Bureau of Mines information [42, 43] for additional information about acceptable ductwork concentrations and lower explosive limits (LELs). The heating value of hydrocarbons at the LEL has been determined to be approximately 2030 kJ/Nm^3 (50 Btu/scf). For catalytic units, the heat of combustion should normally be less than approximately 400 kJ/Nm^3 (10 Btu/scf) so that the catalyst temperature does not become excessive. Regenerative heat recovery would, however, reduce this number further. Similarly, if the oxygen content of the gas stream is lower than approximately 20%, auxiliary air may be required for the combustion process [41].

The inlet temperature to the catalytic bed must be higher than the catalytic ignition temperature to ensure that the reaction occurs [41]. Incomplete combustion can lead to emissions of compounds with more severe health effects than the original hydrocarbon gas stream [1, 44]. Many catalytic incinerators use a preheat burner to raise the temperature of the gas stream higher than that obtained in the heat exchanger. This direct combustion also has the effect of creating a high chemical activity [45].

If the gas stream contains halogenated or sulfur compounds, a scrubber may be required to remove acidic gases after treatment. Depending on the catalyst chosen, these compounds may need to be removed before the catalyst. In either case, the scrubber or removal mechanism should be considered as an integral part of the incinerator system and should not be overlooked even in preparation of rough cost estimates because the cost of the scrubber can be a significant portion of the overall system cost [46].

Procedures for the basic design of the heat exchange and fuel consumption for a catalytic incinerator are given in the *OAQPS Control Cost Manual* [41]. The procedures were developed to ensure that the catalyst temperature remains above the catalytic ignition temperature but below about 920 K (650°C, 1200°F), at which damage to the catalyst may result. The first step is to determine the appropriate outlet temperature of the catalyst bed. Combustion Engineering suggests that this should be between approximately 420 and 760 K (150° and 480°C, 300° and 900°F) to achieve 90–95% destruction [47]. The exact value will, however, depend on the constituents of the gas stream as well as the type of catalyst. Vendors should be consulted to accurately determine an appropriate outlet temperature. The temperature of the polluted gas stream exiting the heat exchanger is then determined from the equation:

$$T_i = T_w + \mathrm{Eff}(T_c - T_w) \tag{8.1}$$

where

T_c = catalyst outlet temperature, °C (or °F)
T_r = reference temperature; 21°C (70°F) is typical
T_i = inlet gas stream temperature, after heat exchanger, °C (or °F); generally 260°–320°C (500°–600°F) [48]
T_w = temperature of inlet gas stream before heat exchanger, °C (or °F)
Eff = anticipated fractional thermal efficiency of the incinerator (usually between 0.5 and 0.95)

Procedures for calculating the total flue gas flow (i.e., the flow rate of gases leaving the incinerator) are given in the *OAQPS Control Cost Manual* [41] and by Vatavuk [49]:

$$Q = Q_w + Q_f \tag{8.2}$$

where

Q = total flue gas flow rate, normal m^3/min, abbreviated Nm^3/min (in English units, standard ft^3/min, abbreviated scfm)
Q_w = waste gas flow rate at inlet to incinerator, Nm^3/min (scfm)
Q_f = flow rate of auxiliary fuel, Nm^3/min (scfm)

The flow rate of auxiliary fuel is calculated from the equation:

$$Q_f = Q_W X/Y \tag{8.3}$$

where

$X = 1.1\,C_{po}(T_c - T_r) - C_{pi}(T_i - T_r) - h_w$
$Y = h_f - 1.1C_{po}(T_c - T_r)$
C_{po} = mean heat capacity of exhaust stream from the incinerator, kJ min/Nm^3 °C (Btu scf^{-1} $°F^{-1}$). The heat capacity of air is typically used because most of the gas stream is air. At approximately 670 K (400°C, 750°F), the heat capacity of air is 1.38 kJ/Nm^3 °C (0.0181 Btu/scf °F)
C_{pi} = mean heat capacity of inlet streams to the incinerator (kJ min/Nm^3 °C Btu scf^{-1} $°F^{-1}$). Again, the heat capacity of air is typically used because most of the incoming stream is air. (At 311 K, 38°C, 100°F), the heat capacity of air is 1.30 kJ/Nm^3 °C (0.017 Btu/scf °F)
h_w = heat content of waste gas stream, kJ/Nm^3 (Btu/scf)
h_f = Heat content of fuel, kJ/Nm^3 (Btu/scf). For natural gas, h_f is approximately 40,200 kJ/Nm^3 (1000 Btu/scf)

h_f may be calculated in SI units from the following equation if the heat content is known in kJ/kg: h_f = heat content in kJ/kg × fuel molecular weight in kg/kg-mol/22.4 Nm^3 per kg-mol. In English units, h_f can be calculated from the known heat content in Btu/lb: h_f = heat content in Btu/lb × fuel molecular weight in lb/lb-mol/386.5 ft^3 per lb-mol.

This must result in a positive value for the fuel consumption, or the amount of heat that is being recovered is too great to maintain the outlet temperature at the desired value. In addition, if there is a preheater combustion chamber, the amount of auxiliary fuel should be at least 5% of the total energy to make sure that the burner flame is stable. If the required fuel consumption is less than 5% of the energy, a lower heat transfer efficiency should be used, and the calculations should be repeated.

If a preheat burner is used, the temperature at the outlet of the preheat burner and the inlet to the catalyst is calculated from the equation:

$$T_{ci} = \frac{\rho_f Q_f[(-\Delta h_f) + 1.1C_{pi}T_r] + \rho_w Q_w C_{pa}(T_i + 0.1T_r)}{1.1C_{pa}(\rho_f Q_f + \rho_w Q_w)} \tag{8.4}$$

where

T_{ci} = catalyst inlet temperature (after the preheater combustion chamber), °C (°F)

ρ_f = density of fuel, 0.6533 kg/m^3 (0.0408 lb/ft^3) for methane at 298 K (25°C, 77°F) and 1 atm

ρ_w = density of polluted gas stream, which can usually be adequately represented by the density of air, 1.183 kg/m^3 (0.0739 lb/scf) for air at 298 K (25°C, 77°F) and 1 atm

The value of T_{ci} that is calculated should be checked against values of catalytic ignition temperatures to ensure that the preheat temperature is high enough to initiate the reaction. The catalytic ignition temperature should be available from the vendor.

The amount of catalyst required will depend on the type of catalyst, the required removal efficiency, the VOCs to be destroyed and the anticipated reaction temperature. If the required residence time and the percentage of void space in the reactor is known, the volume can be estimated. As an approximation, 0.043–0.057 m^3 (1.5–2.0 ft^3) of catalyst is required for every thousand cubic feet of combusted gas to achieve 90–95% destruction [21]. Velocity of the gas is usually approximately 3–6 m/s (10–20 ft/s) through the incinerator [2].

These equations assume a 10% loss of energy from the incinerator. Note that no allowance is included for volume changes due to changes in the total number of moles in the system that may occur during reaction. Additional correction may be needed for gas streams that contain large concentrations of organic compounds, particularly if the organic compounds contain significant amounts of oxygen.

Instrumentation to consider for monitoring the overall performance and temperature history of the catalytic combustor is given in Table 8.2.

8.4 ECONOMIC ESTIMATES

There is typically a trade-off between the higher capital cost associated with a catalytic incinerator and the higher operating cost associated with a thermal

TABLE 8.2 Recommended Instrumentation

Inlet and outlet temperature
Inlet/outlet concentration
Fan speed
Pressure drop
Auxiliary fuel rate

incinerator [41]. Estimates of the capital cost of a fixed bed catalytic incineration system are given in the *OAQPS Manual* [41] according to the following equations: In SI units;

$$CC = a\,(Q/0.026)^{b} \tag{8.5}$$

In English units;

$$CC = a\,Q^{b} \tag{8.6}$$

where

CC = purchased price, in 1988 dollars
Q = flow rate of gases after combustion, Nm^3/min (scfm)
a, b = parameters dependent on the amount of heat recovery required. For 0% heat recovery, a = 1105 and b = 0.5471. For 50% heat recovery, a = 1215 and b = 0.5575. At 70% heat recovery, a = 1443 and b = 0.5527.

These estimates are valid for incinerators between approximately 50 and 1300 Nm^3/min (2000 and 50,000 scfm).

Estimates of the capital cost of a fluid bed catalytic incineration system are calculated from the following equations:
In SI units:

$$CC = c + dQ/0.026 \tag{8.7}$$

In English units:

$$CC = c + dQ \tag{8.8}$$

where

CC = purchased price, in 1988 dollars
Q = flow rate of gases after combustion, Nm^3/min (or standard ft^3/min, scfm) c, d, = parameters dependent on the amount of heat recovery required. For 0% heat recovery, c = 84,800 and d = 13.2. For 50% heat recovery, c = 86,600 and d = 15.8. At 70% heat recovery, c = 83,900 and d = 19.2 [41].

These estimates are valid for incinerators with the flowrates approximately 50 to 650 Nm^3/min (2000 to 25,000 scfm).

In addition to the catalyst and preheat burner, the price includes a heat exchanger, fan, motor, instrumentation and controls, insulation, structural supports, and a short stack. Ductwork, particulate removal devices, scrubbers, and site-specific construction requirements are not included. If the incinerator is to be placed a long distance from the generation site, both the cost of ductwork and the cost of additional booster fans should be considered. As in the other chapters in this book, cost estimates are given to indicate a general order of magnitude and must be scaled up to current dollars to get even an estimated cost. They are not specific enough to use as construction costs and should be confirmed with actual prices from vendors.

To estimate the total installed capital expenditure, the factors in Table 8.3 may be used, which generally follow the methods in the EPA's *OAQPS Cost Control Manual* [41, 50].

Annual costs may be approximated according to the guidelines in the *OAQPS Cost Control Manual* [41], as given in Table 8.4. Site-specific values for unit

TABLE 8.3 Factors for Estimating the Total Installed Capital Cost (TCC) of Incineration Systems

Cost Element	Estimated Cost
Direct Costs	
Purchased equipment cost	
Incinerator and auxiliary equipment	CC, Estimated from equation
Taxes	0.03 CC
Freight	0.05 CC
Total purchased equipment cost (PEC)	1.08 CC
Installation costs	
Foundations and supports	0.08 PEC
Erection and handling	0.14 PEC
Electrical	0.04 PEC
Piping, insulation, and painting	0.04 PEC
Total installation cost	0.30 PEC
Site preparation (SP)	As required for specific site
Buildings (BLDG)	As required for specific site
TOTAL DIRECT COSTS	1.30 PEC + SP + BLDG
Indirect Costs	
Engineering and supervision	0.10 PEC
Construction, field expenses, and fees	0.15 PEC
Start-up and performance test	0.03 PEC
Contingency	0.03 PEC
TOTAL INDIRECT COSTS	0.31 PEC
Total Installed Capital Costs (TCC)	1.61 PEC + SP + BLDG

Data adapted from refs. 41, 50.

TABLE 8.4 Annual Operating Costs for Incinerator Systems

	Unit Cost	Estimated Requirement
Direct Annual Costs		
Utilities		
Electricity	\$0.059/kW-h	See Chapter 4
Natural gas	\$127/1000 Nm^3 \$3.30/1000 scf	Calculated above
Catalyst Replacement	\$23,000/m^3 \$650/ft^3 (metal oxide)	100% replacement every 2 years
Operating labor	\$12.95/h	1/2 h per shift
Maintenance labor	\$14.25/h	1/2 h per shift
Maintenance materials	Equal to maintenance labor	
Supervisory Labor	15% of operating labor	
Indirect Annual Costs		
Overhead	60% of maintenance costs	
Property tax	1% of TCC	
Insurance	1% of TCC	
Administrative costs	2% of TCC	
Capital Recovery (see Chapter 4 for formula to calculate capital recovery factor, CRF)	CRF · (TCC − 1.08 · catalyst cost)	
Total Annual Operating Costs	Direct + indirect costs	

Data adapted from van der Vaart et al. [41].

costs should be used whenever possible to accurately reflect the anticipated costs. Electricity requirements can be estimated from the OAQPS formula given in Chapter 4, assuming a pressure drop of 1.5 kPa (6 in of water column) for a monolith catalyst, 1.5–2.5 kPa (6–10 in of water column) for a fluid bed catalytic chamber. Very high efficiency designs require additional catalyst volume and are likely to increase the pressure drop through the catalyst chamber. The heat exchanger adds another 1–3.7 kPa (4–15 in of water column), corresponding to heat recoveries of 35–70% [41]. Operating and maintenance labor requirements are usually minimal, at approximately one half hour per shift. Dirty gas streams that require frequent catalyst cleaning will, however, increase the amount of maintenance required.

8.5 POTENTIAL PROBLEMS

As with thermal incinerators, catalytic incinerators are best suited to operations that maintain a fairly constant flow rate and emission rate. With catalytic incinerators, however, large fluctuations in concentration or flow may result in overheating of the catalyst, leading to premature catalyst aging. Temperatures above approximately 920 K (650°C, 1200°F) may volatilize the metal catalyst or cause recrystallization of the support or catalyst metal [2]. In addition, high temperature fluctuations increase the stress on seams as the material expands or contracts. For that reason, catalytic incinerators are more appropriate for facilities that operate continuously instead of those that operate for only a few hours each day. The manufacturer's start-up procedure normally specifies a warm-up sequence in which the incinerator is brought to operating temperature gradually. Following this procedure lengthens the life of the refractory but also uses additional fuel and may take longer than is feasible for the process. It is not unusual for maintenance or operating staff to develop a shorter start-up procedure. Such a shortcut lowers the overall gas usage but is likely to result in considerably higher maintenance requirements. For cyclical processes, a regenerative incinerator should be considered that can be maintained at operating temperature with minimal natural gas consumption. Some of these units can even be put on an "idle" mode in which no gas is pulled from the process, but the incinerator recycles exhaust air to the inlet to conserve energy [51].

8.6 INSTALLATION, OPERATION, AND MAINTENANCE CONSIDERATIONS

Particulate concentrations in a gas stream treated by catalytic incineration should be kept below 0.12 grams/Nm^3 (0.05 grains/scf). The concentration of catalytic poisons should not exceed 25 ppm [52].

The main problem with catalytic incinerators is that the catalyst must be kept in relatively good condition to ensure that the reaction proceeds as anticipated. As particulate or other materials build up on the surface, the amount of surface area available for reaction is decreased and the efficiency is reduced. It is generally recommended that the surface of the catalyst be cleaned by blowing jets of air across it, steaming with superheated steam, vacuuming, or cleaning with a nonphosphorus detergent every 3–12 months [13]. For gas streams that contain deactivating compounds, an acid or alkaline wash may be required to restore the catalyst to a normal activity level [4, 53]. As the catalyst ages, pores become permanently blocked from particulate or inorganic compounds, and the metal surfaces tend to erode, vaporize, or be lost by attrition [15, 41, 54, 55], eventually leading to the need to replace the catalyst. This aging process may take many years to occur in a relatively clean constant gas stream, or it may occur within a period of months or even hours if the catalyst temperature limits are exceeded.

It is recommended that the temperature change across the catalyst bed be constantly monitored and recorded because that allows for an assessment of the catalyst activity. A low temperature rise across the catalyst indicates that the catalyst may have become deactivated [19]. Recording the outlet temperature also provides a method for confirming whether the catalyst has been subjected to a period of elevated temperature that may have damaged it. If the catalyst is properly cared for and its activity decreases gradually, the life of the catalyst may be extended somewhat by increasing the preheat temperature so that the average reaction temperature is increased a small amount [2]. This may be particularly helpful in providing for continued VOC reduction while a new catalyst is ordered and delivered.

Catalytic incinerators can be designed to handle a small amount of organic particulate. Operating such an incinerator below the design temperature can, however, result in carbon or organics building up on the surface of the catalyst due to incomplete combustion [2].

Several other process considerations that can help avoid poisoning the catalyst are available [24]. The use of galvanized metal in ovens or ductwork should be avoided because the zinc used for galvanizing is a catalyst poison. Temperature-indicating devices should not include mercury components that could also contaminate the catalyst. Cleaning detergents should not include phosphorus. Lubricants should not contain silicones or other inorganic metal compounds.

The heat exchange surfaces should also be kept clean and free of cracks [56]. In addition to difficulties with the catalytic material in an incinerator, heat stress can take a toll on the heat recovery equipment. At best, heat stress or buildup of material may keep valves from seating properly, a common problem with thermal regenerative units that allows leakage from the contaminated gas stream into the outlet gas stream, decreasing efficiency. At worst, heat stress can crack metal joints in heat exchangers so that gas is allowed to escape from the clean side to the dirty side or vice versa, depending on the position of the fan. Such cracks can be difficult to locate because the metal contracts as it cools to room temperature. The incinerator system should be designed considering the potential leakage of material. By the position of the fan, it can be ensured that any leakage occurs in an acceptable direction.

Any particulate, oily substances, or catalytic inhibitors should be removed before the gas stream enters the incineration system. An overall summary of the advantages and disadvantages of catalytic oxidation systems is given in Table 8.5.

TABLE 8.5 Advantages and Disadvantages of Catalytic Incineration

Advantages	Disadvantages
Reduced fire hazards	High initial cost
Heat recovery for other processes is feasible	No ability to reuse pollutants
High efficiencies possible with proper design	Particulate removal may be necessary
Technology is applicable for a wide variety of gas concentrations and constituents	Disposal of spent catalyst may be difficult
Approximately one-third the fuel usage of thermal incinerators	
Useful for multicomponent streams if the components are all compatible with the catalyst	

Data adapted from refs. 21, 57, 58.

REFERENCES

1. J.J. Spivey, "Complete Catalytic Oxidation of Volatile Organics," *Ind. Eng. Chem. Res.*, **1987**, 26, 2165–2180.
2. "Equipment Characteristics — Catalytic Combustion," in *Air Pollution: The Emissions, The Regulations, and The Controls*, J.P. Tomany, American Elsevier Publishing Company, New York, **1975**, pp. 250–272.
3. M.S. Jennings; N.E. Krohn; R.S. Berry, *Control of Industrial VOC Emissions by Catalytic Incineration: Vol. 1. Assessment of Catalytic Incineration and Competing Controls*, US Environmental Protection Agency Office of Research and Development, Research Triangle Park, NC, **1984**.
4. W. Chu; H. Windawi, "Control VOCs via Catalytic Oxidation," *Chem. Eng. Prog.*, **1996**, March, pp. 37–43.
5. L.D. Pfefferle; W.C. Pfefferle, "Catalysis in Combustion," *Cata. Rev. Sci. Eng.*, **1987**, **29**, (2–3), 219–267.
6. "Catalytic Afterburners," in *Air Pollution Engineering Manual*, 2nd Ed., J.A. Danielson, Ed., United States Environmental Protection Agency, Research Triangle Park, NC, **1973**, pp. 179–181.
7. J.A. Jahnke; D. Beachler, "Stack Gas Afterburners," in *APTI Course 415: Control of Gaseous Emissions, Instructor's Guide*, Report EPA 450/2-81-004, J. Joseph; D. Beachler; J. Jahnke, US Environmental Protection Agency, Air Pollution Training Institute, Research Triangle Park, NC, **1981**, Lesson 5.
8. I.E. Sungkono; H. Kameyama; T. Koya, "Development of Catalytic Combustion Technology of VOC Materials by Anodic Oxidation Catalyst," *Appl. Surf. Sci.*, **1997**, 121/122, pp. 425–428.
9. L. Theodore; A.J. Buonicore; J.D. McKenna; I.J. Kugelman; J.S. Jeris; J.J. Santoleri; T.F. McGowan, "Waste Management," in *Perry's Chemical Engineers' Handbook*, 7th Ed., D.W. Green; J.O. Maloney, Eds., McGraw-Hill, New York, **1997**, Ch. 25, pp. 25-1–25,111.
10. "Incineration," in *Air Pollution Control Equipment, Vol. II: Gases*, L. Theodore; A.J. Buonicore, CRC Press, Boca Raton, 1988, pp. 115–136.
11. J.J. Spivey, "Catalytic Incineration of Gaseous Wastes," in *Thermal Process*, H.M. Freeman, Ed., Technomic Publishing, Cincinnati, OH, **1990**, pp. 95–108.
12. "Incineration," in *Survey of Control Technologies for Low Concentration Organic Vapor Gas Streams*, Report EPA-456/R-95-003, R.A. Zerbonia; J.J. Spivey; S.K. Agarwal; A.S. Damle; C.W. Sanford, Control Technology Center, US Environmental Protection Agency, Research Triangle Park, NC, **1995**, Sect. 2.
13. J.P. Bilotti; T.K. Sutherland, "Incinerators," in *Air Pollution Control Equipment: Selection, Design, Operation, and Maintenance*, L. Theodore, A.J. Buonicore, Eds., ETS, Inc., Roanoke, VA, **1992**, pp. 133–151.
14. "Stack Gas Afterburners," in *APTI Course 415: Control of Gaseous Emissions, Student Workbook*, G.T. Joseph, D.S. Beachler US Environmental Protection Agency, Air Pollution Training Institute, Research Triangle Park, NC, **1981**, p. 4-4–4-5.
15. T.H. Snape, "Catalytic Incineration," in *Air Pollution Control and Design Handbook* (in Two Parts), P.N. Cheremisinoff; R.A. Young, Eds., Marcel Dekker, New York, **1977**, pp. 543–563.

16. C. Libanati; D.A. Ullenius; C.J. Pereira, "Silica Deactivation of Bead VOC Catalysts," *Appl. Catal. B: Environ.*, **1998**, 15, 21–28.

17. N. De Nevers, "Control by Combustion," in *Air Pollution Control Engineering*, McGraw-Hill, New York, **1995**, pp. 300–315.

18. J.R. Kittrell; C.W. Quinlan; J.W. Eldridge, "Direct Catalytic Oxidation of Halogenated Hydrocarbons," *J. Air Waste Manage. Assoc.*, **1991**, 41, (8), 1129–1133.

19. N. Mukhopadhyay; E.C. Moretti, "Catalytic Oxidation (Catalytic Incineration)," in *Current and Potential Future Industrial Practices of Reducing and Controlling Volatile Organic Compounds*, Center for Waste Reduction Technologies, American Institute of Chemical Engineers, New York, **1993**, p. 49.

20. A.T. Bell; L.E. Manzer; N.Y. Chen; V.W. Weekman; L.L. Hegedus; C.J. Pereira, "Protecting the Environment Through Catalysis," *Chem. Eng. Prog.*, February **1995**, pp. 26–34.

21. A.J. Buonicore, "Incineration," in *Air Pollution Engineering Manual*, A.J. Buonicore; W.T. Davis, Eds., Van Nostrand Reinhold, New York, **1992**, pp. 58–62.

22. RY. Purcell; G.S. Shareef, "Catalytic Incineration," in *Control Technologies for Hazardous Air Pollutants*, Air and Energy Engineering Research Laboratory, Office of Research and Development, U.S. Environmental Protection Agency, Research Triangle Park, NC, **1986**, pp. 53–59.

23. H.F. Dey, "Afterburners," in *Air Pollution Engineering Manual*, J.A. Danielson, Ed., U.S. Department of Health, Education, and Welfare, National Center for Air Pollution Control, Cincinnati, Ohio, **1967**, pp. 171–192.

24. "Applications of Catalytic Incinerators," in *Catalytic Incineration for Control of Volatile Organic Compound Emissions*, M.S. Jennings; N.E. Krohn; R.S. Berry, Noyes Publications, Park Ridge, NJ, **1985**, pp. 35–53.

25. A.E.R. Budd, *Environ. Pollut. Manage.*, (January/February) **1978**, 13–14.

26. *Catalytic Oxidation of Discharge from Polymer Process Vessel Vents*; Research Disclosure 14052 from Industrial Opportunities, Ltd., and The Firestone Tire and Rubber Co., Hampshire, UK and Akron, OH, **1975**, as given in J.J. Spivey, "Complete Catalytic Oxidation of Volatile Organics," *Ind. Eng. Chem. Res.*, **1987**, **26**, 2165–2180.

27. J.J. Spivey, "Complete Catalytic Oxidation of Volatile Organics," *Ind. Eng. Chem. Res.*, **26**, **1987**, 2165–2180.

28. J.C. Bonacci; R.M. Heck, "The Platinum Group Metals: An In-Depth View of the Industry," *International Precious Metals Institute Seminar*, Williamsburg, VA, April, **1983**.

29. A. Balough; J. Herendi; J. Bathory, "Catalytic Purification of Gaseous Factory Effluents to Protect the Environment," Report AD-A017807, National Technical Information Service, Springfield, VA, **1975**.

30. R.E. Roberts; J.B. Roberts, *J. Air Pollut. Control Assoc.*, **1976**, **26**, (4), 353–358.

31. J.P. Draves, "Destroying VOCs and Looking to the Future," *Environ. Prot.*, July, 1996, pp. 36–38.

32. "Fume Oxidizers for Destruction of VOCs, Odors, and Smoke," Glenro Inc. sales literature, Glenro Inc., Paterson, NJ, **1998** (http://www.glenro.thomasregister.com).

33. "Anguil Regenerative Catalytic Oxidizer," sales literature from Anguil Environmental Systems, Inc., Milwaukee, WI, **1998**, available at http://www.anguil.com.

34. "Catalytic Oxidation Process Cleans Volatile Organics From Exhaust," *C&EN*, June 27, **1994**, p. 42.
35. R.M. Heck; R.J. Farrauto; M. Durilla, "Employing Metal Catalysts for VOC Emission Control," *Pollut. Eng.*, April, **1998**, pp. 52–55.
36. Chemical Section Engineers, New Source Review Division of Texas Natural Resource Conservation Commission, *Air Permit Technical Guidance for Chemical Sources: Flares and Vapor Oxidizers*, Texas Natural Resources Conservation Commission, **1998**, available at http://www. tnrcc.state.tx.us/air/nsr_permits/guidedoc.html.
37. J.H. Krohling, "Catalytic Oxidation: An Energy Saving Approach to Fume and Odor Control," presented at the Air Pollution Control Association Specialty Conference on State of the Art of Odor Control Technology II, Pittsburgh, PA, March **1977**.
38. M.F.M. Zwinkels; S.G. Järås; P.G. Menon; T.A. Griffin, "Catalytic Materials for High-Temperature Combustion," *Catal. Rev. Sci. Eng.*, **1993**, 35, (3), pp. 319–358.
39. Chris Anguil, Anguil Environmental, Inc. Personal communication.
40. "Anguil Rotor Concentrator/Oxidizer," sales literature (leaflet RCS-3/97-5M), Anguil Environmental Systems, Inc., Milwaukee, WI, **1997**, available at http://www.anguil.com.
41. D.R. van der Vaart; J.J. Spivey; W.M. Vatavuk; A.H. Wehe, "Thermal and Catalytic Incinerators," in *OAQPS Control Cost Manual*, 5th Ed., United States Environmental Protection Agency, Office of Air Quality Planning and Standards, Research Triangle Park, NC, **1995**, Ch. 3, available at http://www.epa.gov/ttn/catc/products.html.
42. U.S. Bureau of Mines, "Limits of Flammability of Gases and Vapors," U.S. Bureau of Mines, Washington, DC, Bulletin Number 503 [no date].
43. R.D. Ross, "Thermal Incineration," in *Air Pollution Control and Design Handbook* (in Two Parts), P.N. Cheremisinoff; R.A. Young, Eds., Marcel Dekker, New York, **1977**, pp. 449–468.
44. R.B. Diemer, Jr; T.D. Ellis; G.D. Silcox; J.S. Lighty; D.W. Pershing "Incinerators for Vapors and Gases," in *Kirk-Othmer Encyclopedia of Chemical Technology, 4th Ed., Vol. 14, Imaging Technology to Lanthanides*, J.I. Kroschwitz; M. Howe-Grant, Eds., John Wiley & Sons, New York, **1995**, pp. 118–120.
45. T. Banaszak; R. Miller; A. Zieba, "Thermal-Catalytic Incineration of Waste Gases," in *Combustion Technologies for a Clean Environment: Selected Papers from the Proceedings of the First International Conference, Vilamoura, Portugal, September* 3–6, *1991*, M. da Graca Carvalho; F.C. Lockwood; W.A. Fiveland; C. Papadopoulos, Eds., Gordon and Breach Publishers, Switzerland, **1995**, pp. 633–642.
46. Personal communication from Andrew Jones to Donald R. van der Vaart, March 28, 1988, as given by D.R. van der Vaart; J.J. Spivey; W.M. Vatavuk; A.H. Wehe, "Thermal and Catalytic Incinerators," in *OAQPS Control Cost Manual*, 5th Edition, United States Environmental Protection Agency, Office of Air Quality Planning and Standards, Research Triangle Park, NC, **1995**, Ch. 3, available at http://www.epa.gov/ttn/catc/products.html.
47. Personal communication from Bill Sheffer to Donald R. van der Vaart, March 30, 1988 as given by D.R. van der Vaart; J.J. Spivey; W.M. Vatavuk; A.H. Wehe, "Thermal and Catalytic Incinerators," in *OAQPS Control Cost Manual*, 5th Edition, United States Environmental Protection Agency, Office of Air Quality Planning and Standards, Research Triangle Park, NC, **1995**, Ch.3, available at http://www.epa.gov/ttn/catc/products.html.

48. V.S. Katari; W.M. Vatavuk; A.H. Wehe, "Incineration Techniques for Control of Volatile Organic Compound Emissions, Part II: Capital and Annual Operating Costs," *J. Air Pollut. Cont. Assoc.*, **1987**, 37 (2), 198–201.

49. "Incinerators," in *Estimating Costs of Air Pollution Control*, W.M. Vatavuk, Lewis Publishers, Boca Raton, **1990**, pp. 141–156.

50. W.M. Vatavuk; R. Neveril, "Estimating Costs of Air Pollution Control Systems, Part II: Factors for Estimating Capital and Operating Costs," *Chem. Eng.*, November 3, **1980**, pp. 157–162.

51. Tellkamp Industries, Personal communication. T. Rowe, Sante Fe Springs, CA.

52. Personal communication between G.I. Madden and N.E. Krohn, January 7, **1982**, as described in "Applications of Catalytic Incinerators," in *Catalytic Incineration for Control of Volatile Organic Compound Emissions*, M.S. Jennings; N.E. Krohn; R.S. Berry, Noyes Publications, Park Ridge, NJ, **1985**, pp. 35–53.

53. N. Mukhopadhyay; E.C. Moretti, "Catalytic Oxidation (Catalytic Incineration)," in *Current and Potential Future Industrial Practices of Reducing and Controlling Volatile Organic Compounds*, Center for Waste Reduction Technologies, American Institute of Chemical Engineers, New York, **1993**, pp. 16–17.

54. RY. Purcell; G.S. Shareef, "Catalytic Incinerators," in *Control Technologies for Hazardous Air Pollutants*, Air and Energy Engineering Research Laboratory, Office of Research and Development, U.S. Environmental Protection Agency, Research Triangle Park, NC, **1986**, p. 25.

55. G.T. Joseph, D.S. Beachler, "Catalytic Oxidation," in *APTI Course 415: Control of Gaseous Emissions*, Student Manual, US Environmental Protection Agency, Air Pollution Training Institute, Research Triangle Park, NC, **1981**, pp. 3-27–3-30.

56. T.T. Shen; C.E. Schmidt; T.R. Card, "Thermal Oxidation," in *Assessment and Control of VOC Emissions from Waste Treatment and Disposal Facilities*, Van Nostrand Reinhold, New York, **1993**, pp. 164–166.

57. L. Theodore; A.J. Buonicore, "Thermal and Catalytic Combustion Reactors and Flares," in *Industrial Control Equipment for Gaseous Pollutants*, Vol. II, CRC Press, Cleveland, OH, **1975**, pp. 75–87.

58. L. Theodore; A.J. Buonicore; J.D. McKenna; I.J. Kugelman; J.S. Jeris; J.J. Santoleri; T.F. McGowan, "Waste Management," in *Perry's Chemical Engineers' Handbook*, 7th Ed., D.W. Green; J.O. Maloney, Eds., McGraw-Hill, New York, **1997**, Ch. 25, pp. 25–1 to 25–111.

9

BIODEGRADATION

9.1 TECHNOLOGY DESCRIPTION

Biodegradation is the process of using microorganisms to convert pollutants into simple compounds. In air pollution control devices, the pollutants are first absorbed into a water film that contains microbes. Then the microbes metabolize the contaminant to obtain energy and nutrients. When VOCs are treated by biodegradation, the principal products of this metabolic process are carbon dioxide, water, and microbe cell mass. This process is the same one that has been used for decades in wastewater treatment plants and landfills to treat organic material [1].

In the most common type of biodegradation system used for air pollution control, gas is passed through a moist solid medium, where pollutants are absorbed into the liquid layer on the surface of the solids. The biological activity results from the high concentration of microorganisms in this layer, or biofilm. These biofilters are common air pollution control devices in Europe, with more than 500 in place [2]; they are used to remove odorous compounds and VOCs from waste gases [3, 4]. These systems are particularly effective for pollutants that are easy to degrade such as toluene, xylene, alcohols, butanol, formaldehyde, trimethylamine, acetaldehyde, and butyric acid [3]. Biofilters can also be combined with carbon adsorption or condensation to treat higher concentrations of gases. Biofilters are also capable of degrading volatile inorganic compounds (VICs) [5].

A typical *biofiltration* system for air pollution consists of a humidification chamber and a reactor. In addition to these components, particulate collectors may be required upstream of the biofilter to remove solid matter from the gas stream that would otherwise plug the bed and hinder gas distribution, in addition to increasing the pressure drop through the bed. Although the humidification system is controlled manually in many simpler units, automatic controls are sometimes used in more sophisticated biofiltration systems designed for low maintenance. The reactor itself contains a solid medium, usually compost, soil, or peat, which provides a surface on which the microorganisms can reside. Organic substrates also provide nutrients to sustain the organisms. Inorganic media may be used, which are more expensive [6] but are also more resistant to compaction and maintain the porosity of the bed for longer bed life. The solids are approximately 1 m (3 ft) deep.

Gas enters the biofilter from either the top or the bottom of the bed. Incoming gas travels through a piping manifold system or molded concrete blocks designed to uniformly distribute the gas across the surface to obtain uniform residence time of the gas stream, as well as to keep portions of the bed with higher gas velocities from drying out. Moisture levels throughout the medium must be maintained at a level adequate to prevent drying and cracking of the bed but low enough to avoid liquid channeling and the development of areas in the bed where oxygen is not present. Any anaerobic zones that develop can result in significant odors being emitted from the bed. The optimum moisture content is generally considered to be approximately 35–60% in compost biofilters for removing H_2S [7] and VOCs [8–11]. For biofilters with inorganic substrate materials, the optimum moisture is approximately 40–50% [10].

Biofiltration is capable of efficiencies ranging from 80 to 99%, depending on the characteristics of the pollutants [12]. The efficiency of removing VOCs from a gas stream by biodegradation depends on several factors: the degree of contact between the gas stream and the medium, the moisture level, the pollutant concentration, the residence time, and the specific physical characteristics of the pollutant. The physical characteristics of importance are the Henry's

TABLE 9.1 Typical Applications of Biodegradation

Pharmaceutical processes
Solid stabilization processes for landfills
Rendering plants
Chemical manufacturing processes
Print shops
Flavors and fragrances
Coffee and cocoa roasting
Sewage treatment odor removal
Coating processes
Composting
Food processing
Livestock farms
Foundries

Data adapted from refs. 3, 4, 17–19.

law constant, the solubility in water, the molecular weight, and the presence of complex bonding structures that prevent microbial attack [1, 10, 13]. Additional information about biodegradation for air pollution control is given by Skladany et al. [14], Bowker [15], and Heinsohn and Kabel [16]. Some typical applications for biodegradation systems are listed in Table 9.1 [3, 4, 17–19].

9.2 TYPES OF BIODEGRADATION SYSTEMS

There are two types of biodegradation systems. In systems with *bio-scrubbers*, the microbes are contained in a wash liquid that is in contact with the contaminated air as in an absorber [19]. This chapter focuses on *biofilters* (a typical diagram of which is shown in Fig. 9.1 [20]), systems in which the microbes are contained on a solid material like compost, soil, granular activated carbon (GAC), diatomaceous earth, or inert synthetic packing material [1, 11, 21]. The microbes live in a water film around the solid particles. Gases are adsorbed into the water film and digested. Trickle bed air biofilters [22], or biotrickling filters [23], are a specific type of biofilter in which a liquid stream of water (with or without additional nutrients) is constantly applied to the solid medium. Recycling the liquid effluent from the bed results in microbe reseeding, better control of pH, and ultimately superior performance, according to several authors [11, 24–29]. In particular, the concentration of styrene or benzene that can be degraded with a 90% removal efficiency in a biotrickling filter has been shown to be more than twice as high as that which can be removed in a conventional biofilter [23, 30].

Although the general configurations of biofiltration systems are similar, each system has specific design considerations. They vary according to the flue gas pretreatment equipment that is required, the number of beds used, the solid medium used, and the gas distribution system. Biofilters may be closed to the

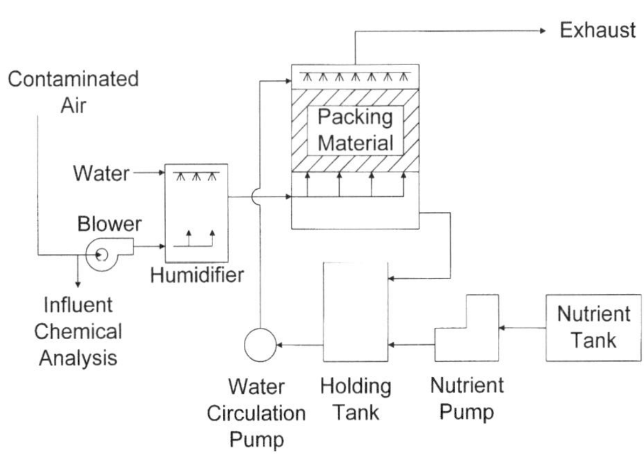

Figure 9.1 Typical flow diagram of a biofilter. (Courtesy of J.B. Quarles.)

Figure 9.2 Open type biofilter [31].

atmosphere with the exception of a vent stack, or they may have open tops so that the gas flows up through the bed and directly into the atmosphere (Fig. 9.2 [31]).

9.3 BIOFILTER SYSTEM DESIGN

Because of the significant effect that the microbe activity can have on the efficiency of the operation, it is generally required that design parameters be developed from previous experience or from pilot testing [3, 6]. This testing can help determine the optimum solid medium for the filter, an estimate of the required moisture addition rate, the appropriate microbes, and the appropriate residence time.

Before the overall size of the biofilter is determined, it is helpful to identify an appropriate solid bed material because the material of choice will affect the overall operating costs of the filter, as well as the required size. As mentioned earlier, the filter medium may be organic or inert synthetic material [11]. Typical organic materials are compost, peat, leaves, soil, bark, wood chips, activated carbon, or clay [1, 32–35]. Inert materials that are used include perlite, pelletized ceramics, ceramic monoliths, diatomaceous earth, and synthetic packing material [6, 21, 36]. Although organic materials provide many nutrients needed by the microbes [6], they are also degraded at a faster rate than the VOCs themselves [37], resulting in additional compaction of the bed and a corresponding increase in pressure drop through the medium [1]. After several years, maintenance is required to remove the material or mix it to reduce the pressure drop and improve overall performance. Cleaning of the air distribution system may also be needed to remove bed material that has collected there. The overall performance of the filter bed can be enhanced by addition of inert solids like polystyrene beads to reduce compaction, extend bed life, and increase porosity [11, 38]. Alternatively, the entire bed can be constructed of an inert medium like pelletized ceramics, activated carbon, or ceramic monoliths [6]. Although an inert medium results in more uniform gas distribution [11] and less maintenance, it is also typically more expensive [6]. Additives such as lime are sometimes used to provide buffering capacity for the bed, particularly if the bed is used to treat chloride or sulfide compounds, which can result in acidic decomposition products [39, 40]. Activated carbon can also be added to improve retention of the pollutants and to maintain a consistent feed source for the microorganisms in cases in which the process does not emit a constant level of pollutants [39, 41, 42].

The medium should provide for even air distribution and pressure drop through the bed, high specific surface area, high porosity, adequate inorganic nutrients, efficient drainage, good mechanical strength to resist attrition, minimal pressure drop, and a surface that accommodates attachment of the microbes [11, 43–45].

Nutrients must be present for the microbes. With organic media, the microbes get many of the inorganic nutrients required from the media itself. Micronutrients are sometimes required, however, and are available in forms that can be dissolved in water and spread onto the surface of the bed. The necessary nutrients include nitrogen, phosphorus, iron, calcium, and magnesium in addition to the carbon, hydrogen, and oxygen [39] that are available in the gas stream. The accepted ratio of carbon to nitrogen and phosphorus is approximately 100 : 10 : 1 [46]. If the essential nutrients and micronutrients are not present, samples of the bed material will have evidence of microbes that have encapsulated themselves [46].

Microorganisms typically found in biofilters are similar to those found in water treatment operations. They include both bacteria and fungi species. Bacterial species may include *Actinomyces Globisporus, Micrococcus albus, Micromonospora vulgarus, Proteus vulgarus, Bacillus cereus*, and *Streptomyces* species. Typical fungi are species of the genuses *Penicillium, Cephalosporium, Mucor, Circinella, Cephalotecium, Ovularia*, and *Stemphilium* [39].

Based on the work of several investigators [39, 47, 48], the reaction rate is generally assumed to follow the Monod (or Michaelis-Menten) relationship:

$$R = R_{max}\, C_l/(C_l + K_m) \tag{9.1}$$

where

R = biodegradation rate, units of mass per length cubed per time ($M\, L^{-3}\, T^{-1}$)
R_{max} = maximum biodegradation rate ($M\, L^{-3}\, T^{-1}$)
C_l = concentration in the liquid biofilm ($M\, L^{-3}$)
K_m = monod (or Michaelis-Menten) constant ($M\, L^{-3}$)

R_{max} can be further described by the relationship:

$$R_{max} = X\ \mu_m / y_i \tag{9.2}$$

where

X = active microorganism cell concentration ($M\ L^{-3}$)
μ_m = maximum rate of growth (T^{-1})
y_i = coefficient of cell yield

The reaction in biofilters is generally considered to be first order at low VOC concentrations in which the biofilm is not completely saturated. Therefore, the rate of VOC destruction is controlled by the mass transfer rate of pollutant into the biofilm [6, 39]. As the concentration in the gas stream increases, the concentration of pollutant in the liquid (C_l) is much larger than the Monod constant. At this point, the system is no longer limited by mass transfer because the biofilm is saturated and the microorganisms have immediate access to the compounds. The point at which this transition from first order to zero order is pollutant specific and is determined empirically.

This biologic model assumes that there is no interaction between multiple pollutants in the gas phase. In addition, it is based on the simplifying assumptions that the gas flow through the biofilter is plug flow and that equilibrium concentrations exist at the boundaries of the liquid and gas that can be described by Henry's law. Furthermore, the biofilm is considered to be adequately represented as flat because the biofilm thickness is small compared with the particle size. It is also assumed that biologic reactions occur only in the liquid phase [39]. Availability of oxygen has not been found to be rate limiting, even in gas streams with oxygen concentrations lower than ambient air [49].

The size of vessel required for a biofiltration system can be estimated from the following equation [50] when the concentration is less than 1 g/m^3:

$$V = 0.0155\ Q\ C^{0.5} \tag{9.3}$$

where

V = required volume of filter bed, m^3
Q = flow rate of gas stream to be treated, m^3/h
C = VOC concentration, g/m^3

For gas streams with concentrations higher than 1 g/m^3, the required volume is predicted by

$$V = 0.01\ Q\ [C + (3 - 0.1 \cdot C^{0.5})] \tag{9.4}$$

An alternative way to estimate the volume required in a biofilter is to predict it on the basis of the removal rates that have been achieved in similar systems. A measure of the effectiveness of a biofilter is the elimination capacity (EC) for a given compound. The EC is the maximum amount of pollutant that the system can degrade, expressed in g/h of the pollutant divided by the volume of the empty reactor (m^3) [43]. ECs of several compounds that have been reported in the literature are given in Table 9.2 [4, 8].

The other important factor in the design of a biofiltration system is adequate, even distribution of moisture. If the medium becomes too dry, it will shrink, causing cracks to develop that allow gas to pass through the bed untreated [59]. If the medium is too wet, gas channeling can result, causing areas of low oxygen supply and creating anaerobic zones in the filter. Moisture is usually controlled by saturating the incoming gas stream to a level of 90–95% relative humidity. This, however, is not adequate to ensure that the bed does not dry out. Because the decomposition of VOCs is an exothermic process, the temperature in the biofilter is usually a few degrees higher than the inlet gas temperature. Therefore, even if the inlet gas stream is completely saturated, it will continue to remove moisture from the bed as the gas warms up in the biofilter and approaches saturation at the new temperature [39]. Approximately 10–20% of the moisture required to replace the water evaporated because of this temperature difference is produced in the biofilter as the VOCs are decomposed [45]. The moisture balance in a biofilter is also affected by the ambient temperature and the amount of heat exchanged with the surroundings. This can cause condensation in the biofilter, particularly as it is started up.

For simple systems with relatively low loadings of VOCs, the moisture balance may be fairly stable so that water can be added manually to the biofilter either with a hose or with a spraying system that is controlled manually. For systems with high amounts of moisture evaporated (such as those with VOC concentrations greater than 0.5 g/m^3 or elimination capacities exceeding 50 g/m^3h) or with variable conditions, automatic monitoring and control of moisture are more critical [45]. The moisture can be monitored by measurement of electric conductivity or capacitance in given spots, but the most common method is to use load cells to determine the moisture content in a portion of the bed or in the entire bed. Load cells are devices that are installed in the support structure for the bed to monitor the "load" or weight of the bed. This type of system cannot be used with open-type biofilters because any additional weight such as growth of vegetation or snow on the surface of the biofilter will affect the reading. The load cells may need to be recalibrated occasionally to compensate for medium losses [45]. With VOC elimination capacities greater than 100 g/m^3h, it may be difficult to maintain a proper moisture level in a large system, even with automated measurement and controls.

Biofilters with gas flowing down through the medium can assist in maintaining proper moisture levels because water can be easily applied to the surface of the filter medium. The majority of the media drying occurs at the entrance of the gas stream to the bed, so a downflow biofilter allows water to be directed more easily to the locations where it is needed. An exception to this is beds for treating feeds with chlorides or sulfides that can contain acidic decomposition products. Upflow filters may be preferable for such beds because the more concentrated acidic byproducts are at the bottom of the bed and can be flushed out more

TABLE 9.2 Maximum Elimination Capacities (EC) of Various Compounds

Compound	Media	EC, $g/m^3/h$	EC, $lb/ft^3/h$	Reference
Methyl formiate	Unspecified	934.4	0.0583	51
Butyl acetate	Peat	64.3	0.00401	47
Butanol	Peat	64.3	0.00401	47
N-Butanol	Compost	64.1	0.00400	52
BTEX (benzene, toluene, ethyl benzene, xylene)	Compost	20–30	0.0012–0.0019	53
BTX (benzene, toluene, xylene)	Pelletized biologically active media	55	0.0034	54
BTX	Peat moss and chicken manure	5.8–63.6	0.00036–0.00397	55
Ethyl acetate	Peat	54.2	0.00338	47
Ethyl benzene	Perlite	85.2	0.00531	36
Ethanol	Peat	120	0.00749	43; inlet loading of 135 $g/m^3/h$
Ethanol	Unspecified	175	0.0109	23
Toluene	Peat	42.2	0.00263	47
Toluene	Peat	70	0.0044	43; inlet loading of 100 $g/m^3/h$
Toluene	Perlite	72.9	0.00455	36
Toluene	Compost and inerts	21	0.0013	48
Simulated JP-4 fuel (85% kerosene, 15% gasoline)	Lawn compost, composted sewage sludge, perlite, gypsum, activated sludge	At loadings of 25–1000 ppm m^3/m^2min and residence times of 1–3 min, >99% removal		56
TEX (toluene, ethyl benzene, *o*-xylene)	Perlite	70	0.0044	36
o-Xylene	Perlite	63.6	0.00397	36
Methanol	Unspecified	36.0	0.00225	51
Methanethiol (methyl mercaptans)	Peat	24	0.0015	57
Styrene	Unspecified	70	0.0044	23
Dimethyl disulfide	Peat	18	0.0011	57
Dimethyl sulfide	Peat	10	0.00063	57
Ammonia	Peat	4.3	0.00027	58
H_2S	Compost	130	0.00811	7

Data are from refs. 4, 8 and listed references.

TABLE 9.3 Recommended Instrumentation

Gas stream temperature and pressure (inlet and outlet)
Inlet gas stream relative humidity
Fan speed
Moisture content of the inlet gas stream
Moisture content of the medium (may be manual sample)
pH of the liquid effluent

Data adapted from van Lith et al. [45].

completely [45]. Gas flow through biofilters is usually between 1 and 5 cubic feet per hour per square foot of medium surface [4].

Other important factors for the system design include adequate oxygen content of the gas stream [4] (although this is not usually a limiting factor in the system performance [49]), removal of decomposition products that inhibit biodegradation, supply of nutrients to the reactor, and adequate buffering capacity.

Studies [22, 60] have shown that supplying nitrogen to the reactor in the form of nitrates instead of ammonia reduces the amount of nitrifying bacteria that are believed to inhibit the effectiveness of the biofilter. Proper biofiltration operation requires the addition of vitamin solutions that contain macronutrients, micronutrients, and buffers [11, 36]. Buffering agents may be used to adjust the pH of the medium or the liquid nutrient mixture.

In reactors used to treat gas streams with acidic decomposition products, providing adequate buffering capacity in the biofilter is important because the biofilter performance may decline significantly once the buffering capacity is exhausted, and the medium must be replaced. Instrumentation that should be considered for biofiltration systems is listed in Table 9.3 [45].

9.4 ECONOMIC ESTIMATES

Estimates of the total installed capital cost of a biofiltration system are given in Figure 9.3 in terms of (1997) dollars per cubic meter of medium. At each filter volume, there is a range of costs. The low end of the range represents open biofilter systems, which frequently require longer residence times (and therefore larger vessels) to achieve a given level of control due to less efficient gas distribution, less precise moisture control, and a nonoptimized filter medium. The upper end of the range represents the cost associated with installing an enclosed, controlled, multilevel biofilter system [45].

Alternatively, the capital cost of the purchased equipment alone can be estimated from the following equations [50], and then the cost can be scaled according to standard factors to arrive at an estimate of the total installed capital cost. For a system that requires less than 200 m^3 of medium:

$$CC = 51{,}315\ V^{0.49} \tag{9.5}$$

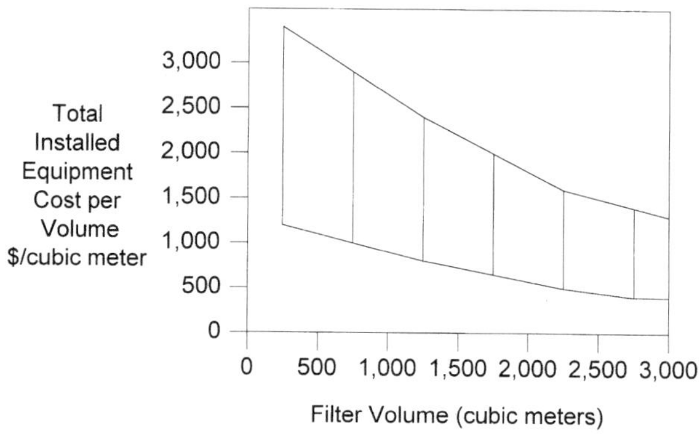

Figure 9.3 Installed capital cost range for biofilter systems greater than 100 m^3 [45].

where CC is capital cost in 1993 dollars. For a system that requires more than 200 m^3 of medium, the appropriate equation is

$$CC = 231{,}254 + 2285\ V \tag{9.6}$$

To estimate the total installed capital expenditure, the factors in Table 9.4 may be used [50], which generally follow the methods in the EPA's *OAQPS Cost Control Manual* [61] and the HAP-PRO software [62] used for other technologies. The *OAQPS Cost Control Manual* and HAP-PRO software do not currently contain information for biofilters.

Annual costs may be approximated with the methods given in the *OAQPS Cost Control Manual* [61] for other systems, as given in Table 9.5. Electricity requirements can be estimated from the formulas given in Chapter 4, assuming a pressure drop of 50–3000 Pa (0.2–12 in of water), depending on the medium chosen, the velocity through the medium, the medium depth, and the number of beds. Water requirements can be estimated from the general guideline of 5–10 gallons of water required for every 100,000 standard cubic feet of gases treated [4]. Operating and maintenance labor requirements are also highly variable, ranging from a few minutes a day for a well-controlled system treating relatively low concentrations of pollutants to a few hours a day for an open, uncontrolled system with high pollutant loadings and high moisture losses [45].

9.5 POTENTIAL PROBLEMS

As the solid medium in a biofilter degrades, the bed compacts and the pressure drop through the filter increases. The pressure drop through the bed should be monitored carefully so that required replacement of medium can be anticipated and the required outage time can be scheduled with the production facility. Monitoring the pressure drop through the bed also enables early detection of cracks in the bed that allow gas to pass through untreated [4].

TABLE 9.4 Factors for Estimating the Total Installed Capital Cost (TCC) of Biofiltration Systems

Cost Element	Estimated Cost
	Direct Costs
Purchased equipment cost (PEC)	
Biofilter and auxiliary equipment	Capital cost (CC), estimated from equation
Taxes	0.03 CC
Freight	0.05 CC
Total purchased equipment cost	1.08 CC
Installation costs	
Foundations and supports	0.08 PEC
Erection and handling	0.14 PEC
Electrical	0.04 PEC
Piping, insulation, and painting	0.04 PEC
Total installation cost	0.30 PEC
Site preparation (SP)	As required for specific site
Buildings (BLDG)	As required for specific site
TOTAL DIRECT COSTS	1.30 PEC + SP + BLDG
	Indirect Costs
Engineering and supervision	0.10 PEC
Construction, field expenses, and fees	0.15 PEC
Start-up and performance test	0.03 PEC
Contingency	0.03 PEC
TOTAL INDIRECT COSTS	0.31 PEC
Total Installed Capital Costs (TCC)	1.61 PEC + SP + BLDG

Data adapted from Mukhopadhyay and Moretti [50].

Optimum temperature for microbial activity is approximately 30°–40°C [4, 7, 45]. At lower or higher temperatures, the removal of VOCs in the biofilter may be significantly reduced (by approximately 50% for each 10°C drop in temperature [63]). Due to the exothermic nature of the metabolic reactions, however, it is possible to maintain adequate VOC removal even in ambient temperatures that are significantly below the optimum [7, 64]. The most prevalent concerns with biofilter operation are maintaining adequate moisture, removing solids that build up in the media, "fluffing" or replacing of the solid media, and maintaining appropriate temperatures.

Biofiltration should be used with some caution on gas streams that contain contaminants that are difficult to degrade because the compound may not be broken down completely. In these cases, it is possible to form more toxic byproducts than the original pollutants [65]. For example, the aerobic degradation of trichloroethylene can form vinyl chloride as a byproduct [66].

TABLE 9.5 Annual Operating Costs for Biofilter Systems

	Unit Cost	Estimated Requirement
	Direct Annual Costs	
Utilities		
Electricity	\$0.05/kW-h	See Chapter 4
Water		5–10 gal/100,000 f^3 [4]
Operating labor	\$12.96/h	1/2 h per shift
Maintenance labor	\$14.26/h	1/2 h per shift
Maintenance materials	2% of TCC	
Supervisory labor	15% of operating labor	
	Indirect Annual Costs	
Overhead	60% of maintenance costs	
Property tax	1% of TCC	
Insurance	1% of TCC	
Administrative costs	2% of TCC	
Capital recovery (see Chapter 4 for formula to calculate capital recovery factor)	CRF · TCC	
Total Annual Operating Costs	Direct + indirect costs	

Data adapted from Mukhopadhyay and Moretti [50].

9.6 INSTALLATION, OPERATION, AND MAINTENANCE CONSIDERATIONS

As discussed earlier, the moisture control in a biofilter is a key parameter in ensuring that the biofilter operates as designed. Therefore, the moisture distribution system, including the humidification chamber, and the water distribution system in the bed must operate properly.

Likewise, particulate that builds up in the gas stream distribution system increases the pressure drop and hinders good distribution, so any particulate collection devices should also be checked periodically to ensure that they are in good operating condition.

Because medium replacement is almost inevitable, the system must be designed and constructed with adequate space and access for the large equipment needed to "fluff" the biofilter contents or replace it. Studies have shown that periodic backflushing of the filter with water may be beneficial in reducing the amount of excess biomass that builds up in the filter over time, increasing the pressure drop [22, 67, 68]. Table 9.6 lists advantages and disadvantages of biofiltration [45, 59].

TABLE 9.6 Advantages and Disadvantages of Biofiltration

Advantages	Disadvantages
Relatively simple equipment and operation	Effectiveness is pollutant specific
Usually no additional pollutants formed other than carbon dioxide, water, biomass, and mineral salts	Potential corrosion of ductwork due to moisture in gas stream
Process is at ambient temperatures and is therefore inherently safer	Sensitivity to temperature, concentration, and moisture
Typically lower capital and operating costs than other technologies	Proper moisture control of streams with high loadings of organic compounds may be difficult
Low energy requirements	

Data adapted from refs. 45, 59.

REFERENCES

1. G. Leson; A.M. Winer, "Biofiltration: An Innovative Air Pollution Control Technology For VOC Emissions," *J. Air Waste Manage. Assoc.*, **1991**, 41(8), 1045–1054.
2. H. Bohn, "Consider Biofiltration for Decontaminating Gases," *Chem. Eng. Prog.*, Apr. **1992**, pp. 34–40.
3. K. Fouhy, "Cleaning Waste Gas, Naturally," *Chem. Eng.*, Dec. **1992**, pp. 43–46.
4. T.O. Williams; F.C. Miller, "Biofilters and Facility Operations," *Biocycle* Nov. **1992**, pp. 75–79.
5. H. Bohn, "Consider Biofiltration for Decontaminating Gases," *Chem. Eng. Prog.*, Apr. **1992**, pp. 34–40.
6. F.G. Edwards; N. Nirmalakhandan, "Biological Treatment of Airstreams Contaminated with VOCs: An Overview," *Water Sci. Technol.*, **1996**, 34(3–4), 565–571.
7. Y. Yang; E.R. Allen, "Biofiltration Control of Hydrogen Sulfide: 1. Design and Operational Parameters," *Air Waste*, **1994**, 44, 863–868.
8. R.L. Corsi; L. Seed, "Biofiltration of BTEX: Media, Substrate, and Loadings Effects," *Environ. Prog.*, **1995**, 14(3), 151–158.
9. J.C. Mueller, "Biofiltration of Gases—A Mature Technology for Control of a Wide Range of Air Pollutants," Project Number 2-51-797, **1988**, Report to the National Research Council of Canada and the British Columbia Ministry of Advanced Education and Job Training.
10. T.S. Webster; J.S. Devinny; E.M. Torres; S.S. Basrai, "Biofiltration of Odors, Toxics, and Volatile Organic Compounds from Publicly Owned Treatment Works," *Environ. Prog.*, 15(3), 141–147.
11. G.A. Sorial; F.L. Smith; M.T. Suidan; P. Biswas; R.C. Brenner, "Evaluation of Trickle Bed Biofilter Media for Toluene Removal," *J. Air Waste Manage. Assoc.*, Oct., **1995**, 45, 801–810.
12. H. Bohn, "Consider Biofiltration for Decontaminating Gases," *Chem. Eng. Prog.*, Apr. **1992**, 34–40.

13. V.F. Medina; T. Webster; M. Ramaratnam; J. Devinny, "Treatment of Gasoline Residuals by Granular Activated Carbon Based Biological Filtration," *J. Environ. Sci. Health*, Feb. **1995**, A30(2), 407–422.
14. G.J. Skladany; M.A. Deshusses; J.S. Devinny; A.P. Togna; T.S. Webster, "Biological Systems: Biofilters," in *Odor and VOC Control Handbook*, H.J. Rafson, Ed., McGraw-Hill, New York, **1998**, pp. 8.150–8.191.
15. R.P.G. Bowker, "Biological Systems: Aeration Basins," in *Odor and VOC Control Handbook*, H.J. Rafson, Ed., McGraw-Hill, New York, **1998**, pp. 8.192–8.202.
16. "Bioscrubbers, Biofilters, and Trickle-Bed Reactors," in *Sources and Control of Air Pollution*, R.J. Heinsohn; R.L. Kabel, Prentice Hall, Upper Saddle River, NJ, **1999**, pp. 488–499.
17. Kessler, Luch, "Biofiltration Makes the Organics Go Down Biofilters Remove Low Level Organics from Coater Exhausts," *Chem. Eng.*, Apr. **1994**, p. 153.
18. R. Kapahi; M. Gross, "Biofiltration for VOC and Ammonia Emissions Control," *Biocycle*, Feb. **1995**, pp. 87–90.
19. B.J. Glowiak; K. Grefen; H. Jüntgen; K. Kirchner; E. Pilarczyk; F. Sporenberg; E. Weber, *Ambient Air Pollutants for Industrial Sources*, M.J. Suess, K. Grefen, D.W. Reinish, Eds., Elsevier, New York, **1985**, pp. 708–713.
20. J.B. Quarles, personal communication, **1997**.
21. D.S. Hodge; J.S. Devinny, "Biofilter Treatment of Ethanol Vapors," *Environ. Prog.*, **1994**, 13(3), pp. 167–173.
22. F.L. Smith; G.A. Sorial; M.T. Suidan; A.W. Breen; P. Biswas, "Development of Two Biomass Control Strategies for Extended, Stable Operation of Highly Efficient Biofilters with High Toluene Loadings," *Environ. Sci. Technol.*, **1996**, 30(5), pp. 1744–1751.
23. A.P. Togna; M. Singh, "Biological Vapor-Phase Treatment Using Biofilter and Biotrickling Filter Reactors: Practical Operating Regimes," *Environ. Prog.*, **1994**, 13(2), pp. 94–97.
24. K. Kirchner; U. Schlachter; H.J. Rehm, "Biological Purification of Exhaust Air Using Fixed Bacterial Monocultures," *Appl. Microbiol. Biotechnol.*, **1989**, 31, 629.
25. R.M.M. Diks; S.P.P. Ottengrapf, "Verification Studies of a Simplified Model for the Removal of Dichloromethane from Waste Gases Using a Biological Trickling Filter (parts I and II), *Bioproc. Eng.*, **1991**, 6, 131.
26. S. Hartmans; J. Tramper, "Dichloromethane Removal from Waste Gases with a Trickle Bed Bioreactor," *Bioproc. Eng.*, **1991**, 6, 83.
27. G.A. Sorial; F.L. Smith; P.J. Smith; M.T. Suidan; P. Biswas; R.C. Brenner, "Development of Aerobic Biofilter Design Criteria for Treating VOCs," Paper Number 93-TP-52A.04, Air and Waste Management Association's 86th Annual Meeting, Denver, Colorado, June 13–18, **1993**.
28. G.A. Sorial; F.L. Smith; P.J. Smith; M.T. Suidan; P. Biswas; R.C. Brenner, "Evaluation of Biofilter Media for Treatment of Air Streams Containing VOCs," Proceedings of the Water Environment Federation's 66th Annual Conference, Vol. 10, Facility Operations, pp. 429–439.
29. H.F. Ockeloen; T.J. Overcamp; C.P.L. Grady, Jr., "A Biological Fixed-Film Simulation Model for the Removal of Volatile Organic Air Pollutants," Proceedings of the 85th Annual Meeting of the Air and Waste Management Association, Kansas City, Missouri, June 21–26, **1992**.

30. A.P. Togna; B.R. Folsom, "Removal of Styrene from Air Using Bench-Scale Biofilter and Biotrickling Filter Reactors," Proceedings of the 85th Annual Air and Waste Management Association Meeting, Kansas City, Missouri, June 21–26, **1992**.
31. Open biofilter at the Advanced Organics facility, Philadelphia, Pennsylvania.
32. M. Rothenbühler, et al., "Biofiltration of Volatile Organic Emissions in Reference to Flexographic Printing Processes," *Water Air Soil Pollut.*, **1995**, 83, pp. 37–50.
33. M.G. Parvatiyar; R. Govind; D.F. Bishop, "Biodegradation of Toluene in a Membrane Biofilter," *J. Membr. Sci.*, **1996**, 119, pp. 17–24.
34. G. Wu, et al., "Biofiltration of Volatile Organic Emissions," Proceedings of the First North America Conference and Exhibition on, Emerging Clean Air Technologies and Business Opportunities, Poster session, Toronto, Canada, September 26–30, **1994**.
35. G. Martin; P. Laffort, *Odeurs et Désodorisation dans l'Environnement, Tech Doc*, Lavoisier, Paris, **1992**.
36. C. Kennes; H.H.J. Cox; H.J. Doddema; W. Harder, "Design and Performance of Biofilters for the Removal of Alkylbenzene Vapors," *J. Chem. Technol. Biotechnol.*, **1996**, 66, pp. 300–304.
37. H.L. Bohn, "Biofiltration: Design Principles and Pitfalls," Paper Number 93-TP-52A.01, 86th Annual Air and Waste Management Association Meeting, Denver, Colorado, June 13–18, **1993**.
38. S.P.P. Ottengraf; J.J.P. Meesters; A.H.C. Van Den Oever; H.R. Rozema, "Biological Elimination of Volatile Xenobiotic Compounds in Biofilters," *Bioproc. Eng.*, **1986**, 1, pp. 66–69.
39. R.F. Michelsen, "Biofiltration," in *Handbook of Air Pollution Control Engineering and Technology*, J.C. Mycock; J.D. McKenna; L. Theodore, Eds., Lewis Publishers, Boca Raton, **1995**, pp. 375–394.
40. H. Bohn; R. Bohn, "Soil Beds Weed Out Air Pollutants," *Chem. Eng.*, **1988**, 95(6), 73–76.
41. S.P.P. Ottengraf, "Method for Biological Treatment of Waste Gases," US Patent Number 4,662,900, issued May 5, **1987.**
42. G. Leson; D.S. Hodge; F. Tabatabai; A.M. Winer, "Biofilter Demonstration Projects for the Control of Ethanol Emissions," presented at the Air and Waste Management Association 86th Meeting and Exhibition, Denver, CO, June 13–18, **1993**.
43. L. Theodore; A.J. Buonicore; J.D. McKenna; I.J. Kugelman; J.S. Jeris; J.J. Santoleri; T.F. McGowan, "Air Pollution Management of Stationary Sources," in *Perry's Chemical Engineers' Handbook*, 7th Ed., D.W. Green; J.O. Maloney, Eds., McGraw-Hill, New York, **1997**, Ch. 25, pp. 41–43.
44. K. Kiared; G. Wu; M. Beerli; M. Rothenbühler; M. Heitz, "Application of Biofiltration to the Control of VOC Emissions," *Environ. Technol.*, **1997**, 18, pp. 55–63.
45. C. van Lith; G. Leson; R. Michelsen, "Evaluating Design Options for Biofilters," *J. Air Waste Manage. Assoc.*, **1997**, 47, pp. 37–48.
46. Steve Leach, personal communication, May **1998.**
47. S.P. Ottengraf, "Exhaust Gas Purification," in *Biotechnology*, Vol. 8, H.J. Rehm, G. Reed, Eds., VCH Verlagsgesellschaft, Weinheim, FRG, **1986**, pp. 427–452.
48. S.P.P. Ottengraf; A.H.C. van den Oever, "Kinetics of Organic Compound Removal from Waste Gases with a Biological Filter," *Biotechnol. Bioeng.*, **1983**, XXV, 3089–3102.

49. M.L. Bradford; R. Krishnamoorthy, "Consider Bioremediation for Waste Site Cleanup," *Chem. Eng. Prog.*, **1991.** 87(2), pp. 80–85.
50. N. Mukhopadhyay; E.C. Moretti, *Current and Potential Future Industrial Practices for Reducing and Controlling Volatile Organic Compounds*, Center for Waste Reduction Technologies, American Institute of Chemical Engineers, New York, **1993**, pp. 61–65.
51. C. van Lith; S.L. David; R. Marsh, "Design Criteria for Biofilters," Chemical Engineering's Symposium on Effluent Treatment and Waste Disposal, March **1990**, University of Leeds, England, pp. 127–281.
52. R. Helmer, "Desodorierung von Geruchsbelander Abluft in Bodenfiltern," *Gesundheits-Ingenieur*, 95(1), pp. 21–26.
53. R. Kamarthi; R.T. Wellington, "Bench-Scale Evaluation of Air Pollution Control Technology Based on Biological Treatment Process," Proceedings of the 87th Annual Meeting of the Air and Waste Management Association, Cincinnati, Ohio, **1994**.
54. B.F. Severin, et al., "Destruction of Gas Industry VOCs in a Biofilter," *Proceedings of the IGT Sixth International Symposium on Gas, Oil, and Environmental Technology*, Colorado Springs, CO, **1994**.
55. K. Tahraoui, et al., "Biodegradation of BTX from Waste Gases in a Biofilter Reactor," Poster session of the 87th Annual Meeting of the Air and Waste Management Association, Cincinnati, OH, **1994**.
56. D.A. Peters, et al., "Laboratory Assessment of Biofiltration for Fuel-Derived VOC Emissions Control," Proceedings of the 86th Annual Meeting of the Air and Waste Management Association, Denver, CO, **1993**.
57. K. Cho; M. Hirai; M. Shoda, "Removal of Dimethyldisulfide by the Peat Seeded with Night Soil Sludge," *J. Ferment. Bioeng.*, **1993**, 71(4), 289–291.
58. M. Shoda, "Methods for the Biological Treatment of Waste Gases," in *Biological Degradation of Wastes*, A.M. Martin, Ed., Elsevier Science Publishers, New York, **1991**, pp. 31–46.
59. D.S. Hodge; V.F. Medina; R.L. Islander; J.S. Devinny, "Treatment of Hydrocarbon Fuel Vapors in Biofilters," *Environ. Technol.*, **1991**, 12(8), 655–662.
60. X. Zhu; M.J. Rihn; M.T. Suidan; B.J. Kim; B.R. Kim, "The Effect of Nitrate on VOC Removal in Trickle Bed Biofilters," *Water Sci. Technol.*, 34(3–4), 573–581.
61. W.M. Vatavuk, *OAQPS Control Cost Manual* 5th Ed., US EPA, Research Triangle Park, NC, **1996**.
62. Pacific Environmental Services, Inc., *HAP-PRO 2.2* (Hazardous Air Pollutant Program), J. Steigerwald, project officer, EPA Control Technology Center, Research Triangle Park, NC, **1996**.
63. H.P. Rothbaum, "Heat Output of Thermophiles Occurring in Wool," *J. Bacteriol.*, 81, 165–171.
64. W.H. Prokop; H.L. Bohn, "Soil Bed Systems for Control of Rendering Plant Odors," *J. Air Poll. Contr. Assoc.*, **1985**, 35(12), 1332–1338.
65. J. Devinny; D. Hodge, "Formation of Acidic and Toxic Intermediates in Overloaded Ethanol Biofilters," *J. Air Waste Manage. Assoc.*, **1995**, 45(2), 125–131.
66. G.E. Speitel; D.S. McClay, "Biofilm Reactors for Treatment of Gas Streams Containing Chlorinated Solvents," *J. Environ. Eng.*, **1993**, 119(4), 658–678.

67. F.L. Smith; G.A. Sorial; M.T. Suidan; P. Biswas; R.C. Brenner, *Proceedings of the EPA Symposium on Bioremediation of Hazardous Wastes: Research, Development, and Field Evaluations*, San Fransisco, California, **1994**, pp. 89–97.

68. G.A. Sorial; F.L. Smith; M.T. Suidan; P. Biswas; R.C. Brenner, *Paper 94-RA115A.05*, Proceedings of the 87th Annual Meeting of the Air and Waste Management Association Meeting, Cincinnati, OH, **1994**.

10

EMERGING TECHNOLOGIES

10.1 INTRODUCTION

The need for better destruction efficiency and lower cost of air pollution control systems drives the development of new methods of VOC elimination [1]. The basic technology for some of these systems, such as membrane filtration and use of ozone for VOC oxidation, has been in existence for many years. As regulations become more stringent and as public awareness increases, however, the technologies are reconsidered and redesigned to address air pollution control issues. In addition, there are new technologies, such as plasma destruction, that have recently been discovered as potential methods for controlling air pollutants. Although many of these systems are available commercially, there is not much information available yet about costs, maintenance, and operational concerns. This chapter describes several of these innovative technologies.

10.2 MEMBRANE SEPARATION

Membrane separation processes are used to remove organic compounds from a polluted gas stream by establishing a partial pressure gradient across a semipermeable glassy or rubbery surface that constitutes the membrane. The membrane is designed to allow either air molecules or pollutant molecules to pass preferentially, resulting in a more concentrated pollutant stream on one side of the membrane. One common system works by pressurizing the incoming gas stream, removing as much of the pollutant as possible by condensation, and then exposing the resulting gas stream to a membrane that allows passage of organic molecules while limiting the passage of air molecules [2]. The organic *permeate* stream

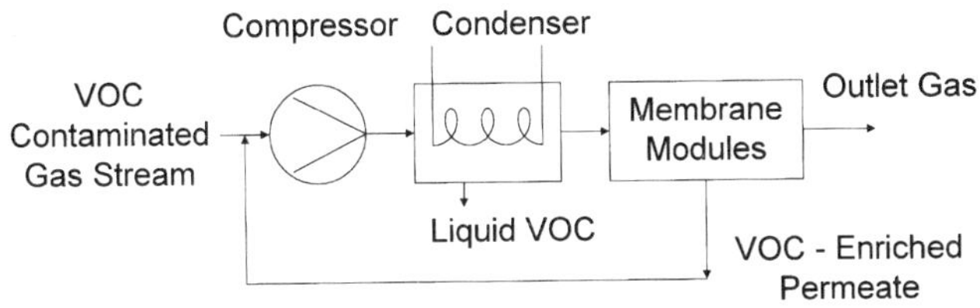

Figure 10.1 Flow diagram for a membrane separation system. (Adapted from Membrane Technology and Research [5].)

may also be under vacuum to increase the driving force for mass transfer. The permeate stream may be recycled to the compressor inlet. Once the concentration builds up sufficiently, the pollutant is removed by the condensation process [3]. Two membranes in series may be used if the gas stream is dilute to reduce the amount of permeate stream that must be recycled [4].

A typical flow diagram for a membrane separation system is shown in Figure 10.1 [5]. The pollutant-laden gas is compressed to 0.31–1.38 MPa gauge pressure (45–200 psig) and then flows into a condenser. From the condenser, the gas flows through the membrane system, where it is separated into an exhaust gas stream that has very low concentrations of organics and a permeate gas stream that is a more concentrated VOC stream [6–8]. The permeate gas stream may be recycled to the compressor. In this system, a condenser is added between the compressor and the membrane system, and the recycled vapor eventually reaches a state in which it is concentrated enough to condense. The permeate stream may also be treated by a conventional technology such as thermal or catalytic incineration. Membrane systems are also available in which the pollutant passes across the membrane into a scrubbing liquor, where it is presumably neutralized or destroyed [9].

As with other technologies, the constituents of the gas stream should be well characterized before a membrane system is employed to ensure that the membrane is chemically compatible with the potential gas stream constituents. Membranes are susceptible to fouling, and it is advisable to remove any particulate or oily compounds prior to the membrane system. Efficiencies of membrane separation systems are reported to be as high as 90–99.99% [10, 11], although it is important to note that greater removal efficiencies result in lower permeate stream concentrations [2].

Membrane separation is perhaps the most developed of the technologies that are covered in this chapter. There are approximately 60 membrane filtration systems that have been installed for VOC recovery [2, 3, 12, 13]. The system is based on technology that was first envisioned in 1950 by Weller and Steiner [14]. The technology has been used for several decades in the chemical process industry and in drinking water treatment plants [4, 15]. Typical and potential applications of membrane systems are listed in Table 10.1 [3, 8, 12, 13, 16–19]. The principles of gas permeation that are the foundation of membrane design work are given in the literature [3, 20–28].

TABLE 10.1 Applications of Membrane Separation Systems for VOC Removal

Vapor recovery from gasoline loading and unloading
Hospital sterilizer vents
Pharmaceutical plants
Polymer production
Industrial chiller purge streams
Film drying
Liquid ring vacuum pump exhaust
Tank and drum filling
Removal of higher hydrocarbons from natural gas

Data are from refs. 3, 8, 12, 13, 16–19.

Membrane systems are convenient in that they are continuous processes [9], they can be used with a large fluctuation in inlet concentration or flow [5], and they are normally modular in design [9], which simplifies installation and increases in capacity. They are usually cost competitive with carbon adsorption when the VOC concentration is above approximately 1000 ppm [2]. They are also applicable in situations in which carbon adsorption is not appropriate, such as with low-molecular-weight compounds [6] or compounds that polymerize on activated carbon.

Membrane systems can be classified according to the type of membrane used. Glassy membranes (such as polyetherimide) are more permeable to air molecules than to organic compounds [3]; rubbery membranes (such as natural rubber) are more permeable to organic compounds. Furthermore, glassy polymers in general allow substantially less material to flow through than do rubbery membranes (about 1/1000 as much). Because glassy membranes are more resistant to flow, a much larger membrane is required than if a rubbery membrane is employed. This is aggravated by the fact that the air must pass through the glassy membrane to get an effective separation. Because the proportion of air is usually much higher than that of the pollutant, glassy membranes result in much larger amounts of material being transferred through the membrane. The amount of material that must pass through the membrane, in combination with the permeability of the membrane, determines the surface area of membrane required. Therefore, rubbery membranes are generally used in VOC separation because the required surface area is significantly less than with glassy membranes [2]. Glassy membranes, though, can provide greater selectivity, resulting in a purer gas stream exiting the system [3]. The remainder of this chapter deals with rubbery membranes because they are generally the most cost effective for VOC removal.

Membrane systems can also be classified according to the design of the membrane unit. As shown in Figure 10.2 [29], plate and frame units are similar to filter presses, where the feed stream travels into a series of plates. The pollutant migrates from the plates through the membrane into a "frame." A frame is a space between the plates that direct the permeate out of the membrane module.

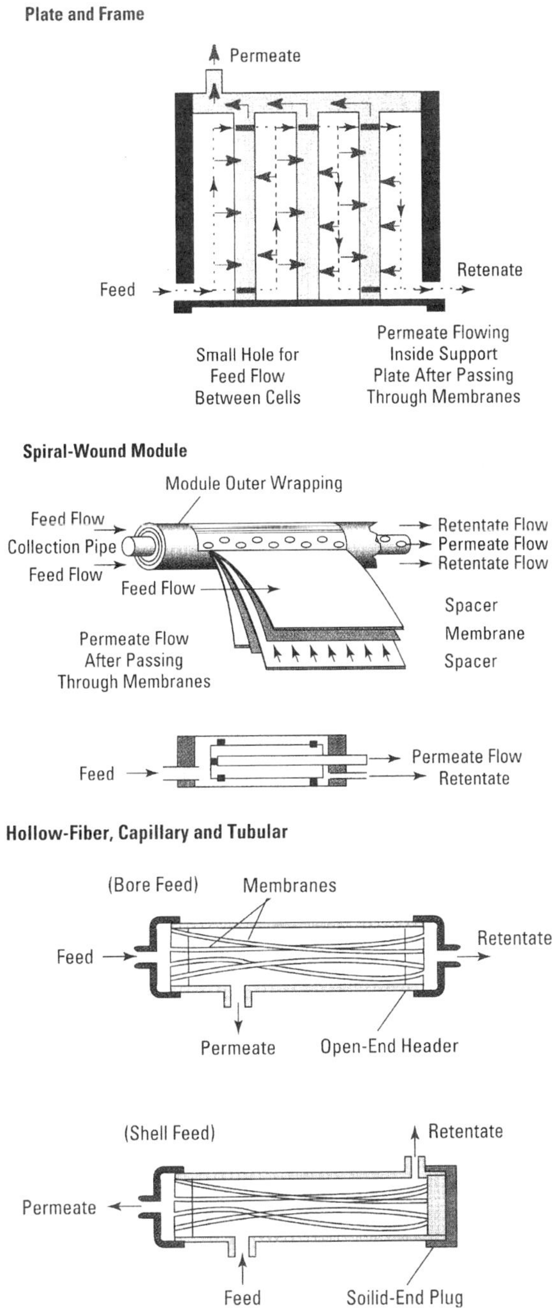

Figure 10.2 Various configurations of membrane separators. (from W.J. Koros, "Membranes: Learning a Lesson from Nature," *Chemical Engineering Progress*, October, **1995**, pp. 68–81. Reproduced with permission of the American Institute of Chemical Engineers.

The spiral-wound module shown is similar to a stack of flat membranes and spacers that have been wound around a hollow center shaft. The feed flows into one end of the space created by the spacers and travels by the membrane, and the permeate goes through the membrane into another cavity that winds to the hollow center shaft. The permeate collects in the center shaft and is removed from the end of the shaft. The feed that does not go through the membrane is collected at the end of the membrane module.

The third type, which is not commonly used for air pollution applications, is similar to a shell-and-tube heat exchanger. There are "heads" on each end that hold the ends of hollow fibers or tubes. The feed stream can flow either in the "shell" side of the membrane module, in which the permeate is removed from the "tube" side, or the feed stream can flow in the "tube" side and the permeate is removed from the "shell" side.

The degree of separation that can be achieved in a membrane module is governed by the selectivity of the membrane to the pollutant over the remaining components of the gas stream. The selectivity of the membrane is given by the formula

$$\alpha_{A/B} = P_A/P_B \tag{10.1}$$

where

$\alpha_{A/B}$ = selectivity of the membrane to pass component A instead of component B
P_A = permeability of gas A through the membrane
P_B = permeability of gas B through the membrane

Usually, membranes have selectivities of approximately 20–100 [3], indicating that approximately 20–100 times as much organic material as air will pass through the membrane. Permeability is a measure of the volume of a compound that moves through a (1 cm thick) membrane at a standard pressure difference of 1 cm Hg. The permeability is usually given in units of Barrers. One Barrer is equal to 1×10^{-10} cm^3 (STP)/(cm^2s cm Hg). The permeability of a compound is related to the diffusion coefficient, a measure of the rate of transport of the compound through the membrane material, and the Henry's law sorption coefficient, a measure of the condensability of the compound [2].

It is important that the relative permeabilities or selectivity be measured in experiments with the simulated gas composition as opposed to measuring the permeabilities of pure gases separately and comparing them. This technique is necessary because rubbery membranes swell in the presence of organic compounds, changing the diffusion of each compound through the membrane. This phenomenon results in a higher selectivity for the organic compound than would be predicted otherwise [3].

Likewise, temperature can affect the selectivity and permeability of the membrane. The selectivity usually improves as the temperature is decreased [3]. Polymer membranes are generally limited to 363 K (194°F, 90°C) [30]. The amount of membrane surface required depends on the type of membrane, the permeability of the pollutant through the membrane, the required removal efficiency, and the difference in pressure across the membrane [3].

Little cost information is available for membrane systems. The cost for membranes that allow passage of organic compounds has been estimated at approximately \$71/1,000 standard m^3 of gas (\$2/1,000 scf gas) [29] and at \$4–11/kg of VOC recovered (\$2–5/lb of VOC recovered) [31]. Most of the annual cost associated with a typical membrane system is the cost to compress the gas stream. Annual energy consumption associated with compressing the gas may be approximated according to the guidelines in the *OAQPS Cost Control Manual* [32]:

$$\text{Power (kW)} = AQ\Delta P/\varepsilon \tag{10.2}$$

where

A = factor, 1.66×10^{-5} for SI units (1.17×10^{-4} for English units, shown in parentheses)

Q = total flue gas flow rate, standard m^3/min or sccm (standard ft^3/min or scfm)

ΔP = pressure change, Pa (inches of water)

ε = combined efficiency of motor and compressor

If a vacuum pump is used on the permeate stream, it would also need to be included in the electrical cost estimate.

There are few moving parts in membrane systems, so they are less likely to require maintenance [9]. Membrane systems are less susceptible to malfunction due to changes in concentration and flow rate than are other pollution control systems, as long as the fluctuations were anticipated as part of the original design. The temperature must, however, be maintained low enough to effectively separate the compounds and to ensure that the membrane is not damaged. Similarly, care must be taken to ensure that the gas stream constituents are compatible with the membrane material and with other materials of construction so that the membrane is not damaged by solvent action of the gas stream. Particulate or oils that could accumulate on the membrane or in the pores of the membrane should also be removed to extend the life of the membrane. High concentrations of organic compounds should also be avoided to minimize the risk of explosions. See the discussion of lower explosive limits in Chapter 6.

An overall summary of the advantages and disadvantages of membrane systems is given in Table 10.2 [5, 33, 34].

10.3 AMBIENT OXIDATION PROCESSES

Ultraviolet (UV) oxidation uses the same types of photochemical reactions that occur in the atmosphere [6, 35] to assist in the oxidation of organic compounds to form water and carbon dioxide. UV light and oxidants such as ozone, peroxide, hydroxyl radicals, or oxygen radicals are used to break down the VOCs into

TABLE 10.2 **Advantages and Disadvantages of Membrane Separation**

Advantages	Disadvantages
Solvent recovery No secondary wastes Compact design High efficiencies possible with proper design Technology is applicable for variable pollutant concentrations.	High energy costs

Data are from refs. 5, 33, 34.

carbon dioxide and water [1]. The wavelength of light used is determined based on the VOCs to be destroyed. In one system, the incoming gas stream passes through a filter to remove particulate and then into a chamber where the reactions occur. Following the main reaction chamber, any residual VOCs may be scrubbed with water or adsorbed onto activated carbon. The scrubbing liquor and carbon regeneration may also include oxidants to further enhance removal efficiency [35].

Photocatalytic oxidation (PCO) or *heterogeneous photocatalysis* [36] is a similar process, but it includes a catalyst surface (usually titanium dioxide, although nickel and mixtures of titanium dioxide and platinum have also been used) [37, 38]. In PCO, the catalyst surface is a semiconductor material that becomes active in UV or near-UV light. The light source may be either the sun or a fluorescent bulb [39, 40]. The process operates by adsorption of the pollutant compound onto the surface of the catalyst and then decomposition of the compound in a manner similar to catalytic oxidation. Ozone, peroxide, hydroxyl radicals, or oxygen radicals may also be added to or generated within the process [41]. The process is applicable for aliphatic hydrocarbons, ethers, ketones, alcohols, and aldehydes [42]. It is particularly well suited to chlorinated hydrocarbons because they are more reactive than nonchlorinated compounds. In conventional technologies, such as thermal incinerators, the chlorinated compounds are more difficult to destroy than the nonchlorinated compounds [39]. Aromatics can also be treated; the addition of ozone or hydrogen peroxide increases the rate of reaction of aromatics substantially [43]. Microbes are also destroyed in the process [44].

One type of PCO reactor is shown in Figure 10.3 [45]. The reactor consists of two concentric tubes. Beads coated with catalyst are packed between the two tubes, and a UV light source is put in the center tube. The polluted gas stream flows through the beads. Alternatively, the reactor can be designed with the catalyst arranged on a series of disks over which the gas flows [46]. The goal of either design is to expose as much of the catalyst surface as possible to the light while maintaining a low pressure drop for the gas stream. The mechanism of the surface reaction is not well understood, although it is recognized that

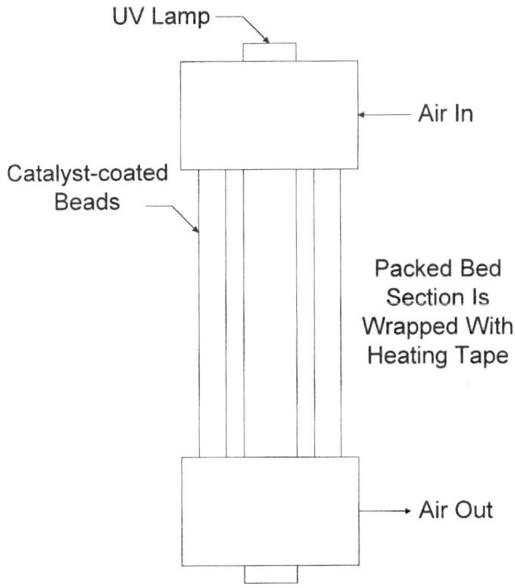

Figure 10.3 Flow diagram of a packed bed PCO reactor [45].

the excitation of the semiconductor catalyst material promotes the formation of electron-hole pairs in the solid [41].

The constituents of the gas stream must be identified in order to determine if a PCO system would be suitable. In addition, particulates, oily compounds, or compounds such as hexamethyldisilazane, which could foul the surface of the catalyst, must be removed prior to the reactor.

Efficiencies of laboratory PCO systems have been reported as high as 99% at high temperatures or residence times [47, 48]. Efficiencies at ambient temperatures and residence times that are more realistic for commercial systems have been reported as high as 95% for a 400 ppm gas stream [43]. Commercial systems are currently being designed to achieve greater than 99% destruction [49].

PCO technology is best suited for low concentration (<1000 ppm) gas streams at low to moderate flow rates (<567 m^3/min or 20,000 cfm). At higher concentrations, a larger catalyst area is required, and the system becomes less cost competitive with other technologies [45]. Normal fluorescent lighting may be an adequate light source for destroying odors and very low concentrations of pollutants using TiO_2 as a photocatalyst [50]. Potential applications of PCO systems are listed in Table 10.3 [40, 45, 49, 51, 52].

For the reaction to occur, the organic species must first be adsorbed onto the catalyst surface. The rate of adsorption is governed by the concentration of the pollutant in the gas stream and the temperature. Increasing temperature generally increases the rate of destruction of the compound. Increased temperature, however, also reduces the speed of the adsorption process, so less compound is available for destruction. Care must be taken to ensure that adequate surface area is available for adsorption. Once the surface of the catalyst is saturated with compound, increases in concentration have little or no effect on the destruction

TABLE 10.3 Applications of PCO Systems for VOC Removal

Soil or groundwater remediation vent streams
Contact lens degreasing
Lithography equipment
Paint spray booths
Cleaning equipment
Indoor air pollution from paint and solvents

Data are from refs. 40, 45, 49, 51, 52.

rate of the pollutant, so the destruction efficiency declines rapidly [45]. It is also important, therefore, to ensure that adsorption sites are not blinded by particulate or nonreactive compounds. The reaction also will not occur without light. The light can, however, be quickly turned off or on for an intermittent pollutant stream, resulting in lower energy consumption [49].

Little cost information is available for these oxidation systems. Cost estimates have been prepared for a photocatalytic oxidation system in the same manner as costs for other controls were estimated by Mukhopadhyay and Moretti [53], which generally follow the methodology of the EPA's *OAQPS Cost Control Manual* [54]. The cost for a PCO system has been estimated at approximately \$883–1069/standard m^3 per min of gas (\$25–30/scfm of gas), although the final cost will depend on the relative reactivity of the compound to be treated. Unlike other control systems, the cost increases with the concentration of organic compound, but remains relatively stable with increasing flow rates [39].

There are not many moving parts in PCO systems, so it is anticipated that annual maintenance costs would be relatively low. Because of the low temperatures, simple materials of construction can be used such as plastic piping [55], which reduce both capital and maintenance costs. An overall summary of the advantages and disadvantages of PCO systems is given in Table 10.4 [49].

TABLE 10.4 Advantages and Disadvantages of PCO

Advantages	Disadvantages
Can be turned off with a switch to save energy between batches	May form additional pollutants if process is not properly designed
Has a compact design	Solvent recovery is not feasible
High efficiencies possible with proper design	
Technology is applicable for small or short fluctuations in pollutant concentrations	

Data are from Kittrell et al. [49].

10.4 CORONA DESTRUCTION

The *corona destruction* process also operates at ambient temperature. A high-voltage/low-current electrical discharge is used to produce high energy electrons that collide with VOCs and decompose them into carbon dioxide and water [1, 56]. The basic process is described by Yamamoto et al. [57]. It is believed that the technology may eventually be applicable for low concentration VOC streams; however, available information indicates that the energy requirements are currently too high to be competitive with other technologies. Packed bed systems that use a bed of dielectric beads (perovskite $BaTiO_3$) with an electrode connected on either end are being investigated. The beads store electrical energy and discharge it out of phase with the applied electric field, producing electrons that destroy the VOC. High destruction efficiencies are possible with the corona discharge processes; however, it has been demonstrated that ozone is formed in substantial concentrations (approximately 500 ppm; see Chapter 11). In addition, NO_x and incomplete products of combustion may also be formed [56]. The potential benefits of corona destruction are that it would not require auxilliary fuel, would operate at ambient temperatures, and would be able to treat both halogenated and nonhalogenated compounds [1].

10.5 PLASMA DESTRUCTION

Several types of plasma destruction are also being investigated to determine if they are applicable for controlling chlorinated or nonchlorinated compounds [6]. Plasma technologies can be divided into high and low temperature processes. In high temperature processes, the contaminated gas stream is mixed with oxygen or water vapor and passes into the reactor. A plasma or high temperature ionized gas state is maintained in the reactor by an electric discharge (arc) or by radiofrequency energy. The plasma is extremely reactive, and the gas reacts with the water or oxygen to form carbon dioxide [6]. One such system is used in Japan to destroy chlorofluorocarbons and hydrochlorofluorocarbons. Destruction efficiencies have been measured at 99.99% and higher using 35–138 kW of electricity for a gas stream containing 36–100 kg/h [58].

Cold plasma techniques heat only the electrons in the plasma by a series of micro-arcs similar to tiny lightening bolts. The electron temperatures are typically 3 eV, approximately equal to 30,000 K (29,700°C or 54,000°F). The remainder of the gas is, however, not heated. The reactions that occur are similar to the free radical reactions that occur in thermal oxidation, but the free radicals are produced by the collisions of the electrons and the gas molecules, not by the temperature. Efficiencies in the range of 90–99% have been demonstrated on laboratory scale units. Ozone, carbon monoxide, nitrous oxides, and nitric acid, however, may also be formed as byproducts of the reactions. Preliminary cost estimates are approximately $1.18/scmm ($33.3/1000 scfm) [59].

10.6 OZONE CATALYTIC OXIDATION

Catalytic oxidation of VOCs with ozone as the primary oxidant offers the advantage of low temperatures of reaction. This topic is covered in Chapters 11 and 12.

REFERENCES

1. E.C. Moretti; N.K. Mukhopadhyay, "VOC Control: Current Practices and Future Trends," *Chem. Eng. Prog.*, **1993**, 7, 20–26.
2. R.W. Baker; J. Kaschemekat; J.G. Wijmans, "Membrane Systems for Profitable VOC Recovery," *Chemtech*, July **1996**, pp. 37–43.
3. R.W. Baker; J.G. Wijmans, "Membrane Separation of Organic Vapors from Gas Streams," in *Polymeric Gas Separation Membranes*, D.R. Paul; Y.P. Yampolskii, Eds., CRC Press, Boca Raton, **1994**, pp. 353–397.
4. N. Mukhopadhyay; E.C. Moretti, "Membrane Technology," in *Current and Potential Future Industrial Practices for Reducing and Controlling Volatile Organic Compounds*, Center for Waste Reduction Technologies, American Institute of Chemical Engineers, New York, **1993**, pp. 22–23.
5. "VaporSep®: Membrane Systems For VOC Recovery," Membrane Technology and Research, Menlo Park, CA, **1998**.
6. R.W. Baker; M. Jacobs, "Improve Monomer Recovery from Polyolefin Resin Degassing," *Hydrocarbon Processing*, March **1996**.
7. R.G. McInnes, "Explore New Options for Hazardous Air Pollutant Control," *Chem. Eng. Prog.*, **1995**, 11, 36–47.
8. V. Simmons; J. Kaschemekat; M.L. Jacobs; D.D. Dortmundt, "Membrane Systems Offer a New Way to Recover Volatile Organic Air Pollutants," *Chem. Eng.*, Sep. **1994**.
9. C.M. Caruana, "Membranes Vie for Pollution Cleanup Role," *Chem. Eng. Prog.*, Oct. **1993**, pp. 11–14.
10. R.W. Baker; V. Simmons; H. Wijmans, "Membrane Recovery of Volatile Organic Compounds (VOCs) in the Pharmaceutical Industry," *Pharmaceutical Eng.*, **1993** 13 (5).
11. J. McCallion, "Membrane Process Captures Vinyl Chloride, Other VOCs: Recovers Values, Eases Incineration Burdens," *Chem. Proc.*, Sep. **1994**.
12. K. Ohlrogge; K.V. Peinemann; J. Wind; R.D. Beling, *Sep. Sci. Technol.*, **1990**, 25, 13–15.
13. K. Matsumoto; K. Ishii; T. Kuroda; K. Inoue; A. Iwama, *Polym. J.*, **1991**, 23 (5).
14. R.R. Zolandz; G.K. Fleming, "Applications," in *Membrane Handbook*, W.S. Winston Ho; K.K. Sirkar, Eds., Van Nostrand Reinhold, New York, **1992**, Ch. 5, pp. 91–93.
15. L. Theodore; A.J. Buonicore; J.D. McKenna; I.J. Kugelman; J.S. Jeris; J.J. Santoleri; T.F. McGowan, "Waste Management," in *Perry's Chemical Engineers' Handbook*, 7th Ed., D.W. Green; J.O. Maloney, Eds., McGraw-Hill, New York, **1997**, Ch. 25, pp. 25-1–25-111.
16. K. Ohlrogge, presentation at the Ninth Annual Membrane Technology Planning Conference, Newton, MA, Nov. **1991**.

17. J. Wijmans, U.S. Patent 5,071,451, December 19, **1991**, as given by R.W. Baker; J.G. Wijmans, "Membrane Separation of Organic Vapors from Gas Streams," in *Polymeric Gas Separation Membranes*, D.R. Paul; Y.P. Yampolskii, Eds., CRC Press, Boca Raton, **1994**, pp. 353–397.
18. K. Watler, U.S. Patent 4,857,078, August 15, **1989**, as given by R.W. Baker; J.G. Wijmans, "Membrane Separation of Organic Vapors from Gas Streams," in *Polymeric Gas Separation Membranes*, D.R. Paul; Y.P. Yampolskii, Eds., CRC Press, Boca Raton, **1994**, pp. 353–397.
19. N. Mukhopadhyay; E.C. Moretti, "Membrane Technology," in *Current and Potential Future Industrial Practices for Reducing and Controlling Volatile Organic Compounds*, Center for Waste Reduction Technologies, American Institute of Chemical Engineers, New York, **1993**, pp. 50–51.
20. R.M. Barrer, *Diffusion in and Through Solids*, MacMillan, New York, **1941**.
21. G.J. van Amerongen, *J. Appl. Phys.*, **1946**, 17, 972.
22. R.M. Barrer; G. Skinner, *J. Polym. Sci.*, **1948**, 3, 549.
23. A. Aitken; R.M. Barrer, *Trans. Faraday Soc.*, **1948**, 51, 549.
24. G.J. van Amerongen, *J. Polym. Sci.*, **1950**, 5, 307.
25. G.J. van Amerongen, *Rubber Chem. Technol.*, **1964**, 37, 1065.
26. P. Meares, *J. Appl. Polym. Sci.*, **1965**, 9, 917.
27. S.A. Stern, in *Gas Permeation Process in Industrial Processing with Membranes*, R.B. Lacey; S. Loeb, Eds., Wiley-Interscience, New York, **1972**.
28. R.M. Baker; J.G. Wijmans, U.S. Patent 5,129,921, July 14, **1992**.
29. W.J. Koros, "Membranes: Learning a Lesson from Nature," *Chem. Eng. Prog.*, **1995**, 10, 68–81.
30. "Development of New Membranes," Fraunhofer Institut Grenzflächen-und Bioverfahrenstechnik (available at http://www.igb.fhg.de/MuPT/home.en.html) **1998**.
31. "VOC Offgas Membrane Separation" in *The DOE Industrial Projects Locator* (available at http://www.oit.doe.gov/Access/locator/2114.htm), July **1998**.
32. D.R. van der Vaart; J.J. Spivey; W.M. Vatavuk; A.H. Wehe, "Thermal and Catalytic Incinerators," in *OAQPS Control Cost Manual*, 5th Ed., United States Environmental Protection Agency, Office of Air Quality Planning and Standards, Research Triangle Park, NC, **1995**, ch. 3 (available at http://www.epa.gov/ttn/catc/products.html).
33. "Ethylene, Propylene Recovery System Spotlighted," Press Release from Membrane Technology and Research, Menlo Park, CA, **1995**.
34. "Novel Membranes for Natural Gas Separation," Department of Energy Tech Brief, U.S. Department of Energy, Washington, DC, Nov. **1994**.
35. N. Mukhopadhyay; E.C. Moretti "Ultraviolet Oxidation Technology," in *Current and Potential Future Industrial Practices for Reducing and Controlling Volatile Organic Compounds*, Center for Waste Reduction Technologies, American Institute of Chemical Engineers, New York, **1993**, p. 23.
36. R.A. Zerbonia; J.J. Spivey; S.K. Agarwal; A.S. Damle; C.W. Sanford, "Heterogeneous Catalysis," in *Survey of Control Technologies for Low Concentration Organic Vapor Gas Streams*, Control Technology Center, U.S. Environmental Protection Agency, Research Triangle Park, NC, **1995**, Sect. 6.2.
37. T.T. Shen; C.E. Schmidt; T.R. Card, "Treatment Process Selection/Modification Control Technologies," in *Assessment and Control of VOC Emissions from Waste*

Treatment and Disposal Facilities, Van Nostrand Reinhold, New York, **1993**, pp. 157–177.

38. E.J. Wolfrum; R. Rabago; A.S. Jassal, *Destruction of Volatile Organic Compound (VOC) Emissions by Photocatalytic Oxidation (PCO): Final Report* (ESHC003), Technology Transfer Document #97013236A-ENG, SEMATECH, Inc., Austin, TX, **1997** (available at http://www.sematech.org).
39. C.S. Turchi; E.J. Wolfrum; R.A. Miller, "Gas-Phase Photocatalytic Oxidation: Cost Comparison with Other Air Pollution Control Technologies," Presented at Advanced Oxidation Technologies for Water and Air Remediation, London, Ontario, Canada, June 1994, NREL/TP-41-7014, National Renewable Energy Laboratory, Golden, CO, **1994**.
40. Arizona State University/Zentox Corporation: "Photocatalytic Oxidation with Air Stripping," Arizona State University, Tempe, AZ, **1999** (available at http://www.clu-in.org/remed1.html).
41. D.A. Gratson; M.R. Nimlos; E.J. Wolfrum, "Photocatalytic Oxidation of Gas-Phase BTEX-Contaminated Waste Streams," prepared for the Air & Waste Management Association, 88th Annual Meeting and Exhibition, San Antonio, TX, June 1995, Report NREL/TP-473-7575, National Renewable Energy Laboratory, Golden, CO, **1995**.
42. "KSE, Inc, Adsorption-Integrated-Reaction Process," KSE, Inc., Amherst, MA, **1998** (available at http://www.clu-in.org/remed1.html).
43. C.S. Turchi; R. Rabago, *Benchscale Testing of Photocatalytic Oxidation (PCO) to Destroy Volatile Organic Compound (VOC) Emissions*, Technology Transfer Document #95042791A-ENG, SEMATECH, Inc., Austin, TX, **1995**, (available at http://www.sematech.org).
44. *Sun-Air*^{TM} *300: with Phototech*^{TM} *Photocatalytic Air Purification and Detoxification Technology*, Universal Air Technology, Inc., Lake Hopatcong, NJ, **1998** (available at http://www.universalair.com).
45. C.S. Turchi; R. Rabago; A.S. Jassal, *Destruction of Volatile Organic Compound (VOC) Emissions by Photocatalytic Oxidation (PCO): Field Test Results*, Technology Transfer Document #96023084A-ENG, SEMATECH, Inc., Austin, TX, **1996** (available at http://www.sematech.org).
46. *Radiant Decomposition of Airborne VOCs*, U.S. Navy, Indian Head Division, Naval Surface Warfare Center, Indian Head, MD, **1999** (available at http://www.fsec.ucf.edu//~at/detox.html).
47. J. Blanco; P. Avila; A. Bahamonde; E. Alvarez; B. Sánchez; M. Romero, "Photocatalytic Destruction of Toluene and Xylene at Gas Phase on a Titania Based Monolithic Catalyst," *Catalysis Today* **1996** 29, 437–442.
48. A.V. Vorontsov; E.N. Savinov; G.B. Barannik; V.N. Troitsky; V.N. Parmon, "Quantitative Studies on the Heterogeneous Gas-Phase Photooxidation of CO and Simple VOCs by Air Over TiO_2," *Catalysis Today*, **1997**, 39, 207–218.
49. J.R. Kittrell; C.W. Quinlan; J.W. Shepanzyk, "Photocatalytic Destruction of Hexane Eliminates Emissions in Contact Lens Manufacture," in *Emerging Solutions to VOC and Air Toxics Control: Proceedings of A Specialty Conference*, Clearwater Beach, FL, February–March, 1996, Air and Waste Management Association, Pittsburgh, PA, **1996**, pp. 389–409.
50. A. Fujishima, "The Raising of The Titanium," *Look Japan*, June **1995**, 26–27.

51. H.M. Liljestrand; M.L. Sattler, "Photocatalytic Oxidation Processes for In-Door Air Pollution," paper 96-WA61.07, presented at the Air and Waste Management Association 89th Annual Meeting, Nashville, TN, 1996, Air and Waste Management Association, Pittsburgh, PA, **1996**.

52. "Destruction Technology for Air Stripper and Vapor Extraction Off-Gas," **1999** (available at http://www.bwonotes5.wdc.pnl.gov/iof.nsf). Department of Energy, NREL.

53. N. Mukhopadhyay; E.C. Moretti, "Economic Analysis of Selected VOC Abatement Technologies," in *Current and Potential Future Industrial Practices for Reducing and Controlling Volatile Organic Compounds*, Center for Waste Reduction Technologies, American Institute of Chemical Engineers, New York, **1993**, pp. 57–65.

54. W.M. Vatavuk, *OAQPS Control Cost Manual*, 5th Ed., G. Moyer, Ed., United States Environmental Protection Agency, Office of Air Quality Planning and Standards, Research Triangle Park, NC, **1995**, ch. 3 (available at http://www.epa.gov/ttn/catc/products.html).

55. "Photocatalytic Oxidation: Low Temperature Oxidation," Zentox (available at http://www.zentox.com/fpco.html). Ocala, FL, **1996**.

56. R.A. Zerbonia; J.J. Spivey; S.K. Agarwal; A.S. Damle; C.W. Sanford, "Corona Discharge Processes," in *Survey of Control Technologies for Low Concentration Organic Vapor Gas Streams*, Control Technology Center, U.S. Environmental Protection Agency, Research Triangle Park, NC, **1995**, Sect. 6.1 (available at http://www.epa.gov/ttn/catc/products.html).

57. T. Yamamoto; K. Ramanathen; M.K. Owen; D.S. Ensor; J.J. Spivey, *Corona Destruction and Other Technologies for VOC Control*, report to the U.S. EPA: AEERL, Cooperative Agreement CR815169, Nov. **1990**.

58. "Induction-Coupled Plasma Used to Destroy CFCs in Japan," *Air Pollut. Consultant*, V.A. Dean, Ed. **1997** 7 (4), pp. 1.11–1.12.

59. J.J. Coogan; A.S. Jassal, *Silent Discharge Plasma for Point-of-Use (POU) Abatement of Volatile Organic Compound (VOC) Emissions*, final report (ESHC003), Technology Transfer Document #97023244A-ENG, SEMATECH, Inc., Austin, TX, **1997** (available at http://www.sematech.org).

11

OZONE PROPERTIES, HANDLING, AND PRODUCTION

11.1 PROPERTIES

In ordinary conditions, ozone is a light blue gas with a pungent odor. It is a thermally unstable powerful oxidant, but its high reactivity makes it useful in synthetic applications and as a disinfectant. In the upper atmosphere ozone is beneficial, absorbing dangerous ultraviolet light and contributing to the thermal equilibrium of the planet. Close to the ground, however, ozone is harmful, damaging plant and animal tissue and contributing to smog formation. The general properties and chemistry of ozone are covered in a comprehensive encyclopedia article [1] and in several books [2–4].

Ozone was first recognized as a distinct chemical species by Schönbein in 1839 [5], but its formula, based on three oxygen atoms, was first suggested by Soret in 1865 [6]. A reliable synthesis of high concentration ozone was first developed by von Siemens [7] using an electrical discharge.

11.1.1 Structure and Bonding

Ozone is an allotrope of oxygen consisting of three, rather than two, atoms. It has a bent structure of C_{2v} symmetry with an angle of 116°49′ and equal oxygen–oxygen bond distances of 0.128 nm. This distance is closer to that of O_2, with 0.121 nm, than of H_2O_2, with 0.149 nm, and hence the bonds in O_3 have considerable double bond character with a bond order of 1.7 [8]. The resonance structures [9] of ozone are shown in Figure 11.1.

The contributions of forms **1** and **4**, which have a positively charged terminal oxygen with only six electrons, have been used to account for the electrophilic properties of ozone [10]. These forms are consistent with the 1,3-addition

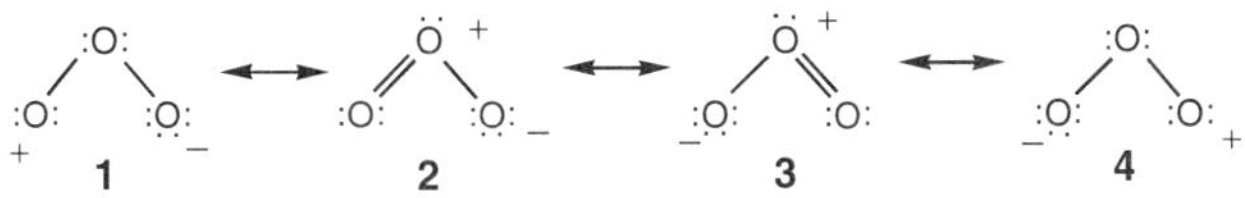

Figure 11.1 Resonance structures of ozone.

chemistry of ozone to double bonds. The resonance forms contain no unpaired electrons, and ozone is diamagnetic, with a low dipole moment of 0.53 debye [9] and a low magnetic susceptibility of 2.00×10^{-9} cgs units [11].

Bonding in ozone [12] can be viewed as occurring through σ and π interactions between sp^2-hybridized atoms (Fig. 11.2). The nonbonded sp^2 orbitals contain two electrons, and the entire molecule shares a π cloud containing four electrons. The low dipole moment of ozone can be explained by the balancing effect of the unshared electron pair in the apical atom.

Molecular orbital and valence bond calculations [13–15] suggest that a 1,3-singlet diradical structure, •O–O–O•, may contribute significantly to the ground state structure of ozone. This could explain the reaction of ozone with saturated C–H bonds. The diradical can still have considerable dipolar character because the electrons remain paired [16]. Other studies [17, 18] favor a hypervalent structure, with bonding similar to resonance structures **2** and **3**, and negligible diradical character. An *ab initio* Gaussian calculation also suggests that those resonance forms are the major contributors, with the terminal oxygen atoms being negatively charged [19].

11.1.2 Thermodynamic Properties

Ozone is a stronger oxidizing agent than molecular oxygen or hydrogen peroxide and reacts with most substances at room temperature. In acid solution, the oxidizing power of ozone is exceeded only by fluorine, the perxenate ion, atomic oxygen, OH radicals, and a few other such species [8]. The potentials in aqueous

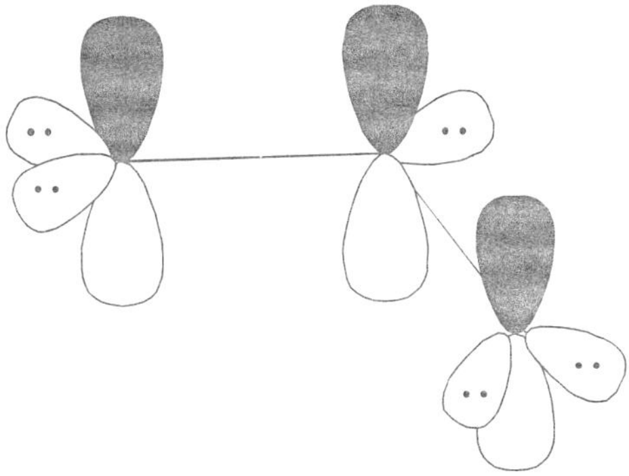

Figure 11.2 Orbitals in ozone.

solution are [11]

$$O_2 + 2H^+ + 2e^- \longrightarrow H_2O_2 \qquad E^\circ = 0.68 \text{ eV} \tag{11.1}$$

$$H_2O_2 + 2H^+ + 2e^- \longrightarrow 2\ H_2O \qquad E^\circ = 1.78 \text{ eV} \tag{11.2}$$

$$O_3 + 2H^+ + 2e^- \longrightarrow O_2 + H_2O \quad E^\circ = 2.07 \text{ eV} \tag{11.3}$$

Ozone is thermodynamically unstable, and its decomposition is exothermic.

$$O_3 \longrightarrow 3/2\ O_2 \tag{11.4}$$

The standard heat of reaction is $\Delta H^\circ_{298} = -142$ kJ mol^{-1} (-34.0 kcal mol^{-1}), the standard free energy of reaction is $\Delta G^\circ_{298} = -163$ kJ mol^{-1} (39.0 kcal mol^{-1}) [20], and the entropy is $S^\circ_{298} = 239$ J K^{-1}mol^{-1} (57.1 cal mol^{-1}K^{-1}) [21]. The heat capacity (J K^{-1}mol^{-1}) of ozone is [20]:

$$c_p = 54.258 + 2.004 \times 10^{-3}T - 1.556 \times 10^6 T^{-2} \tag{11.5}$$

This corresponds to a value of $c_P = 39.2$ J K^{-1}mol^{-1} (9.37 kcal K^{-1}mol^{-1}) at room temperature. The melting point is 80.45 K (-192.7°C); the boiling point is 161.25 K (-111.9°C); the critical point is at 242.31 K (-12.1°C) and 67.9 bar (54.6 atm) [11].

11.1.3 Spectral Properties

The three principal infrared absorption bands of ozone are $\omega_1 = 1135$ cm^{-1} (a_1), $\omega_2 = 716$ cm^{-1} (a_1), and $\omega_3 = 1089$ cm^{-1} (b_2) [15]. The corresponding Raman lines are $\omega_1 = 1103$ cm^{-1} and $\omega_2 = 702$ cm^{-1} [22]. The modes are indicated in Fig. 11.3.

The absorption cross sections [23] and absorption spectra [24, 25] of ozone in the visible and ultraviolet (UV) regions have been reported. There are three major regions in the absorption spectrum: the strong Hartley band at 200–300 nm, the Huggins band at 300–350 nm, and the weak Chappuis band at 400–750 nm. In the visible region (Chappuis band), ozone displays a diffuse absorption band in the visible region centered at ~600 nm and an absorption cross section, $\sigma = 489 \times 10^{-20}$ cm^2. The molar absorptivity is 2.94 L mol^{-1}cm^{-1} [26]. In the UV region (Hartley band) the spectrum has a single broad absorption centered at

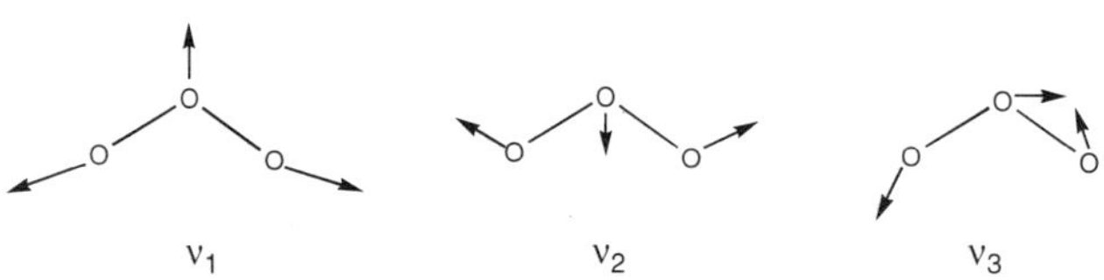

Figure 11.3 Vibrational modes of ozone.

TABLE 11.1 Comparison of Oxygen, Hydrogen Peroxide, and Ozone

Properties	Oxygen	Hydrogen peroxide	Ozone
Bond distance, nm	0.121	0.149	0.128
Bond order	2	1	1.7
Dipole moment, debye	0	2.2	0.53
Oxidation potential, eV	0.68	1.78	2.07
Magn. susceptibility, cgs	3.45×10^{-5}	-1.77×10^{-7}	2.00×10^{-9}

Data are from Weast and Astle [11].

~254 nm. At this wavelength and 273 K, the absorption cross section is $\sigma = 1147 \times 10^{-20}$ cm^2. The molar absorptivity at 260 nm and 298 K is 2930 L mol^{-1} cm^{-1} [26]. The Hartley and Chappuis bands are symmetric, while the Huggins band is a decreasing shoulder region of the Hartley band.

11.1.4 General and Physical Properties

Ozone is a pale blue gas (d = 2.1415 g/L at 0°C and 1 atm [101.3 kPa]), which condenses to an indigo blue liquid (d = 1.354 kg/L at −112°C (161 K); bp = −119.6°C [153.6 K]) and freezes to a deep blue-violet solid (d = 1.728 kg/L at −195.8°C (77.4 K); mp = −195.2°C [78 K]) [1, 27]. The critical parameters are $T_c = -12.1$°C (261.1 K), $P_c = 5532.8$ kPa, and $\rho_c = 0.436$ kg/L. The solubility of ozone in water at 1 atm and 0°C is 1.1 g/L H_2O; that in organic solvents is given elsewhere [28].

Table 11.1 summarizes and compares the properties of oxygen, hydrogen peroxide, and ozone [11]. These species constitute oxidizing forms of the element oxygen.

11.2 STABILITY AND REACTIVITY

11.2.1 Gas Phase

Ozone decomposes only slowly in the absence of UV radiation or catalysts below 523 K (250°C) [8]. The thermal gas phase decomposition of ozone is described by a simple mechanism first proposed by Jahn [29] and later adapted [30, 31].

$$O_3 + M \underset{k_2}{\overset{k_1}{\rightleftharpoons}} O + O_2 + M \tag{11.6}$$

$$O_3 + O \xrightarrow{k_3} 2O_2 \tag{11.7}$$

where O is an oxygen atom, and M is a third body weighted according to $[M] = [O_3] + 0.44[O_2] + 0.41[N_2] + 1.06[CO_2] + 0.34[He]$[30], where the brackets represent concentrations. Most bimolecular gas phase reactions reported in this chapter require a third body to take away the energy of reaction. The third body will not be explicitly shown in these reactions. The rate constants k_1 [32, 33], k_2 [26], and k_3 [26] are

$$k_1(M = O_3) = 7.7 \times 10^{-9} \exp(-12{,}000/T) \quad cm^3 molec^{-1} s^{-1} \qquad (11.8)$$

$$= 4.6 \times 10^{12} \exp(-12{,}000/T) \quad L\ mol^{-1} s^{-1}$$

$$k_2(M = O_2) = 6.0 \times 10^{-34} (T/300)^{-2.8} \quad cm^3 molec^{-1} s^{-1} \qquad (11.9)$$

$$= 2.2 \times 10^{8} \quad (T/300)^{-2.8} \quad L^2 mol^{-2} s^{-1}$$

$$k_3 = 8.0 \times 10^{-12} \exp(-2{,}060/T) \quad cm^3 molec^{-1} s^{-1} \qquad (11.10)$$

$$= 4.8 \times 10^{9} \quad \exp(-2{,}060/T) \quad L\ mol^{-1} s^{-1}$$

The rate of ozone decomposition is given by

$$-\frac{d(O_3)}{dt} = \frac{2k_1k_3(O_3)^2}{k_2(O_2)\left[1 + \dfrac{k_3(O_2)}{k_2(O_2)[M]}\right]} \qquad (11.11)$$

At high ozone concentrations this reduces to $-d(O_3)/dt = 2\ k_1(O_3)[M]$. The half-life of ozone is then $t_{1/2} = \ln 2/k_1[M]$, which gives values of $t_{1/2}$ for ozone in oxygen at atmospheric pressure of 25°, 100°, and 250°C of 160 h, 210 s, and 0.03 s, respectively. Thus, ozone mixtures in oxygen are stable at room temperature; however, impurities produced during ozone generation (e.g., nitrogen oxides) can lead to decomposition. A 1 mol% ozone mixture produced by silent discharge in air will be 30% decomposed in 12 h [1].

11.2.2 Detonation

The detonation properties of ozone have been summarized elsewhere [4]. There are a number of general theories for gas phase detonations. In one theory [34–36], detonation occurs by the rise in temperature and pressure accompanying the detonation shock wave. In another [37, 38], detonation occurs because of the existence of branching reactions that increase the net number of chain carriers (radicals or excited species). A chain reaction for ozone has been proposed [39, 40]. The detonation properties of ozone in oxygen have been studied [41], and it has been found that detonation can be induced by spark at high concentrations (60–100 mol%, ~0.1 bar) or by detonation wave at lower concentrations (9–60 mol%, ~0.6–0.2 bar). For liquid ozone in oxygen at −183°C (90 K), the concentration threshold for spark-initiated explosion is reported to be 19 mol% [27]. Ozone has been reported to be produced routinely at concentrations as high as 20% [42], while acknowledging that the detonation limit was 13–17%. Ozone is thermodynamically unstable, and explosions can be initiated by not only spark or shock wave but also by heat, flashes of intense light, mechanical shock, impurities, or sudden changes in pressure or temperature [1]. It should be handled with extreme care.

11.2.3 Liquid Phase

The decomposition of ozone in pure water occurs through a complex radical chain mechanism [1]. The sequence is initiated by OH^-.

$$O_3 + HO^- \longrightarrow HO_2\bullet + O_2 \qquad (11.12)$$

The propagation steps involve O_2^- radical ions and HO• radicals [43–46]. In this sequence, O_3^- is a radical ion.

$$HO_2\bullet \rightleftharpoons O_2^- + H^+ \qquad (11.13)$$

$$O_2^- + O_3 \longrightarrow O_3^- + O_2 \qquad (11.14)$$

$$O_3^- + H^+ \rightleftharpoons HO_3\bullet \qquad (11.15)$$

$$HO_3\bullet \longrightarrow HO\bullet + O_2 \qquad (11.16)$$

$$HO\bullet + O_3 \rightleftharpoons HO_4\bullet \qquad (11.17)$$

$$HO_4\bullet \longrightarrow HO_2\bullet + O_2 \qquad (11.18)$$

$$\text{NET:} \quad 2O_3 \longrightarrow 3O_2 \qquad (11.19)$$

As can be seen, two ozone molecules are destroyed each time the sequence turns over once. The termination steps are [1]

$$\bullet HO_4 + \bullet HO_4 \longrightarrow H_2O_2 + 2O_3 \qquad (11.20)$$

$$\bullet HO_4 + \bullet HO_3 \longrightarrow H_2O_2 + O_3 + O_2 \qquad (11.21)$$

The half-life of ozone in distilled water is 7 min [1]. The decomposition increases rapidly with increasing alkalinity [47–49], but is inhibited by HO• (radical scavengers (e.g., HCO_3^-, CO_3^{2-}, HPO_4^{2-}) [43, 50]. In the presence of scavengers, the pseudo-first-order initiation rate constant at 293 K is $175 \times 10^{pH-14}\ s^{-1}$, yielding a half life of 66 min at pH 8 . The decomposition can be also be initiated by metal ions (e.g., Fe^{2+}) and accelerated by hydrogen peroxide, which forms HO_2^- in alkaline solutions [1]. The latter reacts rapidly with ozone to form the radical ion O_2^- [43].

11.3 PHOTOCHEMISTRY

Ozone occurs naturally in the atmosphere, constituting up to 27% by weight at altitudes between 15 and 25 km [51]. It is formed from molecular oxygen by the absorption of solar UV radiation in the range 240–300 nm via the reactions

$$O_2 + h\nu \longrightarrow 2\ O \qquad (11.22)$$

$$O + O_2 + M \longrightarrow O_3 + M \qquad (11.23)$$

The oxygen atoms generated by radiation are mostly in the ^3P ground state, with some in the ^1D excited state [52], but there is little difference between the two for the reaction with oxygen molecules to produce ozone [53]. Oxygen atoms can also be generated from vibrationally excited oxygen molecules at wavelengths of 207 nm [53] and 226 nm [54].

$$O_2 + h\nu \longrightarrow O_2{}^*(X^3\Sigma_g{}^-) \qquad (11.24)$$

$$O_2 + O_2{}^*(X^3\Sigma_g{}^-) \longrightarrow O_2 + 2\ O \qquad (11.25)$$

These excited oxygen molecules can also react to form ozone in the energy range 175-200 nm [53].

$$O_2 + O_2{}^*(X^3\Sigma_g{}^-) \longrightarrow O_3 + O \qquad (11.26)$$

Ozone itself absorbs UV radiation in the range of 200–600 nm. This leads to a reversal of the formation reaction and to the establishment of a steady-state concentration. A consequence of these processes is absorption of considerable UV radiation and its conversion into heat. This protects the earth's surface from the harmful effects of UV light on living plant and animal tissue.

The reactions that occur depend on the wavelength of light. Above 400 nm, only ground state oxygen molecules ($^3\Sigma_g{}^-$) and oxygen atoms (^3P) are formed [55].

$$O_3 + h\nu(>400\ \text{nm}) \longrightarrow O_2(^3\Sigma_g{}^-) + O(^3P) \qquad (11.27)$$

This is followed by the ozone decomposition reaction [23], which results in a maximum quantum yield, Φ, of 2.0 mol/hν [1]. With 334 nm light, excited singlet oxygen molecules ($^1\Delta_g$) and excited atoms (^1D) are formed [56].

$$O_3 + h\nu(<400\ \text{nm}) \longrightarrow O_2(^1\Delta_g) + O(^1D) \qquad (11.28)$$

The reaction to produce excited oxygen atoms, $O(^1D)$, and ground state molecules, $O_2(^3\Sigma_g{}^-)$, was believed to be unimportant because it is spin forbidden, but recent measurements show that it occurs between 300 and 340 nm [57]. The ^1D oxygen atoms return to ground state by collision with oxygen molecules and the generation of singlet sigma oxygen molecules ($^1\Sigma_g{}^+$) [1].

$$O_2 + O(^1D) \longrightarrow O_2(^1\Sigma_g{}^+) + O(^3P) \qquad (11.29)$$

The electronically excited singlet oxygen molecules go on to decompose ozone [58].

$$O_2(^1\Sigma_g^- + {}^1\Delta_g) + O_3 \longrightarrow 2O_2 + O \tag{11.30}$$

The net result of reactions in Eqs. 11.28–11.30 together with Eq. 11.23 is the chain decomposition of ozone with a maximum $\Phi = 6$ [59, 60]. The larger quantum yield, coupled with the greater absorptivity of ozone, results in higher rates of decomposition in the UV region than in the visible region.

11.3.1 Reactions in the Stratosphere

Photochemical synthesis and decomposition in the stratosphere were covered in the previous section. In addition to these reactions, there are a number of decomposition reactions catalyzed by extraneous species. The third body (M) required in these reactions will not be explicitly shown. Supersonic aircraft in the ozone layer release NO and NO_2 [61]. NO can participate in the following cycle [62].

$$NO + O_3 \longrightarrow NO_2 + O_2 \tag{11.31}$$

$$NO_2 + O \longrightarrow NO + O_2 \tag{11.32}$$

NO_2 can react in the following manner [62]:

$$NO_2 + O_3 \longrightarrow NO_3 + O_2 \tag{11.33}$$

$$NO_3 + h\nu \longrightarrow NO + O_2 \tag{11.34}$$

Chlorofluorocarbons like $CFCl_3$ and CF_2Cl_2, used as foam-blowing agents, aerosol propellants, and refrigerants, are transported to the upper atmosphere where they are photochemically decomposed to produce Cl atoms [8]. These catalyze ozone decomposition as follows [63]:

$$Cl + O_3 \longrightarrow ClO + O_2 \tag{11.35}$$

$$ClO + O \longrightarrow Cl + O_2 \tag{11.36}$$

11.3.2 Reactions in the Troposphere

Ozone close to the ground constitutes a serious air pollution problem. At ground level, ozone is not emitted directly, but is formed from the photochemical interactions of VOCs and nitrogen oxides. The reactions involved are summarized below [64, 65]. They are initiated by the photolysis of NO_2.

$$NO_2 + h\nu \longrightarrow NO + O(^3P) \tag{11.37}$$

$$O(^3P) + O_2 \longrightarrow O_3 \tag{11.38}$$

The ozone formation reaction is balanced by the reaction in Eq. 11.31 in which ozone oxidizes NO to NO_2, resulting normally in low levels of ozone. In the presence of VOCs, however, radicals are formed that consume NO or convert NO to NO_2, and this leads to an increase in O_3 levels. The major processes can be summarized as follows:

$$\mathrm{VOC + OH \longrightarrow RO_2 + products} \tag{11.39}$$

$$\mathrm{RO_2 + NO \longrightarrow NO_2 + radicals} \tag{11.40}$$

$$\mathrm{Radicals \longrightarrow OH + products} \tag{11.41}$$

Many types of VOCs are released into the atmosphere that react in the above manner at different rates by a variety of mechanisms. The relative reactivities of these VOCs have been summarized elsewhere [65], as well as the reactivities of alkenes [66]. The reactions of alkenes with ozone are particularly important in the formation of photochemical oxidants [67], which have a significant impact on urban air pollution and regional air quality [68].

11.4 REACTIONS WITH ORGANIC MOLECULES

Ozone reacts with most elements of the periodic table except for the noble metals, fluorine, and the inert gases. A broad summary of the reactions of ozone and their rate constants has been compiled [1]. The reactions of ozone with organic compounds have been particularly well documented [2, 3]. Ozone reacts with practically all organic and organometallic functional groups, including C–H, C–O, C–N, C–S, and C–M bonds. It acts primarily as an electrophile, reacting at points of high electron density in organic substrates. Its reactivity is enhanced by the presence of electron-donating groups and suppressed by the effect of electron-withdrawing substituents.

In general, ozone does not completely oxidize organic compounds to CO_2 and H_2O. This is because low-molecular-weight partial oxidation products such as acetic acid and oxalic acid are relatively unreactive. Considerable work with organic compounds has been carried out both in solution and in the gas phase because of applications in water treatment and atmospheric chemistry, respectively. The results in both media are discussed in parallel below. As before, rate constants in aqueous phase will be second-order constants with units of $L\ mol^{-1}s^{-1}$, refered to the consumption of ozone. For the gas phase, second-order rate constants will be reported in units of $cm^3 molec^{-1}s^{-1}$. Units of $L\ mol^{-1}s^{-1}$ may be converted to $cm^3 molec^{-1}s^{-1}$ by multiplication by 1.66×10^{-21}.

Close to room temperature in aqueous solutions below pH 7, ozone reacts directly with organic compounds. Rate constants vary over a large range, $k \sim 10^{-5}$ for acetic acid and $k \sim 10^9$ for phenolate ion [69]. For pH > 7, ozone can also react indirectly through generation of OH radicals. These radicals are kinetically more reactive than pure ozone, being able to activate organic species through radical addition, hydrogen abstraction, and electron transfer. Rate constants vary over a more narrow range $k = 10^7–10^{10}$ in alkaline solutions.

Table 11.2, based on a comprehensive critical review, [70] shows the orders of magnitude, at room temperature, of several gas phase rate constants. The same review [70] reports the lifetimes, τ, of organic compounds due to reaction with ozone [$\tau = (k[O_3])^{-1}$] (Table 11.3). These are for an ozone concentration of 1×10^{12} molec cm^{-3} (~40 ppb at ground level).

The data in Tables 11.2 and 11.3 make clear why homogeneous conversion of nonalkene hydrocarbons with ozone cannot be practically accomplished at close to room temperature (25°–100°C, with normal activation energies). The rate constants are too low. They also explain the observation that gas phase oxidation of the most reactive hydrocarbons, the alkenes, results in partial oxidation products. Although the rate constants for the alkenes are high, those for the C_1–C_3 oxygenate products are much lower.

The best known reaction of ozone is the oxidation of olefins to give aldehydes by the Criegee mechanism [71, 72]. Considerable work has been carried out on the reaction [73–76], and a modified, three-step version is shown below. Ozone first forms a π complex (R is hydrogen or alkyl) (**5**) [77] and then is believed to add across the double bond in a 1,3-dipolar cycloaddition, involving the ozone resonance structures (**1**) and (**4**). The reaction is stereospecific [78, 79] and likely concerted (no rotation around the C–C axis). The initial ozonide, a 1,2,3-trioxolane or molozonide (**6**), is unstable and undergoes reversion, fragmenting into a carbonyl species (**7**) and a peroxidic dipolar ion or zwitterion (**8**).

TABLE 11.2 Rate Constants for Reactions of Organic Molecules with Ozone at Room Temperature

Reactant	$k/cm^3 molec^{-1} s^{-1}$	Reactant	$k/cm^3 molec^{-1} s^{-1}$
Alkane	10^{-23}–10^{-24}	Haloalkanes	10^{-19}–10^{-20}
Alkene	10^{-16}–10^{-18}	Polyalkenes	10^{-16}–10^{-18}
Cycloalkenes	10^{-14}–10^{-17}	Alkynes	10^{-20}–10^{-21}
Aromatics	10^{-20}–10^{-22}	Polyaromatics	10^{-19}–10^{-20}
C_1-C_3 oxygenates	10^{-19}–10^{-21}	C_{4+} oxygenates	10^{-17}–10^{-18}
N-containing compounds	10^{-17}–10^{-20}	S-containing compounds	10^{-19}–10^{-20}

TABLE 11.3 Lifetimes of Organic Molecules Exposed to Ozone (1×10^{12} molec cm^{-3})

$k/cm^3 molec^{-1} s^{-1}$	τ	$k/cm^3 molec^{-1} s^{-1}$	τ
1×10^{-15}	0.3 h	1×10^{-19}	0.3 year
1×10^{-16}	3 h	1×10^{-20}	3 year
1×10^{-17}	1 day	1×10^{-21}	30 year
1×10^{-18}	10 day		

The dipolar ion and its alternative resonance form, $R_2C = O^+{-}O^-$, can be formed in syn- and anti-configurations, depending on the nature of the substituents. There is some controversy as to whether the peroxidic species, also known as a *carbonyl oxide*, is actually a diradical, $R_2C{\cdot}{-}O{-}O{\cdot}$ [70]. Several molecular orbital calculations indicate that the species can be characterized by a singlet diradical wave function [80–82] in a similar manner as ozone itself. Nevertheless, isotopic labeling experiments and calculations [83] suggest that the addition reactions occur in a concerted manner more characteristic of the reactivity pattern of a dipolar form than a diradical. In fact, though, depending on the polarity of the solvent and on the nature of the substituents, the diradical will behave to a great extent as a dipolar ion. Thus, the distinction may be quite subtle. The electronic state of ozone and the peroxidic species is a superposition of the dipolar and diradical states, and its nature will depend on its surroundings [84]. In Scheme 1, diradical species can be substituted for the dipolar forms depicted.

Following its formation by a cycloreversion, the further reaction of the dipolar ion is influenced by the presence of a solvent and the structure of the olefin. In the absence of a solvent, the ion will be formed in a high energy state, which can lead to its decomposition [70, 80, 85]. If an inert solvent is present and the

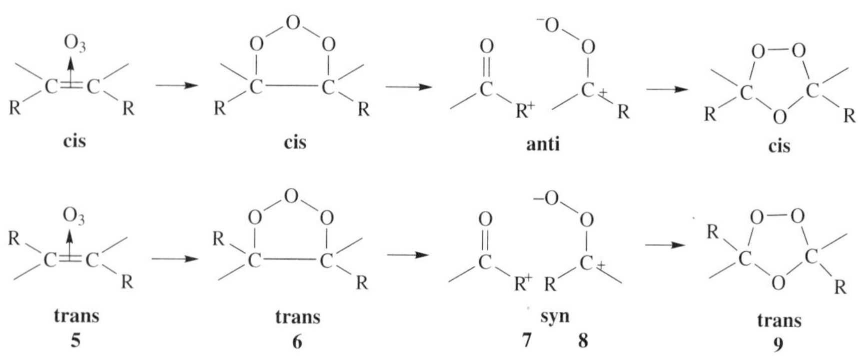

Scheme 1

TABLE 11.4 Rate Constants for the Reaction of Olefins with Ozone

Olefin	A/cm^3 molec^{-1}s^{-1}	E/kJ mol^{-1}	k at 298 K/cm^3 molec^{-1}s^{-1}
Ethene	1.2×10^{-14}	21.9	1.8×10^{-18}
Propene	1.3×10^{-14}	17.5	1.1×10^{-17}
Isobutene	3.6×10^{-15}	14.1	1.2×10^{-17}
cis-2-Butene	3.5×10^{-15}	8.1	1.3×10^{-16}
trans-2-Butene	9.1×10^{-15}	9.4	2.0×10^{-16}
2,3-Dimethylbutene	3.7×10^{-15}	2.9	1.2×10^{-15}
1,3-Butadiene	8.8×10^{-14}	23.0	8.1×10^{-18}
2-Methylbutadiene	1.2×10^{-14}	16.7	1.4×10^{-17}

carbonyl compound is highly reactive (e.g., an aldehyde), the dipolar ion can add to the carbonyl to give the standard ozonide, a 1,2,4-trioxolane (**9**). Thus, 1,1- and 1,2-dialkylethylenes form the ozonide, while tri- or tetraalkyl-substituted alkenes hardly produce it. The ozonide that is formed can be explosive and is generally not isolated. Instead, it can be reduced to form alcohols or oxidized or decomposed [86] to form carbonyl compounds or carboxylic acids.

Typical values of rate constants for the ozonation of olefins are summarized in Table 11.4 [70]. Values for other unsaturated compounds [87] and for cycloalkenes [88] are summarized elsewhere.

11.5 PRODUCTION

The synthesis of ozone is described in a number of publications [e.g., 4], including a comprehensive encyclopedia article [1]. Ozone is generated commercially from air or oxygen in two major ways, silent electric discharge or UV irradiation of oxygen. Substantial progress has been made in the technology and economics of ozone production due to the demand of the chemical in water purification and the paper and pulp industries.

11.5.1 Preparation by Electric Discharge

The production of ozone in electrical discharges has been reviewed elsewhere [89]. In generators operating by electric discharge, ozone is produced from oxygen atoms formed by the collision of electrons with oxygen molecules [90, 91].

$$O_2 + e^- \longrightarrow O_2(A^3\Sigma_u{}^+) + e^- \longrightarrow 2O(^3P) + e^- \qquad (11.42)$$

$$O_2 + e^- \longrightarrow O_2(B^3\Sigma_u{}^-) + e^- \longrightarrow O(^3P) + O(^1D) + e^- \qquad (11.43)$$

The first reaction has a threshold energy of 6 eV, while the second has a threshold of 8.4 eV [92]. Although oxygen atoms are formed in nanoseconds, their subsequent reactions involve microseconds [1]. The generated oxygen atoms react with oxygen molecules in the presence of a third body, M, to form ozone. ($\Delta H^\circ = -106.5$ kJ mol^{-1}[25.5 kcal mol^{-1}])

$$O + O_2 + M \longrightarrow O_3{}^* + M \qquad (11.44)$$

The ozone formed by the above reaction is in a vibrationally excited state, $O_3{}^*$ but undergoes quenching by collisions with other molecules. The principal competing reaction is the recombination of oxygen atoms in the gas phase or at walls. In the gas phase this reaction ($\Delta H^\circ = -498.3$ kJ mol^{-1}[-119.1 kcal mol^{-1}]) has a rate constant, $k = \sim 1 \times 10^{11} L^2 mol^{-2} s^{-1}$ [92].

$$2O + M \longrightarrow O_2 + M \qquad (11.45)$$

The recombination of oxygen atoms is unimportant until the ratio $O/O_2 >$ 0.01 [1]. In silent discharges, an optimal ratio for ozone production is about $O/O_2 \sim 0.002$ [92]. Processes other than recombination of oxygen atoms can however, result in less ozone formation. Ozone can be decomposed directly by thermal means or through electron impact, reaction with oxygen atoms, and reaction with electronically or vibrationally excited oxygen molecules [91]. Rate constants for these processes are available elsewhere [93, 94]. The magnitude of these constants for reactions involving ions such as O_2^-, O_2^+, O^-, O^+ and O_3^- indicate that these are not important. Overall, side reactions can be minimized by the use of low temperatures and short residence times, and for this reason ozone generators generally operate at close to room temperature, with contact times of a few seconds. Ozone production increases with pressure up to about 1 atm, above which it starts to decline. This is apparently due to the pressure dependency of the reactions of O_2 and O_3 with energetic electrons [1].

Production of ozone by electrical discharge can be carried out with pure oxygen or air. Although the use of oxygen requires an air separation step, it is advantageous because yields of ozone are higher and no nitrogen oxides are formed. The chemistry of nitrogen oxides in air-fed ozone generators is complex, and the reactions have been described in a number of publications [95–99]. Electron impact can activate nitrogen molecules in air to excited states ($N_2^* = N_2[A^3\Sigma_u^+]$, $N_2[B^3\Sigma_u^-]$) or dissociatiate them into atoms, and these species can form oxygen atoms by the following reactions [100, 101].

$$N + O_2 \longrightarrow NO + O \tag{11.46}$$

$$N + NO \longrightarrow N_2 + O \tag{11.47}$$

$$N_2^* + O_2 \longrightarrow N_2 + 2O \tag{11.48}$$

At low levels of nitrogen (~4%), these reactions increase ozone production. At higher levels, however, nitrogen oxides are involved in reactions that lead to a decrease in ozone concentration.

$$NO + O_3 \longrightarrow NO_2 + O_2 \tag{11.49}$$

$$NO_2 + O \longrightarrow NO + O_2 \tag{11.50}$$

$$NO_2 + O_3 \longrightarrow NO_3 + O_2 \tag{11.51}$$

Nitrogen atoms themselves are ineffective in ozone decomposition [1].

$$N + O_3 \longrightarrow NO + O_2 \tag{11.52}$$

Nitrogen oxides also react with each other and molecular oxygen.

$$NO + NO_3 \longrightarrow 2NO_2 \tag{11.53}$$

$$NO_2 + NO_3 \rightleftharpoons N_2O_5 \tag{11.54}$$

Nitrous oxide is produced by the following reactions [102].

$$N_2(A^3\Sigma_u{}^+) + O_2 \longrightarrow N_2O + O \tag{11.55}$$

$$N + NO \longrightarrow N_2O \tag{11.56}$$

N_2O_5 and N_2O are relatively inert toward ozone [1]. Although not specified above, all reactions involving binary collisions require a third body. The complex reaction network has been modeled [103, 104]. A simulation [102] with 143 reactions among 30 reacting species provided a good description of the outlet ozone, N_2O_5, and N_2O concentrations, which are the only products present at high concentrations in the effluent of the discharge.

11.5.2 Technology of Electric Discharge Ozone Generators

Chemical processing in electrical discharges has been comprehensively reviewed [92]. Electrical discharges are essentially nonequilibrium plasmas, a partly ionized state of matter consisting of atoms, molecules, ions, electrons, and radicals. The discharges may be classified according to temporal behavior, pressure, appearance (Fig. 11.4), as well as mechanism of generation [92]. Nonequilibrium refers to the fact that the electron energy (and, hence, temperature) is substantially higher than that of the neutral species. The most important discharge for ozone production technology is the silent discharge. Other types of discharges are briefly described below to distinguish them from silent discharges.

An important parameter describing nonequilibrium plasmas is the reduced field E/n, that is, the electric field divided by the neutral gas density, n, which has units of Townsend (Td), with 1 Td $= 10^{-17}$ V cm^2. The reduced field is a measure of the electron energy in a plasma and describes the breakdown strength of a plasma. Typical gases (N_2, O_2) break down at around 100 Td [92].

The glow discharge [92, 105] operates at low pressures (<1 kPa) and is usually formed between flat electrodes in a containment vessel (Fig. 11.5a). A typical electrical field is 10 V cm^{-1}, giving rise to reduced fields of 50 Td and

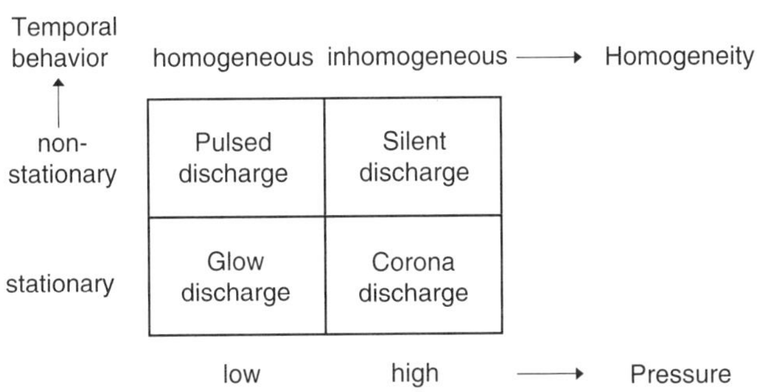

Figure 11.4 Various discharge types (Adapted from Eliasson et al. [92].)

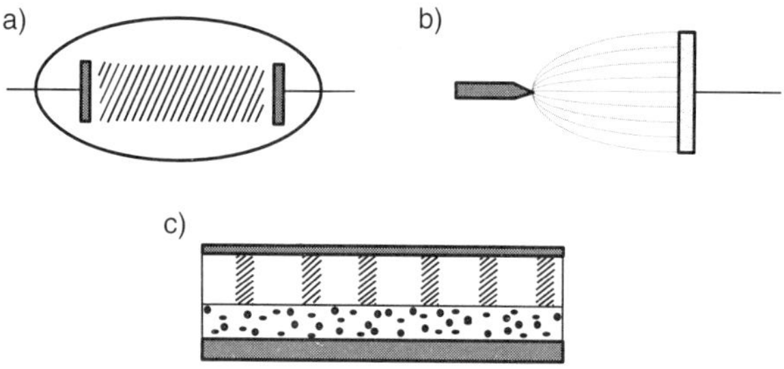

Figure 11.5 Various types of discharges. **a**, The glow discharge. **b**, The corona discharge. **c**, The silent discharge. (Adapted from Eliasson et al. [92].)

imparting electron energies of 0.5–2 eV. Common applications are neon and fluorescent tubes.

The corona discharge [92, 106, 107] is able to operate at higher pressure (1 bar) through the use of inhomogeneous electrode geometries, for example, a pointed electrode or a thin wire (Fig. 11.5b). Typical electric fields are 0.5–50 kV cm^{-1}, giving rise to reduced fields of 2–200 Td and variable electron energies of 5–10 eV. Common uses of corona discharges are in electrostatic precipitators and copying machines, where they produce charged particles, or in laser printers and radiation detectors. Ozone is inadvertently produced in such devices operating in air.

The silent discharge [92] operates at pressures up to 1 bar between two electrodes, importantly one of which (the anode) is covered with a dielectric (insulator) (Fig. 11.5c). For this reason the silent discharge is also known as the "dielectric-barrier discharge." Electric fields are in the range of 0.1–100 kV cm^{-1}, giving rise to reduced fields of 1–500 Td and imparting electron energies of 1–10 eV. This energy range is ideal for exciting atomic and molecular species and breaking chemical bonds.

In silent discharges, current flow occurs in the gap between electrodes as a consequence of ionization, but the presence of the dielectric causes the transported charge to accumulate on its surface and to halt the discharge. There have been numerous variations, such as pulsed high pressure discharges [108, 109], special double discharge techniques [110], electron beam controlled discharges [111], surface discharges [112], and cryogenically operated discharges [112, 113].

In standard silent discharges, the current flows in channels or filaments that are columnar in shape with a diameter of the order of 100 μm and lasting 100 ns or less [91, 114] (Fig. 11.6). Their characteristics have been studied by fast image intensifier recordings [115] and detailed current and charge measurements [116]. The current density in the microdischarge channels is of the order of 100–1000 A cm^{-2}, with the transported charge about $10^{-10} - 10^{-9}$ C and an energy density of 10 mJ cm^{-3}. The microdischarges can be considered to be transient glow discharges with electron temperatures of 50,000 K and gas temperatures close to room temperature [90]. They occur randomly in position

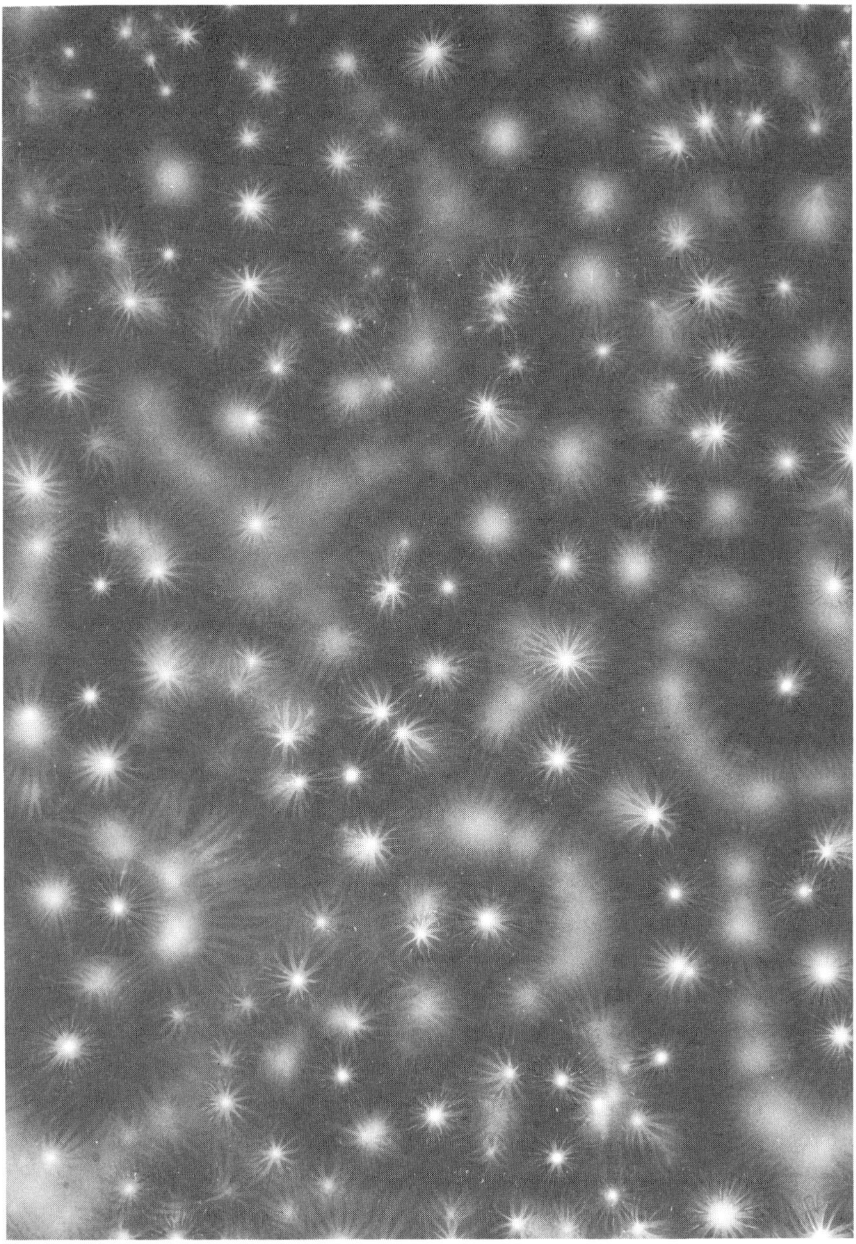

Figure 11.6 Lichtenberg figure showing footprints of individual microdischarges. Original size: 21 × 30 cm. (Courtesy of U. Kogelschatz, Asea Brown Boveri, Baden, Switzerland.)

and time over the entire surface of the electrodes. The physical situation is actually quite complex [104]. Ionization of atoms and molecules occurs by both electron impact and absorption of UV radiation generated by excited species. The dielectric carries out two important functions: It limits the amount of charge conducted by a single microdischarge, and it distributes the microdischarges over the whole electrode area.

In a homogeneous electric field, the condition for discharge initiation is solely a function of the product nd, where n is the gas density and d the gap width, and this defines the Paschen field. In a silent discharge the effective breakdown field is assumed to be 50% higher than the Paschen field. A considerable advantage of silent discharges over other discharges is that the average energy of electrons can be controlled by changing nd (Fig. 11.7) [104].

In most commercial ozone generators, the electrodes are in the form of concentric cylinders, which makes sealing easier than if they were in the shape of plates (Fig. 11.8). The feed gas is oxygen or air that has been dehumidified to prevent inefficiencies in the discharge process. The gas passes through a narrow discharge gap, 1–3 mm in width. The most common dielectric is glass, although other dielectric materials such as ceramics and enamel coatings with optimized geometries and segmentation have been reported. These are claimed to give rise to more effective energy utilization [117]. Most commercial ozone generators use glass tubes of about 1–2 m length and 1.5–2.0 mm wall thickness coated on the inside with metal to serve as the high voltage electrode. These glass tubes are mounted in stainless steel tubes of slightly larger diameter, which are housed in a steel tank in an arrangement similar to that of a crossflow heat exchanger [92] (Fig. 11.9). Cooling water passes between the steel tubes, which are welded to

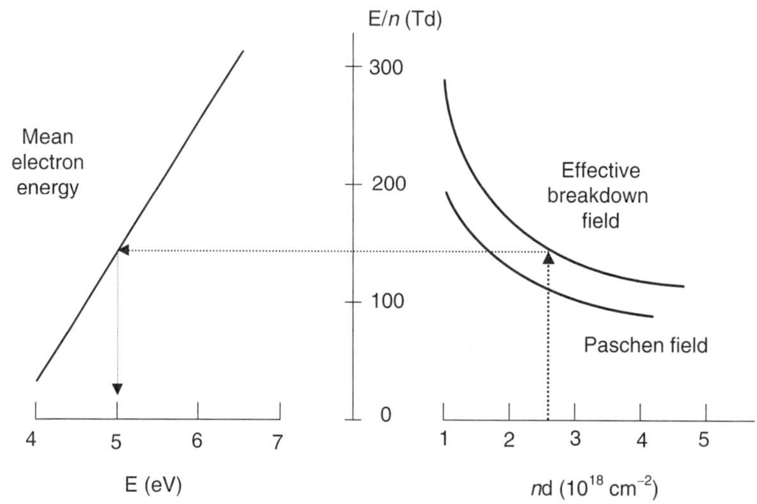

Figure 11.7 Dependence of the mean electron energy, E, on the product nd in xenon. (Adapted from Eliasson and Kogelschatz [104].)

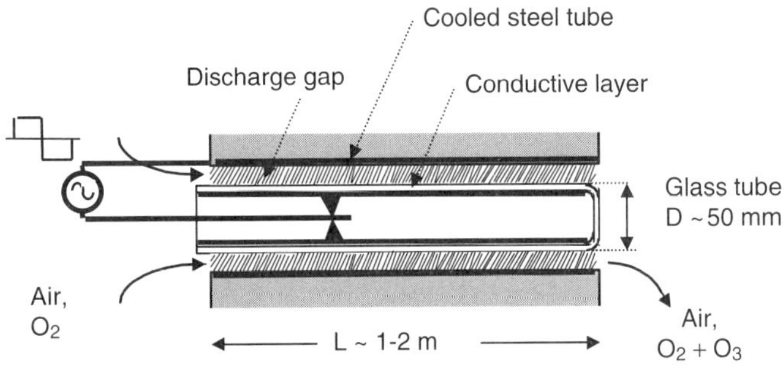

Figure 11.8 Silent discharge tube. (Adapted from Eliasson and Kogelschatz [104].)

two end flanges. In some designs the inner tube is also cooled. Large ozone generators can contain up to 1000 discharge tubes and produce typically 10–50 kg of ozone per hour at a power consumption of 0.1–0.5 MW [104] (Fig. 11.10).

Silent discharges offer the advantage that many elements can be connected in parallel using a common power supply. Unlike other discharges, there is no need to control the current to each individual tube [92]. Because current can only pass through the dielectric in the form of a displacement current, the discharge is driven by an alternating voltage. Originally, ozonizers employed line frequencies of 50/60 Hz, but in the late 1970s generators using medium frequency were introduced that increased efficiency [118]. In the 1980s, thyristor-controlled frequency converters were developed that operate between 500 Hz and a few kiloherz and were able to produce a square-wave current. These further increased efficiency and also raised power density by an order of magnitude, giving rise to much more compact ozone installations [92, 119]. Power density is the energy supplied per unit of discharge surface and is important for the economic production of high ozone concentrations. The absorbed power of an ozone generator is given by the following equation [120]:

$$P = 4fC_D U_B[U - (C_D + C_G)U_B/C_C] \tag{11.57}$$

where

f = frequency of applied alternating current
C_D = electric capacitance of the dielectric
C_C = electric capacitance of the discharge gap
U_B = gap breakdown voltage
U = applied peak voltage

According to this relation, the power density increases with increasing peak voltage and frequency. Peak voltage is limited by sparking and breakdown of the dielectric to values not exceeding 20 kV, corresponding to 14 kV rms

Figure 11.9 Arrangement of silent discharge tubes in steel tank. (Courtesy of U. Kogelschatz, Asea Brown Boveri, Baden, Switzerland.)

for a sinusoidal voltage. Thus, the power density depends on the attainable frequency [120].

With frequencies of 600–1000 Hz, a standard ozone generator achieves a power density of 0.2–0.25 kg $h^{-1}m^{-2}$ in air and 0.35–0.45 kg $h^{-1}m^{-2}$ in oxygen with an electrode voltage of not more than 10 kV rms. As a result, ozone

Figure 11.10 Series of large-scale ozone generators at the Los Angeles Aqueduct Filtration Plant. Capacity is 187.5 kg h^{-1}. (Courtesy of U. Kogelschatz, Asea Brown Boveri, Baden, Switzerland.)

concentrations of up to 3.5 wt% (42 g Nm^{-3}) in air and 6 wt% (79 g Nm^{-3}) in oxygen can be obtained with good yield. The corresponding specific energy consumptions are 17 kW-h kg^{-1} and 9.5 kW-h kg^{-1} [120]. These figures are largerly confirmed: Another study reports that production of 1.5–3 wt% ozone in air requires approximately 20 kW-h kg^{-1}, whereas production of 7–14 wt% consumes 7–13 kW-h kg^{-1} [121]. The very highest concentrations obtainable in ozone generators are reported to be 6 wt% in air and over 20 wt% in oxygen [90]. (See section on explosion limits, Section 11.2.2.)

11.5.3 Economics of Silent Discharge Ozone Generation

Ozone generation equipment is produced by many companies throughout the world [122]. An early pioneer of ozone production using oxygen was the Swiss firm Brown Boveri Corporation (now Asea Brown Boveri). In the late 1980s, the ozone technology was purchased by Ozonia, a joint venture between the French companies l'Air Liquide and Degrémont S.A. with expertise in oxygen and water treatment. In the early 1980s, the German company Messer Griesheim also developed oxygen-fed generation equipment based on small diameter dielectric tubes. These developments stimulated the company Trailigaz, long the leader in conventional production methods, to develop oxygen technology.

In the United States, some major suppliers of ozone equipment are OREC (Phoenix, AZ), PCI-WEDECO, Environmental Technologies, Inc. (W. Caldwell, New Jersey, http://www.pci-ozone.com/pciwedeco/index.html), Praxair-Trailigaz Ozone Co. (Cincinnati, OH), Hess Machine International (Ephrata, PA, http://www3.thomasregister.com/ss/.1963866674/cowebfrm.cgi?Product.17.Coweb.899594217), UVOCS (Montgomeryville, PA), and Ozone Pure Water, Inc.(Sarasota, FL, http://www3.thomasregister.com/ss/.1963866674/cowebfrm.cgi?Product.12.Coweb.899594217). In Canada, suppliers include Ozone Services, Yanco Industries (Burton, British Columbia, http://www.o3zone.com/ozoneser/index.htm), and AZCO Industries, Ltd. (Surrey, British Columbia, http://www.azco.bc.ca/). In Italy there is HRS Engineering, and in Germany, Fischer (Bonn).

Comparisons of the cost of producing ozone under various conditions have been carried out [121, 123]. The main conclusion is that the use of oxygen is more cost effective than the use of air. In an early study [123], the production costs were presented for a capacity range of 1–1000 kg h^{-1} for various plant configurations. The parameters and assumptions used in the calculations are summarized in Table 11.5. Capital costs were based on turnkey plants based on Asea Brown Boveri (Zurich, Switzerland) technology and on cost information from major suppliers of oxygen. Using an interest rate of 12.5% and depreciating the plant over 10 years resulted in an annuity of 18% per year of the invested capital. The power cost included the power consumed by the ozone generation process, compressors, dryers, and waste ozone destruction system within battery limits. Cooling water was assumed available at 20°C (68°F), and the temperature rise in the heat exchangers was set at 5°C (9°F); but the economics were not affected significantly by these values. Oxygen was assumed to be available as liquid oxygen (LOX), and its cost included rental of the storage tank, associated equipment, and maintenance. Labor and maintenance were set at 1% of investment per year.

Several scenarios were treated, including different product ozone concentrations, different on-stream times, the inclusion of air separation plants, the

TABLE 11.5 Operating Parameters and Cost Assumptions

Cooling water	Temperature: 20°C (68°F)			
	Temperature rise: 5°C (9°F)			
Power	Cost: 6.5¢/kW-h			
Capital	Interest: 12.5%/year			
Depreciation	Period: 10 years			
Capacity	On-stream time: 25%			
Labor, maintenance	Percent of capital: 1%			
Oxygen cost	Delivery (st[a]/day):	10	25	50
	Cost ($/st):	80	65	55
	Delivery (st/day):	100	250	500
	Cost ($/st):	42	40	36

[a] 1 st = 1 short ton = 2000 lb = 908 kg.

Adapted from Merz and Gaia [123], with permission.

recycling of oxygen, and the cryogenic concentration of exit ozone [123]. The two main cases treated were the synthesis of ozone with air and with oxygen.

The production of ozone using air was analyzed considering the plant in Figure 11.11 [123]. A minimum number of medium frequency ozone generators for the plant capacity were utilized (the maximum productivity per reactor was 100 kg O_3/h).

The air-utilizing plant consisted of the following units: (1) an air preparation section including compressor, coolers, and desiccant dryer; (2) the ozone generation units; (3) an ozone contacting system including piping; and (4) a catalytic waste ozone destruction system. The main results for producing 2% ozone from air with a time-on-stream of 2190 h year^{-1} (25% capacity utilization) are presented in Figure 11.12.

The production of ozone using oxygen was also analyzed considering the plant in Figure. 11.11. Liquid oxygen was assumed to be available from a LOX tank located outside battery limits. The oxygen plant consisted of the following units: (1) the ozone generators (maximum capacity, 100 kg O_3/h); (2) an ozone contacting system, including piping but excluding civil work; and (3) a catalytic vent ozone decomposition system. The main results for producing 6% ozone from oxygen with a time-on-stream of 2190 h year^{-1} (25% capacity utilization) are presented in Figure. 11.13.

Comparing the results of the above cases of operation at 2190 h per year for ozone production shows that the production cost at 6 wt% ozone from LOX are 16% lower than the production cost of 2.5 wt% from air. Thus, oxygen use not only results in lower costs, but it produces a product that is of higher concentration, which is generally desirable.

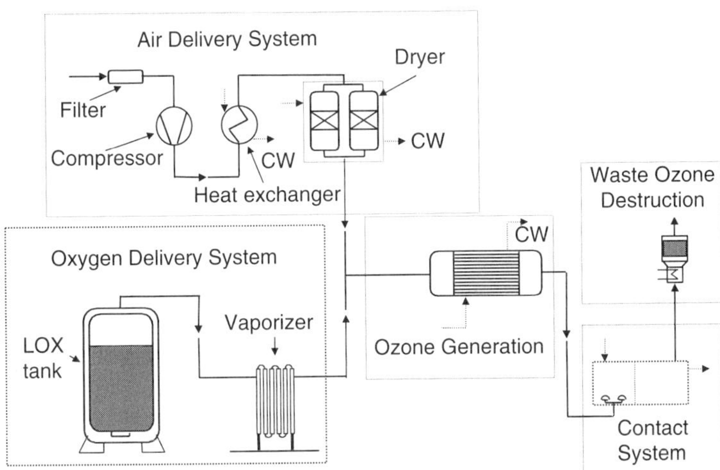

Figure 11.11 Schematic of ozone delivery system showing air and oxygen alternatives. (Adapted from Merz and Gaia [123].)

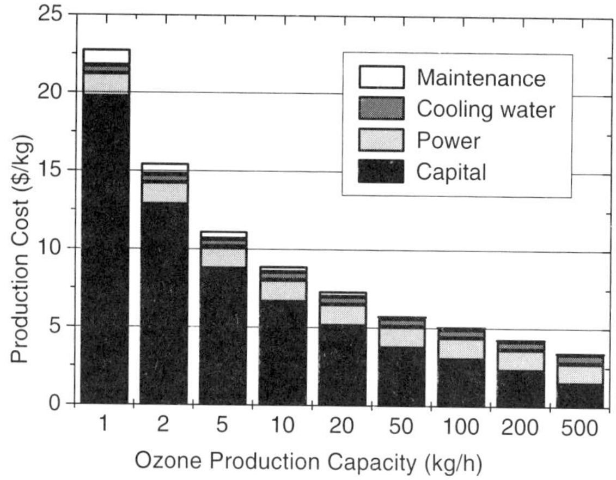

Figure 11.12 Economics of ozone production from air. (Adapted from Merz and Gaia [123].)

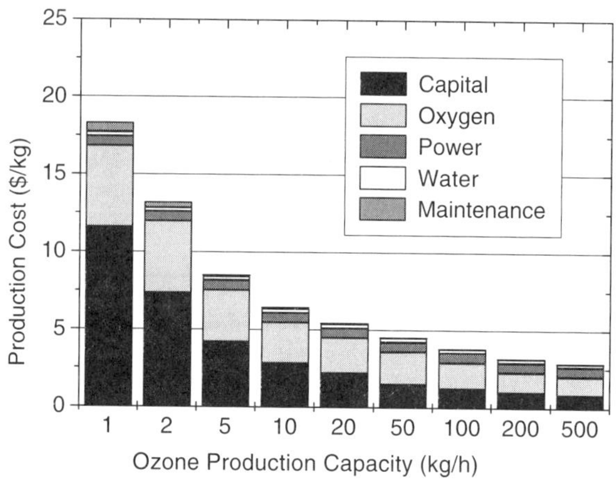

Figure 11.13 Economics of ozone production from oxygen. (Adapted from Merz and Gaia [123].)

The results of this study [123] were largely confirmed in a subsequent analysis (Schmidding-Werke, Cologne, Germany) [121]. Two plants utilizing air and oxygen were considered that essentially had the same design as previously (Figure. 11.11). The operating costs of the two plants producing ozone at a rate of 25 kg h^{-1} were compared based on the assumptions summarized in Table 11.6.

The results of the analysis are summarized in Table 11.7. The capital cost was obtained using an assumed plant investment of DM 2,000,000, interest costs of 15% $year^{-1}$, maintenance costs of 1% $year^{-1}$, and SUM costs of 16% $year^{-1}$.

TABLE 11.6 Operating Parameters and Cost Assumptions

Item	Cost (DM)[a]	Cost (US$)
Capacity	100%	8400 h/year
Power	DM 0.25/kW-h	15¢/kW-h
Cooling water	DM 1.00/m^3	59¢/m^3
Liquid oxygen	DM 0.25/m^3	15¢/m^3
Capital	15%/year	15%/year
Maintenance, spare parts	1%	1%

[a] DM = Deutsche Mark (assumes a conversion of 1.7 DM/$, June 1997).
Adapted from Shadiakhy and Blanken feld [21].

TABLE 11.7 Production of Ozone Production Costs for Air and Oxygen Plants (Adapted from ref. 121)

Items	Air		Oxygen	
	Consumption	Costs ($/kg)	Consumption	Costs ($/kg)
Air treatment	4 kW-h kg^{-1}	0.59	—	—
Ozone production	19 kW-h kg^{-1}	2.79	10 kW-h kg^{-1}	1.47
Contacting system	1 kW-h kg^{-1}	0.15	0.5 kW-h kg^{-1}	0.08
Offgas	1 kW-h kg^{-1}	0.15	0.3 kW-h kg^{-1}	0.05
Cooling water	3 m^3 kg^{-1}	1.76	1.5 m^3 kg^{-1}	0.88
Oxygen	—	—	8 m^3 kg^{-1}	1.18
Capital	DM 2,000,000	1.33	DM 2,000,000	0.89
Total		6.77		4.55

The values for the costs of ozone production from air and oxygen from the two studies [121, 123] are in good agreement with each other. There are minor differences, in particular in the capital costs, but these may be due to different allocations of costs among the various components.

11.5.4 Preparation by Ultraviolet Light Absorption

In the UV light method, UV light absorption causes dissociation of oxygen molecules into atoms that then react with other oxygen molecules to form ozone [124]. The reactions are similar to those that occur in electrical discharges except that they are driven by photons [58].

$$O_2 + h\nu \longrightarrow 2O(^3P) \tag{11.58}$$

$$O_2 + h\nu \longrightarrow O(^3P) + O(^1D) \tag{11.59}$$

The so-formed oxygen atoms react with oxygen molecules by the reaction in Eq. 11.59 to form ozone. Because each photon produces two oxygen atoms and, hence, two ozone molecules, the theoretical quantum efficiency is 2. Because of the photodissociation of ozone itself [58], however, the actual quantum efficiency is reduced to 0.5 (with mercury lamps). Overall, the photochemical method of ozone production gives lower yields based on energy consumed than the methods of silent electric discharge or electrolysis of sulfuric acid solutions [125].

$$O_3 + h\nu \longrightarrow O(^3P) + O_2(X^3\Sigma_g) \tag{11.60}$$

$$O_3 + h\nu \longrightarrow O(^1D) + O_2(a^1\Delta_g) \tag{11.61}$$

$$O_3 + h\nu \longrightarrow O(^1D) + O_2(b^1\Sigma_g{}^+) \tag{11.62}$$

The spectral range that is effective for ozone production has been studied [126–128]. The UV radiation must be energetic enough to dissociate $O_2(\lambda < 242$ nm), but not energetic enough to cause ionization ($\lambda > 103$ nm). In practice, the need for path lengths of a few centimeters lowers the upper wavelength limit to below 190 nm. This makes the 185 nm line of low pressure mercury lamps suitable for ozone production [129]. There are several drawbacks, however, to the use of these lamps. First, the UV efficiency of the 185 nm lamps is low. Second, the absorption coefficient of oxygen at 185 nm is also small. Finally, the 185 nm line is always accompanied by the much more intense 254 nm line, which is strongly adsorbed by ozone, and results in its destruction [129]. Absorption cross sections for ozone have been reported [128]. In the wavelength region 254–280 nm, ozone has an almost symmetric absorption peak in the Hartley band with a maximum cross section of $\sigma \sim 10^{-17} \pm 10\%$ cm^2 for $\lambda = 254$ nm [130].

To overcome the deficiencies of the mercury lamp, better UV sources have been investigated, such as coherent and incoherent excimer sources. A fluorine excimer laser of radiation 158 nm used to treat low pressure oxygen [131] and an incoherent xenon excimer source with wavelengths peaking at 172 nm (half-width 12–14 nm) [129] have been employed. In the latter study, the UV efficiency (UV output/electrical input) was determined and used to relate it to the ozone production. It was calculated that an excimer source ozone generator of 14% efficiency would match the performance of a mercury lamp of 30% efficiency for the 254 nm line. The best value obtained at the time was a 9% efficiency for the excimer.

11.5.5 Preparation by Electrochemical Oxidation

Ozone can be prepared electrochemically in acidic electrolytes [132–134] such as phosphoric acid or sulfuric acid. High current density electrolysis of 68 wt% sulfuric acid can produce 18–25 wt% ozone in oxygen in the anodic gas [1]. Although this method can produce high concentrations of ozone, the output is low, and the cost is several times higher than that of electric discharge processes [135]. Only limited commercial applications have been realized: It has been used to prepare ultrapure water for manufacturing processes in the microelectronic and pharmaceutical industries [136].

Lead dioxide deposited on titania or solid polymers has been explored as an anode material because of its low cost and its stability under high positive potentials [137, 138]. Ozone is suggested to form on the surface of the lead oxide by reaction of adsorbed oxygen atoms and oxygen molecules [139]. Lead dioxide has also been used with iron(III) in anodes to generate O_3 simultaneously with O_2 in a system using an aqueous electrolyte with a phosphate buffer containing potassium fluoride [140].

A proton-exchange-membrane electrochemical flow reactor has been reported that oxidizes water at the anode to form ozone and reduces oxygen to hydrogen peroxide at the cathode [141, 142]. Lead dioxide was used as the electrode material for ozone evolution, while gold, carbon, or graphite were used for peroxide evolution. Low ozone current efficiencies ($\sim$4.5%, at a current density of 2 A cm^{-2}) and high operating potentials (4.5 V) make the process uneconomical [143]. Another proton-exchange gas diffusion system has been described utilizing a cathode impregnated with the proton exchange polymer polytetrafluoroethylene and an anode consisting of glassy carbon and platinum black [144].

11.5.6 Preparation by Chemical Reaction

The generation of ozone by purely chemical reactions is rare [4]. Ozone is formed from the interaction of fluorine and water through the generation of HF and oxygen atoms [4]. Ozone is reported to be formed from the reaction of an iron(III) porphyrin with nitrite ion and oxygen [145]. The reaction is believed to be driven by the stability of the resulting prophyrin iron(II)–NO adduct [146].

$$\text{(Por)Fe}^{\text{III}} + \text{NO}_2^- \rightleftharpoons \text{(Por)FeNO}_2 \xrightarrow{\text{O}_2} \text{(Por)Fe}^{\text{II}}\text{NO} + \text{O}_3 \qquad (11.63)$$

The formation of ozone as a by-product of oxidation of phosphorus has also been reported [147, 148].

$$\text{P}_4 + 20\ \text{O}_2 \longrightarrow \text{P}_4\text{O}_{10} + 10\ \text{O}_3 \qquad (11.64)$$

The reaction is the basis of the POZONE process and has been used for the bleaching of pulp [149]. The reaction occurs between yellow phosphorus (P_4) and oxygen molecules in moist air and produces oxygen atoms and ozone as well as phosphorus oxide radicals such as PO and PO_2. These radicals are also chemically active and could contribute to the destruction of lignin. Using a cost of P_4 of \$2.2/kg (\$0.98/lb) and assuming an additional 50% of phosphorus is needed to account for process inefficiency and losses [150], a reagent cost of \$0.84/kg (\$0.76/lb) is calculated [149]. The POZONE process for ozone production is more cumbersome than conventional electrical discharge technology as solids have to be handled, and the reaction is highly exothermic and difficult to control. The phosphorous oxide is hydrolyzed to phosphoric acid, which can generate a by-product credit.

11.6 HANDLING

11.6.1 Experimental Techniques

General procedures for handling ozone have been summarized [4]. Pyrex glass is a satisfactory material for construction of a small laboratory apparatus for the handling of ozone [151]. Use of stopcock grease should be avoided, and, instead, lubricants such as Fluorolube LG (polymer of trifluorovinyl chloride) should be used. An alternative material is stainless steel, and valves can be of the diaphragm or bellows type. Materials compatible with ozone are listed in Table 11.8 [152].

Considerable care should be taken to protect researchers in case an explosion occurs. Shields of 0.5 cm Plexiglass can prevent injury by flying glass. Valves and stopcocks can be turned by extension rods inserted through holes in the Plexiglass shields. Detonations can occur at concentrations higher than 9 mol%, for any number of reasons, including variations in temperature or pressure, sudden mechanical agitation or shock, and exposure to intense light or sound (see Section 11.2.2). The purity of the oxygen employed in the preparation can be a factor in the stability of the resulting ozone.

TABLE 11.8 Materials Resistant to Ozone

Ozone gas
Glass
Teflon[a]
Stainless steel (300 series)[b]
Aluminum
Ozone in solution
Stainless steel (300 series)
PVC
Concrete[c]
FRP (with Dekarane[d] or Hetron[e] 197 resin)
Plexiglass
Teflon
Glass
Gaskets, fittings, and sealing materials
Stainless steel (300 series)
Kalrez[a]
Kynar[f]
Teflon
Viton[a]

[a] Trademarks of Dupont, Co.
[b] Seamless tubing or pipe is prefered over welded.
[c] Teflon or stain. steel skirt at water line.
[d] Trademark of Dow Chemical USA.
[e] Trademark of Ashland Chemical Inc.
[f] Trademark of Pen Walt.
Adapted from OREC (Phoenix, AZ) brochure.

Mixtures of ozone in oxygen produced in the laboratory can be concentrated by condensing the mixture and pumping off the oxygen. A technique employing two storage bulbs and an evaporation vessel has been described [153]. A safer method involves the adsorption of ozone on refrigerated silica gel, followed by its desorption either in pure form at reduced pressure or diluted by air, nitrogen, argon, or other gases not strongly adsorbed by the silica gel [154]. The amount of ozone adsorbed increases substantially with decreasing temperature; at 298 K, 263 K, 195 K, 183 K (25°C, −10°C, −78.5°C, and −90°C), the amounts adsorbed at 1.37 kPa are 0.042, 0.13, 3.0, and 6.0 kg O_3/100 kg SiO_2, respectively. The process has been piloted at an adsorption temperature of 263 K (−10°C).

11.6.2 Health Aspects

Although its presence in the lower and upper stratosphere protects the Earth's surface from overexposure to UV radiation, in the troposphere ozone is a pollutant and is harmful to many living organisms. Plants undergo damage to its exposed surfaces. It has been found that hydroperoxides (ROOH) are produced on the leaves of plants exposed to ozone [155] and that damage to lipids via ROOH formation would possibly be more extensive than that caused by direct reaction with ozone [156].

Because of ozone's toxicity, the Occupational Safety and Health Administration (OSHA) has set the maximum human allowable exposure to ozone for an 8 h period at 0.08 ppm (0.2 mg/m^3) [157]. (All ppm values here are volume based.) The allowable concentration in the working environment is 0.1 ppm in Japan [158]. The threshold for odor detection is about 0.01–0.05 ppm for most individuals. In the range of 0.1–1 ppm, ozone causes headaches, throat dryness, and damage to mucous membranes [157]. At concentrations of 1–100 ppm, ozone can cause asthma-like symptoms, including tiredness and lack of appetite [1]. Higher concentrations can cause throat irritations, hemorrhaging, pulmonary edema, and even death [159]. Figure 11.14 summarizes the effects on humans [160].

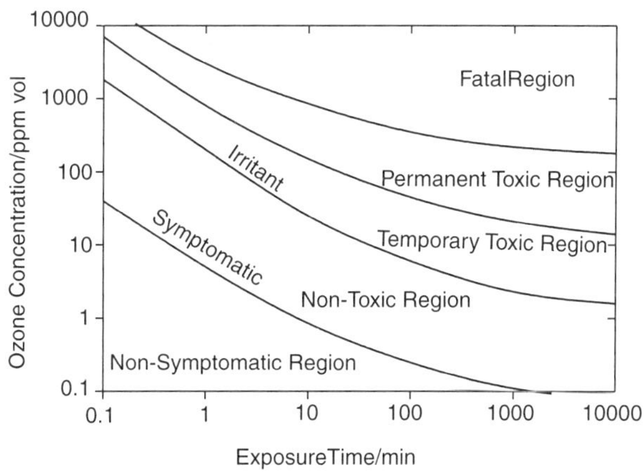

Figure 11.14 Human tolerance for ozone. (Adapted from the U.S. EPA [160].)

11.7 APPLICATIONS

11.7.1 Conventional Uses

The applications of ozone have been comprehensively summarized elsewhere [1]. Ozone is used in disinfection [161–163], waste water control [164, 165], odor control [166], process water treatment [167], and organic synthesis [168].

Ozone is used extensively in water purification [1], and the chemistry, equipment, and processing has been reviewed [169]. The chemical reactions of ozone in aqueous phase have been the subject of in-depth studies that point toward the participation of hydroxyl radicals [170–172]. Ozonation results in the reduction of turbidity, the removal of taste, color, odor, and dissolved manganese and iron [173–176], and the killing of bacteria and viruses [177]. It also prevents the growth of algae, which can clog filters. In addition, ozone supports microflocculation [178], an important effect that reduces the required amount of flocculating agent by as much as 50% [179, 180]. The technology dates back to 1903 and in 1980 was reported to be employed in over 1100 water-treatment installations [157]. Extensive work continues in this area [181–183]. A review [184] of ozone use in industrial wastewater treatment summarizes recent applications, which are extensive. For example, ozone is used to recycle marine aquaria waters [185], treat wastewater from zoo exhibits [186], remove cyanide from electroplating wastewaters [187–189], decolorize and destroy surfactants in textile industry effluents [190–193], destroy organics in petroleum refinery process water [184], prevent fouling in closed-circuit water cooling systems, and purify ultrapure water delivery systems [194–196].

A problem with the use of ozone to treat drinking water is that ozone oxidizes bromine to bromates, which go on to react with organic compounds to form an array of potential carcinogens [197]. The first step is the formation of hypobromous acid (HOBr) and hypobromide ion (BrO^-) [198]. The World Health Organization has set a guideline bromate concentration in drinking water of 0.5 μg/L at a risk level of 10^{-5}, but this may be exceeded in practice [199]. Possible means of dealing with this problem are lowering the ozone dose, decreasing the pH, decomposing the bromate with ammonia or hydrogen peroxide, reducing the bromate catalytically, or separating the bromide precursor or bromate product by membrane filtration [199].

Another major use of ozone is bleaching and delignification of pulp in the paper industry. Ozone is an attractive replacement for chlorine and chlorine dioxide, which can produce carcinogenic dioxins. Lignin, a high molecular weight phenolic material, and cellulose, a β-glycoside, are effectively degraded by ozone [1]. There has been considerable work on the reactions of ozone in the paper-making process [200–202], and the subject has been summarized [203]. A number of plants throughout the world utilize ozone as primary oxidant, and for environmental reasons these are likely to grow in number. A review and historical perspective is available [122]. In 1996, 14 full-scale ozone-utilizing pulp bleaching plants were in operation in Scandinavia, Austria, South Africa,

Brazil, and the United States, and several additional plants were scheduled to be started in 1997.

Ozone is used in a few commercial processes for the production of high-value-added products [1]. In Japan, ozonolysis and reduction of isosafrole yields piperonal (3,4-methylenedioxybenzaldehyde) in 87% yield at a scale of several metric tons per day. Oleic acid, $CH_3(CH_2)_7CH{=}CH(CH_2)_7CO_2H$, is ozonated to produce pelargonic, $C_8H_{17}CO_2H$, and azelaic, $HO_2C(CH_2)_7CO_2H$, acids. Oleic acid is also ozonated to the azealic half-aldehyde that is reduced with hydrogen and ammonia to ω-aminononanoic acid. In Austria, Chemie Linz produces commercial quantities of phthalaldehyde by the ozonation of naphthalene followed by reduction with hydrogen [204]. Phthalaldehyde is an intermediate that is useful in amino acid analysis and organic synthesis. Ozonation of undesirable residual double bonds in cross-linked elastomeric polymers can be used to produce soluble elastomeric latexes [205]. Ozone is used in the pharmaceutical and fine chemicals industries to produce peroxides as germicides in skin lotions, in the synthesis of steroids, and the preparation of vitamin E [1].

The oxidizing properties of ozone are being exploited in a number of processes for materials and device preparation. Ozone has been used for removing photoresists in the flat-panel display and the micromachining industries [206] and to oxidize $YBa_2Cu_3O_{7-x}$ superconducting thin films [207]. In the semiconductor industry, it has been used to remove organic materials from semiconductor wafers [208] and to form controlled oxide layers on the surface of silicon [209–213], silicon-germanium [214], gallium arsenide, and indium phosphide [215, 216]. This is an important application because, as device scales decrease, the thickness of gate oxides in MOSFETs needs to be reduced, and ozone can provide thin oxide layers of high integrity and high charge to breakdown [217]. It has been utilized in catalyst preparation, for example, the oxidation of organic ligands in a supported rhodium catalyst to form $Rh(O_2)$ complexes [218]. Ozone is also more effective than molecular oxygen for the deposition of surface oxygen on metals like Pt, Pd, and Au [219].

There are only limited reports on the use of ozone for the oxidation of gas-phase pollutants. In part, this may be due to the low rates of the uncatalyzed reactions or to concerns about unreacted ozone emissions [220]. In the absence of catalysts, ozone produces mainly partly oxidized products [221]. Hydrocarbons have been oxidized directly in laboratory plasma discharge [222, 223]. The use of such nonequilibrium plasmas for treatment of effluent streams is reasonable because many toxic substances (PCBs, dioxins, solvent vapors, nerve gases, pesticides) are subject to attack by charged particles, and the use of silent discharges does not require heating of the entire gas [92]. Chapter 12 directly addresses the issues of reactivity and ozone release in catalytic reactions.

Ozone is used in disinfection for the killing of bacteria and viruses. It has been utilized in sterilizing devices [224] and for the preservation of fruits and vegetables [225, 226]. In the latter application, ozone not only destroys extraneous organisms but also serves to oxidize ethylene released by fruits in the maturation process. Ethylene is a plant hormone that accelerates ripening.

A new process for the removal of nitrogen and sulfur oxides from the flue gas of coal-burning plants involves the generation of ozone and other reactive species by corona discharge. The complex network of reactions of NO and organics with ozone occurring in a flue gas stream have been modeled [227]. For pulsed coronas, decreasing the voltage pulse width and increasing the rate of rise of the front increases the dissociation rate of gases [228–230]. The reactive species result in oxidation of nitrogen- and sulfur-containing compounds to nitric and sulfuric acids, which are then precipitated out as ammonium salts by reaction with ammonia [231–233]. The ammonium salts $(NH_4)_2SO_4$ and NH_4NO_3 are collected in electrostatic precipitators and are recycled as fertilizers. The performance and economics of this process appear promising [234]. The technology has been piloted in an installation in Italy treating 1000 $m^3\ h^{-1}$ [235].

11.7.2 Advanced Oxidation Processes

Ozone is also used in a number of processes in combination with UV light (Rayox), hydrogen peroxide (Peroxone), solid adsorbents, TiO_2 catalysts (Catazone), or combinations of these. These are often referred to as *advanced oxidation processes*, and there are currently over 100 plants using these technologies [236]. A number of companies in North America, like Ultrox and Solar Chem, are marketing processes for oxidation of organic matter. Catalytic oxidation is described in Chapter 12.

In aqueous phase, the application of UV radiation promotes the decomposition of ozone and the subsequent formation of hydroxyl radicals, which are powerful oxidizing agents [237, 238]. It has been suggested that ozone is first split into an oxygen molecule and an oxygen atom, with the latter reacting with water to form two hydroxyl radicals [239]. The observation of hydrogen peroxide suggests, however, that ozone can react photochemically with water to form H_2O_2, with the latter split by UV light to form the hydroxyl radicals [240]. The dependence of the decomposition rate on ozone concentration, pH, light intensity, and the presence of radical scavengers has been studied in a number of investigations [241, 242]. A number of situations have been described in which the combination of ozone and UV radiation is effective in decomposing refractory substances. For example, cyanide ions in metal finishing wastes can combine with iron to form complexes that are stable toward ozone but that can be attacked by hydroxyl radicals [243]. In the Rayox process, UV irradiation and ozone are used to destroy ordnance and propellant compounds in waste water [244]. Protocatechuic acid formed in the production of olive oil has been oxidized by ozone/UV treatment [245]. Various polycyclic aromatics and alkyl- and halogen-substituted phenols in wastewater have been oxidized by ozone/UV and ozone/UV/H_2O_2 [246]. Aromatic compounds are relatively unreactive, but their ozonation rates can be increased by UV radiation [247].

Ozone has also been used in conjunction with UV irradiation and TiO_2 photocatalysts to degrade organic pollutants in water [248]. UV/ozone treatment has also been applied for the oxidation of organic contaminants on solid surfaces [249]. Oxidation is enhanced not only by the generation of oxygen atoms but by

TABLE 11.9 Relative Costs of Oil Shale Wastewater Purification

Purification Process	Relative Costs at Initial Concentration of Phenols	
	500 mg/L	50 mg/L
Aerobic biooxidation	1.00	1.00
Anaerobic biooxidation	0.38	0.68
Ozonation	0.98	1.13
Hydrogen peroxide	6.30	3.85
Activated carbon adsorption	8.00	1.25

the UV excitation of the organic substances to excited species. A related technique uses microwave discharges to generate ozone decomposition products to oxidize surfaces at low temperatures [250].

Hydrogen peroxide enhances ozone oxidation in a similar manner to UV light by releasing hydroxyl radicals. The kinetics for the oxidation of a variety of organic compounds have been studied in a semibatch reactor [251]. The Peroxone process utilizing H_2O_2 and O_3 has been piloted in drinking water purification and has been found to be more effective than ozone alone in the control of taste and odor compounds and the destruction of microorganisms [252, 253]. Similarly, the decomposition of the stable compound tetrachloroethylene is enhanced [251].

Ozone has also been used in conjunction with sonolysis to decolorize polluted groundwater [254]. The reaction appears to involve the generation of radicals, as scavengers reduce the rate.

Ozone can also be utilized in combination with solid adsorbents. A synergistic effect in utilizing ozonation followed by granular activated carbon (GAC) adsorption has been described in the treatment of dyeing wastewaters from textile mills [255]. A similar enhacement in the absorption of phenol ozonation products by GAC has been reported [256]. Ozonation of GAC beds is also reported to be effective for treatment of effluents from petroleum storage facilities [257].

A comparison of the effectiveness of various treatment processes for purifying phenol laden wastewater was conducted at a pilot plant scale [258]. The results are reported in Table 11.9.

The ozonation process was an advanced oxidation process combining UV radiation, TiO_2 photocatalysts, and ozone. Ozonation had similar costs to aerobic biooxidation, while anaerobic biooxidation was close to 60% cheaper. Because biological methods are slow, purification levels require impractically long residence times. Thus, for effective removal of the phenols, a combination of anaerobic biooxidation and ozonation is deemed appropriate [184].

REFERENCES

1. J.A. Wojtowicz, "Ozone," in *Kirk-Othmer Encyclopedia of Chemical Technology*, 4th Ed., John Wiley & Sons, New York, **1996**, pp. 953–994.

2. P.S. Bailey, *Ozonation in Chemistry*, Vol. 1, Academic Press, New York, **1978**.
3. P.S. Bailey, *Ozonation in Chemistry*, Vol. 2, Academic Press, New York, **1982**.
4. M. Horvárth; L. Bilitzky; J. Hüttner, *Ozone*, Elsevier, Amsterdam, **1985**.
5. C.F. Schönbein, "Beobachtungen über den bei der Elektrolysation des Wassers und dem Ausströmen der gewöhnlichen Electricität aus Spitzen sich Enwickelnden Geruchs," *Poggendors Ann. Phys. Chem.*, **1840**, 50, 616.
6. J.L. Soret, "Recherches sur la Densité de l'Ozone," *Ann. Chim. Phys. (Paris)*, **1865**, 7, 113.
7. W. von Siemens, "Uber die Elektrostatische Induction und die Verzögerund des Stroms in Flaschendrähten," *Poggendors Ann. Phys. Chem.*, **1857**, 102, 66.
8. F.A. Cotton; G. Wilkinson, *Advanced Inorganic Chemistry*, 5th Ed., John Wiley & Sons, New York, **1988**, pp. 452–454.
9. R. Trambarulo; S.N. Ghosh; C.A. Burrus, Jr.; W. Gordy, "The Molecular Structure, Dipole Moment, and *g* Factor of Ozone from its Microwave Spectrum," *J. Chem. Phys.*, **1953**, 21, 851.
10. C. Nebel, "Ozone," in *Kirk-Othmer Encyclopedia of Chemical Technology*, 3rd Ed., M. Gayson, D. Eckroth, Eds., John Wiley & Sons, New York, **1981**, pp. 683.
11. R.C. Weast; M.J. Astle, *CRC Handbook of Chemistry and Physics*, 62nd Ed., CRC Press, Boca Raton, **1981**.
12. W.J. Orville-Thomas, "A Bond-Order/Bond-Length Relation for Oxygen-Oxygen Bonds," *J. Mol. Spectrosc.*, **1958**, 3, 588.
13. P.C. Hiberty, *Isr. J. Chem.*, "The Diradical Character of 1,3-Dipoles," **1983**, 23, 10.
14. R.C. Harcourt; F.L. Skrezenek; R.M. Wilson; R.H. Flegg, "*Ab Initio* Valence Bond Calculations for the Ground State of Ozone," *J. Chem. Soc. Faraday Trans. II*, **1986**, 82, 495.
15. P. Borowski; K. Anderson; P.-Å. Malmqvist; B.O. Roos, "Vibrational Frequencies of Ozone: A Multiconfigurational Approach," *J. Chem. Phys.*, **1992**, 97, 5568.
16. L. Salem; C. Rowland, "The Electronic Properties of Diradicals," *Angew. Chem., Int. Ed. Engl.*, **1972**, 11, 92.
17. D.L. Cooper; J. Gerratt; M. Raimondi, "The Electronic Structure of 1,3 Dipoles: Hypervalent Atoms," *J. Chem. Soc. Perkin Trans. II*, **1989**, 1187.
18. G.P. Kostikova; Y.P. Kostikov; D.V. Korolkov, "Hypervalent π Bonds in the Electronic Structure of N_3^{-1} and O_3," *Dokl. Akad. Nauk. SSSR* (Engl. transl.), **1987**, 296, 963.
19. G.V. Gibbs, personal communication, 1998. Virginia Tech, Blacksburg VA.
20. O. Knacke; O. Kubaschewski; K. Hesselmann, *Thermochemical Properties of Inorganic Substances*, Vol. 2, 2nd Ed., Springer Verlag, Berlin, **1991**, pp. 1495.
21. E.A.V. Ebsworth; J.A. Connor; J.J. Turner, "Oxygen," in *Comprehensive Inorganic Chemistry*, A.F. Trotman-Dickerson, Ed., Pergamon Press, Oxford, 1973.
22. H. Selig; H.H. Claassen, "Raman Spectrum of Ozone", *Isr. J. Chem.*, **1968**, 6, 499.
23. L.T. Molina; M.J. Molina, "Absolute Absorption Cross-Section of Ozone in the 185 to 350 nm Wavelength Range," *J. Geophys. Res.*, **1986**, 91, 14501.
24. E. Vigroux, *Ann. Phys.*, **1953**, 8, 709.

25. E.C.Y. Inn; Y. Tanaka, "Ozone Absorption Coefficients in the Visible and Ultraviolet Regions," in *Ozone Chemistry and Technology*, Advances in Chemistry, Series 21, American Chemical Society, Washington, D.C., **1959**, pp. 263–268.
26. R. Atkinson; D.L. Baulch; R.A. Cox; R.F. Hampson, Jr.; J.A. Kerr; J. Troe, *J. Phys. Chem. Ref. Data*, **1992**, 21, 1125.
27. A.G. Streng, *J. Chem. Eng. Data*, **1961**, 6, 431.
28. R. Battino, Ed., *Oxygen and Ozone, IUPAC Solubility Series*, Vol. 7, Pergamon, Oxford, U.K., 1981.
29. S. Jahn, "Table of Ozone Properties," *Z. Anorg. Chem.*, **1906**, 48, 260.
30. J.A. Zaslowsky; H.B. Urbach; F. Leighton; R.J. Wnuk; J.A. Wojtowicz, "The Kinetics of the Homogeneous Gas Phase Thermal Decomposition of Ozone," *J. Am. Chem. Soc.*, **1960**, 82, 2682.
31. S.W. Benson; A.E. Axworthy, Jr., "Implications of Data on the Gas Phase Decomposition of Ozone," in *Ozone Chemistry and Technology*, Advances in Chemistry, Series 21, American Chemical Society, Washington, D.C., **1959**, pp. 398–404.
32. A.E. Axworthy, Jr.; S.W. Benson, "Mechanism of Gas Phase Decomposition of Ozone. Thermal and Photochemical Reactions," in *Ozone Chemistry and Technology*, Advances in Chemistry, Series 21, American Chemical Society, Washington, D.C., **1959**, pp. 388–397.
33. S.W. Benson; A.E. Axworthy, Jr., *J. Chem. Phys.*, **1957**, 26, 1718.
34. J. von Neuman, *OSRD Report 549*, Ballistic Research Laboratory File No. X-122, **1942**.
35. W. Doring, "Uber den Detonations vorgang in Gasen," *Ann. Physik*, **1943**, 43, 421.
36. V.B. Zeldovich, *J. Exp. Theor. Phys (U.S.S.R.)*, **1940**, 10, 542.
37. N. Semenoff, "Die Oxydation des Phosphordampfes bei Niedrigen Drucken," *Z. Physik*, **1927**, 46, 109.
38. N. Semenoff, "Zur Theorie des Verbrennungs prozesses," *Z. Physik*, **1928**, 48, 571.
39. H.J. Schumacher; G. Sprenger, *Z. Physik. Chem.*, **1930**, B6, 446.
40. H.J. Schumacher; G. Sprenger, *Z. Physik. Chem.*, **1930**, B11, 38.
41. S.A. Harper; W.E. Gordon, "Detonation Properties of Ozone," in *Ozone Chemistry and Technology*, Advances in Chemistry, Series 21, American Chemical Society, Washington, D.C., **1959**, pp. 28–37.
42. U. Kogelschatz; B. Eliasson; M. Hirth, "Ozone Generation from Oxygen and Air: Discharge Physics and Reaction Mechnisms," *Ozone Sci. Eng.*, **1988**, 10, 367.
43. J. Staehelin; J. Hoigné, "Decomposition of Ozone in Water: Rate of Initiation by Hydroxide Ions and Hydrogen Peroxide," *Environ. Sci. Technol.*, **1982**, 16, 676.
44. R.E. Bühler, J. Staehelin; J. Hoigné, "Ozone Decomposition in Water Studied by Pulse Radiolysis. 1. HO_2/O_2^- and HO_3/O_3^- as Intermediates," *J. Phys. Chem.*, **1984**, 88, 2560.
45. J. Staehelin; R.E. Bühler; J. Hoigné, "Ozone Decomposition in Water Studied by Pulse Radiolysis. 2. OH and HO_4 as Chain Intermediates," *J. Phys. Chem.*, **1984**, 88, 5999.
46. H. Tomiyasu; H. Fukutomi; G. Gordon, "Kinetics and Mechanisms of Ozone Decomposition in Basic Aqueous Solution," *Inorg. Chem.*, **1985**, 24, 2962.

47. M.D. Gurol, "Kinetics of Ozone Decomposition: A Dynamic Approach," *Environ. Sci. Technol.*, **1982**, 16, 337.

48. L. Forni; D. Bahnemann; E.J. Hart, "Mechanism of the Hydroxide Ion Initiated Decomposition of Ozone in Aqueous Solution," *J. Phys. Chem.*, **1982**, 86, 255.

49. S. Morooka; K. Ikemizu; Y. Kato, "The Decomposition of Ozone in Aqueous Solution," *Int. Chem. Eng.*, **1979**, 19, 355.

50. C. Yurteri; M. Gurol, *Ozone Sci. Eng.*, **1988**, 10, 277.

51. A.K. Biswas, Ed., *The Ozone Layer*, Pergamon Press, New York, **1979**.

52. M. Horváth; L. Bilitzky; J. Hüttner, *Ozone*, Elsevier, Amsterdam, **1985**.

53. *Gmelins Handbuch der Anorganischen Chemie*, 8th Ed., E.H.E. Pietsch, A. Kotowski, Eds., Sauerstoff, Liefereung 4, System-Nummer 3, Verlag Chemie, Weinheim, **1960**, pp. 984–1184.

54. D.L. Baulch; R.A. Cox; P.J. Crutzen; R.F. Hampton, Jr.; J.A. Kerr; J. Troe; R.T. Watson, "Evaluated Kinetic and Photochemical Data for Atmospheric Chemistry: Supplement I," *J. Phys. Chem. Ref. Data*, **1982**, 11, 327.

55. E. Castellano; H.J. Schumacher, *J. Chem. Phys.*, **1962**, 36, 2238.

56. E. Castellano; H.J. Schumacher, "Die Kinetik des Photochemischen Ozonzerfalles in Licht der Wellenlänge 313 mμ," *Z. Phys. Chem.*, **1971**, 76, 258.

57. A.R. Ravishankara; G. Hancock; M. Kawasaki; Y. Matsumi, "Photochemistry of Ozone: Surprises and Recent Lessons," *Science*, **1998**, 280, 60.

58. D.L. Baulch; R.A. Cox; P.J. Crutzen; R.F. Hampson; J.A. Kerr; J. Troe; R.T. Watson, "Evaluated Kinetic and Photochemical Data for Atmospheric Chemistry: Supplement I," *J. Phys. Chem. Ref. Data*, **1982**, 11, 327.

59. E. Castellano; H.J. Schumacher, "The Kinetics and the Mechanism of the Photochemical Decomposition of Ozone with light of 3340 Å Wavelength," *Chem. Phys. Lett.*, **1972**, 13, 625,

60. C. Cobos; E. Castellano; H.J. Schumacher, "The Kinetics and the Mechanism of Ozone Photolysis at 253.7 nm," *J. Photochem.*, **1983**, 21, 291.

61. H.S. Johnston, *Supersonic Aircraft and the Ozone Layer*, Report LBL-2748, University of California at Berkeley, Lawrence Berkeley Laboratory, May **1974**.

62. H.S. Johnston, "Catalytic Reduction of Stratospheric Ozone by Nitrogen Oxides," *Adv. Environ. Sci. Technol.*, **1973**, 4, 263.

63. S. Elliot; F.S. Rowland, "Chlorofluorocarbons and Stratospheric Ozone," *J. Chem. Educ.*, **1987**, 64, 387.

64. A. Russell; J. Milford; M.S. Bergin; S. McBride; L. McNair; Y. Yang; W.R. Stockwell; B. Coer, "Urban Ozone Control and Atmospheric Reactivity of Organic Gases," *Science*, **1995**, 269, 491.

65. W.P.L. Carter, "Development of Ozone Reactivity Scales for Volatile Organic Compounds," *J. Air Waste Manage. Assoc.*, **1994**, 44, 881.

66. E. Grosjean; D. Grosjean, "Gas Phase Reactions of Alkenes with Ozone: Formation Yields of Primary Carbonyls and Biradicals," *Environ. Sci. Technol.*, **1997**, 31, 2421.

67. W.P.L. Carter, *Atmos. Environ.*, **1990**, 34A, 481.

68. National Research Council, *Rethinking the Ozone Problem in Urban and Regional Air Pollution*, National Academy Press, Washington, D.C., 1991.

69. J. Hoigné; H. Bader, "Rate Constants of Reactions of Ozone with Organic and Inorganic Compounds in Water-II Dissociating Organic Compounds," *Water Res.*, **1983**, 17, 185.

70. R. Atkinson; W.P.L. Carter, "Kinetics and Mechanisms of the Gas-Phase Reactions of Ozone with Organic Compounds under Atmospheric Conditions," *Chem. Rev.*, **1984**, 84, 437.

71. R. Criegee, "Mechanism of Ozonolysis," *Angew. Chem., Int. Ed. Engl.*, **1975**, 14, 745.

72. R.L. Kuczkowski, "Formation and Structure of Ozonides," *Acc. Chem. Res.*, **1983**, 16, 42.

73. R.P. Lattimer; R.L. Kuczkowski; C.W. Gillies, "Mechanism of Ozonolysis. (a) Microwave Spectra, Structures, and Dipole Moments of Propylene and *trans*-2-Butene Ozonides. (b) Orbital Symmetry Analysis," *J. Am. Chem. Soc.*, **1974**, 96, 348.

74. D. Cremer, "Theoretical Determination of Molecular Structure and Conformation. 8. Energetics of the Ozonolysis Reaction. Primary Ozonide vs. Carbonyl Oxide Control of Stereochemistry," *J. Am. Chem. Soc.*, **1981**, 103, 3627.

75. H. Niki; P.D. Maker; C.M. Savage; L.P. Breitenbach, "A FTIR Study of a Transitory Product in the Gas-Phase Ozone-Ethylene Reaction," *J. Phys. Chem.*, **1981**, 85, 1024.

76. R.I. Martinez; J.T. Herron; R.E. Huie, "The Mechanism of Ozone–Alkene Reactions in the Gas Phase. A Mass Spectrometric Study of the Reactions of Eight Linear and Branched-Chain Alkenes," *J. Am. Chem. Soc.*, **1981**, 103, 3807.

77. P.S. Bailey; J.W. Ward; T.P. Carter, Jr.; E. Nieh; C.M. Fischer; A.-I.Y. Khashab, "Studies Concerning Complexes of Ozone with Carbon π Systems," *J. Am. Chem. Soc.*, **1974**, 96, 6136.

78. N.L. Bauld; J.A. Thompson; C.E. Hudson; P.S. Bailey, "Stereospecificity in Ozonide and Cross-Ozonide Formation," *J. Am. Chem. Soc.*, **1968**, 90, 1822.

79. D. Cremer, "Theoretical Determination of Molecular Structure and Conformation. 7. Steroselectivity of the Ozonolysis Reaction", *J. Am. Chem. Soc.*, **1981**, 103, 3619.

80. L.B. Harding; W.A. Goddard, III, "Mechanisms of Gas-Phase and Liquid-Phase Ozonolysis," *J. Am. Chem. Soc.*, **1978**, 100, 7180.

81. G. Karlström; S. Engström; B. Jönnson, Milticontigurational SCF and CI calculations on CH_2O_2, an intermediate in the Ozonolysis of Ethylene," *Chem. Phys. Lett.*, **1979**, 67, 343.

82. K. Yamaguchi; S. Yabushita; T. Fueno; S. Kato; K. Morokuma; S. Iwata, "Ab Initio UHF and UHF NOCl Approaches for Quasi-Degenerate Systems: Methylene Peroxide(CH_2OO)," *Chem. Phys. Lett.*, **1980**, 71, 563.

83. P.S. Nangia; S.W. Benson, "Thermochemistry and Kinetics of Ozonation Reactions," *J. Am. Chem. Soc.*, **1980**, 102, 3105.

84. D. Cremer, "Theoretical Determination of Molecular Structure and Conformation. 6. The Criegee Intermediate. Evidence for a Stabilization of its Syn Form by Alkyl Substituents," *J. Am. Chem. Soc.*, **1979**, 101, 7199.

85. F. Su; J.G. Calvert; J.H. Shaw, "A FTIR Spectroscopic Study of the Ozone-Ethene Reaction Mechanism in O_2-Rich Mixtures," *J. Phys. Chem.*, **1980**, 84, 239.

86. L. Rebrovic, "Catalyzed Process for Oxidation of Ozonides of Unsaturates to Carboxylic Acids," *World Patent WO9305007*, March **1993**, Assigned to Henkel Corp.
87. J. Treacy; M. El Hag; D. O'Farrell; H. Sidebottom, "Reactions of Ozone with Unsaturated Organic Compounds," *Ber. Bunsenges. Phys. Chem.*, **1992**, 96, 422.
88. J. Treacy; M. Curley; J. Wenger; H. Sidebottom, "Determination of Arrhenius Parameters for the Reactions of Ozone with Cycloalkanes," *J. Chem. Soc. Faraday Trans.*, **1997**, 93, 2877.
89. U. Kogelschatz; B. Eliasson, "Ozone Generation and Applications," in *Handbook of Electrostatic Processes*, J.S. Chang, A.J. Kelly, J.M. Crowley, Eds., Marcel Dekker, New York, **1995**, Ch. 26, p. 581.
90. U. Kogelschatz; B. Eliasson; M. Hirth, "Ozone Generation from Oxygen and Air: Discharge Physics and Reaction Mechnisms," *Ozone Sci. Eng.*, *1988*, 10, 367.
91. B. Eliasson; M. Hirth; U. Kogelschatz, "Ozone Synthesis from Oxygen in Dielectric Barrier Discharges," *J. Phys. D Appl. Sci.*, **1987**, 20, 1421.
92. B. Eliasson; U. Kogelschatz, "Nonequilibrium Volume Plasma Processing," *IEEE Trans. Plasma Sci.*, **1991**, 19, 1063.
93. B. Eliasson; U. Kogelschatz, *Basic Data for Modelling of Electrical Discharges in Gases: Oxygen*, Asea Brown Boveri, Baden, Switzerland, **1986.**,
94. R. Atkinson, et al., *J. Phys. Chem. Ref. Data*, **1992**, 21, 1125.
95. V.I. Gibalov; V.G. Samoilovich; M. Wronski, "Electrosynthesis of Nitrogen Oxides and Ozone in Ozonizer," in *Proc. 7th Int. Symp. Plasma Chem.*, Eindhoven, **1985**, pp. 401–406.
96. U. Kogelschatz; P. Baessler, "Determination of Nitrous Oxide and Dinitrogen Peroxide Concentrations in the Output of Air-Fed High Power Density Ozone Generators," in *Proc. Symp. Ozone Ultraviolet Water Treatment, Aquatech, Amsterdam*, **1986**, A.5.1–A.5.13, *Ozone Sci. Eng.*, **1987**, 9, 195.
97. V.G. Samailovich; V.I. Gibalov, "Kinetics of the Synthesis of Ozone an Nitrogen Oxides in a Barrier Discharge," *Russ. J. Phys. Chem.*, **1986**, 60, 1107.
98. B. Elliasson; U. Kogelschatz, "Nitrogen Oxide Formation in Ozonizers," *Proc. 8th Int. Symp. Plasma Chemistry*, Tokyo, **1987**, p. 736.
99. S. Yagi; M. Tanaka, "Mechanism of Ozone Generation in Air-Fed Ozonizers," *J. Phys. D Appl. Phys.*, **1979**, 12, 1509.
100. K. Yoshida; H. Tagashira, "Computer Simulation of Ozone Electrosynthesis in an N_2/O_2 Mixture-Fed Ozonizer," *Memoirs Kitami Institute Technol.*, **1986**, 18, 11.
101. B. Eliasson; U. Kogelschatz; P. Baessler, "Dissociation of O_2 in N_2/O_2 Mixtures," *J. Phys. B Atom Mol. Phys.*, **1984**, 17, L797.
102. B. Eliasson; U. Kogelschatz, "N_2O Formation in Ozonizers," *J. Chim. Phys.*, **1986**, 83, 279.
103. V.I. Gibalov; V.G. Samailovich; Y.V. Filippov, "Physical Chemistry of the Electrosynthesis of Ozone: The Results of Numerical Experiments," *Sov. J. Phys. Chem.*, **1981**, 55, 471.
104. B. Eliasson; U. Kogelschatz, "Modeling and applications of silent Discharge Plasmas," *IEEE Trans. Plasma sci.*, **1991**, 19, 309.
105. H. Sabadil; P. Bachmann; H. Kastelewicz, "Reaktionskinetik der Ozonbildung in der Sauerstoffglimmentladung," *Beitr. Plasmaphys.*, **1980**, 20, 283.

106. M. Lécuiller; M. Goldman, "Analyse des Règimes et des Zones de la Décharge Couronne en Termes de Production d'Ozone," *J. Phys. D Appl. Phys.*, **1988**, 21, 51.

107. R. Peyrous; R.-M. Millot, "Ozone Generation in Oxygen Discharges in Point-to-Plane Gap Subjected to a Chopped DC Positive Voltage," *J. Phys. D Appl. Phys.*, **1981**, 14, 2237.

108. L.M.L.F. Hosselet, "Increased Efficiency of Ozone Production by Electric Discharges," *Electrochim. Acta*, **1973**, 18, 1033.

109. L.A. Rosocha; W.A. Fitzsimmons, "Criteria for the Generation of Homogeneous Oxygen Plasmas Suitable for Ozone Synthesis," *Proc. 5th Int. Symp. Plasma Chem.*, Edinburgh, **1981**, pp. 421–426.

110. C. Yamabe; M. Hayashi; Y. Tachioka; K. Horii, "Ozone Generation Characteristics by New Type Double Discharge Ozonizer," *Proc. 8th Int. Symp. Plasma Chemistry*, Tokyo, **1987**, pp. 742–747.

111. G. Fournier; J. Bonnet; J. Fort; D. Pigache; M. Lécuiller, "Towards a Possible Industrial Production of Ozone with an Electron-Beam Controlled Discharge," *Proc. 4th Int. Symp. Plasma Chemistry*, Zürich, **1979**, pp. 742–747.

112. S. Masuda; S. Koizumi; J. Inoue; H. Araki, "Production of Ozone by Surface and Glow Discharge at Cryogenic Temperatures," *IEEE Trans. Ind. Appl.*, **1988**, 24, 298.

113. J.S. Chang; S. Masuda, "Mechanism of the Ozone Formation in a Near Liquid Nitrogen Temperature Medium Pressure Glow Discharge Positive Column," *Pure Appl. Chem.*, **1988**, 60, 645.

114. U. Kogelschatz, "Ozone Synthesis in Gas Discharges," in *Proc. XVI Int. Conf. Phenomena Ionized Gases (ICPIG XVI)*, Düsseldorf, Germany, Aug. **1983**, pp. 240–250.

115. M. Tanaka; S. Yagi; N. Tabata, "The Observation of Silent Discharge by Image Intensifier," *Trans. IEE Jpn.*, **1978**, 98A, 57.

116. M. Hirth; U. Kogelschatz; B. Eliasson, "The Structure of Microdischarges in Ozonizers and Their Influence on the Reaction Kinetics," in *Proc. 6th Int. Symp. Plasma Chem.*, Montreal, **1983**, pp. 663–668.

117. P. Erni; P.-A. Liechti; A. Poisson, "Ozone Production Today," *Water Supply*, **1995**, 13, 291.

118. K.H. Elsenhans, "More Than 5 Years Experience with High-Frequency Ozone Generators with a Capacity of 10 kg/h," in *Proc. 8th Ozone World Congress*, Vol. 1, Zurich, **1987**, pp. A62–A71.

119. U. Kogelschatz, "Advanced Ozone Generation," in *Process Technologies for Water Treatment*, S. Stucki, Ed., Plenum, New York, **1988**, pp. 87–120.

120. M. Fischer; H.-P. Klein; P.-A. Liechti; P. Dyer-Smith, "Technical and Economic Advantages of Producing and Applying Ozone at High Concentration," *Ozone Sci. Eng.*, **1987**, 9, 93.

121. A. Shadiakhy; D. Blankenfeld, "Potable Water Treatment with Ozone of High Concentration," *Water Supply*, **1995**, 13, 231.

122. J. Rounsaville; R.G. Rice, "Evolution of Ozone for the Bleaching of Paper Pulps," *Ozone Sci. Eng.*, **1997**, 18, 549.

123. E. Merz; F. Gaia, "Comparison of Economics of Various Ozone Generation Systems," *Ozone Sci. Eng.*, **1990**, 12, 401.

124. J.M. Dohan; W.J. Masschelein, "The Photochemical Generation of Ozone: Present State-of-the-Art," *Ozone Sci. Eng.*, **1987**, 9, 315.

125. E. Briner, "Photochemical Production of Ozone," in *Ozone Chemistry and Technology*, Advances in Chemistry, Series 21, American Chemical Society, Washington, D.C., **1959**, pp. 1–6.

126. E. Briner; H. Karbassi, "Production de l'Ozone au Moyen d'une Lumière Ultra-Violetle Intermittente: Lampe à Vapeur de Mercure avec Dispositif d'Interruption; Etincelles entre Electrodes detinc," *Helv. Chim. Acta*, **1945**, 28, 1014.

127. E. Briner; A. Munzhuber, "Researches sur la Production de l'Ozone par les lampes à Vapour de Mercure," *Helv. Chim. Acta*, **1955**, 38, 1994.

128. L.T. Molina; M.J. Molina, "Absolute Absorption Cross-Section of Ozone in the 185 to 350 nm Wavelength Range," *J. Geophys. Res.*, **1986**, 91, 14501.

129. B. Eliasson; U. Kogelschatz, "Ozone Generation with Narrow-Band UV Radiation," *Ozone Sci. Eng.*, **1991**, 13, 365.

130. D. Mor, *Environ. Sci. Technol.*, **1976**, 10, 897.

131. B. Freisinger; U. Kogelschatz; J.H. Schaefer; J. Uhlenbusch; W. Violel, "Ozone Production in Oxygen by Means of F_2-Laser Irradiation at $\lambda = 157.6$ nm," *Appl. Phys. B* **1989**, 49, 121.

132. P.C. Foller; C.W. Tobias, "The Anodic Evolution of Ozone," *J. Electrochem. Soc.*, **1982**, 129, 506.

133. P.C. Foller; C.W. Tobias, "The Effect of Electrolytic Anion Adsorption on current Efficiencies for the Evolution of Ozone," *J. Phys. Chem.*, **1981**, 85, 3228.

134. P. Kötz; S. Stucki, "Ozone and Oxygen Evolution on PbO_2 Electrodes in Acid solution," *J. Electroanal. Chem.*, **1987**, 228, 407.

135. C. Foller, in *Handbook of Ozone Technology and Applications*, Vol. 1, R.G. Rice and A. Netzer, Eds., Ann Arbor Science Publishers, Ann Arbor, MI, **1982**, p. 89.

136. S. Stucki; H. Baumann, "Application of Electrochemical Ozone Generators in Ultrapure Water Systems," in *Process Technologies for Water Treatment*, S. Stucki, Ed., Plenum, New York, **1988**, pp. 191–203.

137. S. Stucki; H. Baumann, H.J. Christen; R. Kötz, "Performance of a Pressurized Electrochemical Ozone Generator," *J. Appl. Electrochem.*, **1987**, 17, 773.

138. S. Stucki; G. Theis; R. Kötz; H. Devantay; H.J. Christen, "In situ Production of Ozone in Water using a Membrel Electrolyzer," *J. Electrochem. Soc.*, **1985**, 132, 367.

139. A.A. Babak; R. Amadelli; A. De Battisti; V.N. Fateev, "Influence of Anions on Oxygen/Ozone Evolution on PbO_2/*spe* and PbO_2/Ti Electrodes in Neutral pH Media," *Electrochim. Acta*, **1994**, 39, 1597.

140. J. Feng; D.C. Johnson; S.N. Lowery; J.J. Carey, "Electrocatalysis of Anodic Oxygen-Transfer Reactions Evolution of Ozone," *J. Electrochem. Soc.*, **1994**, 141, 2708.

141. P. Tatapudi; J.M. Fenton, *J. Electrochem. Soc.*, "Synthesis of Ozone in a Proton Exchange Membrane Electrochemical Reactor," **1993**, 140, 3527.

142. P. Tatapudi; J.M. Fenton, "Simultaneous Synthesis of Ozone and Hydrogen Peroxide in a Proton-Exchange-Membrane Electrochemical Reactor," *J. Electrochem. Soc.*, **1994**, 141, 1174.

143. P. Tatapudi, Ph.D. Thesis, University of Connecticut, Storrs, **1993**.

144. O.J. Murphy; G.D. Hitchens, "Method and Apparatus for Electrochemical Production of Ozone," *US Patent 5,460,705*, October **1995**, Assigned to Lynntech, Inc.

145. C.E. Castro, "Ozone from Iron(III) Porphyrin, Nitrite Ion, and Oxygen," *J. Am. Chem. Soc.*, **1996**, 118, 3984.

146. K.A. Jongeward; D. Magde; D.J. Taube; J.C. Marsters; T.G. Traylor; V. S. Sharma, "Picosecond and Nanosecond Geminate Recombination of Myoglobin with CO, O_2, NO and Isocyanides," *J. Am. Chem. Soc.*, **1988**, 110, 380.

147. *Gmelins Handbuch der Anorganischen Chemie*, E.H.E. Pietsch, A. Kotowski, Eds., 8th Ed., Sauerstoff, Lieferung 4, System-Nummer 3, Verlag Chemie, Weinheim, 1960, pp. 984–1184.

148. S.G. Chang; K. Hu; Y. Wang, "The Use of Yellow Phosphorus to Destroy Toxic Organic Compounds," *J. Environ. Sci.*, **1994**, 6, 1.

149. H. Wang; Y. Shi; L. Le; S.-M. Wang; J. Wei; S. -G. Chang, "POZONE Technology to Bleach Pulp," *Ind. Eng. Chem. Res.*, **1997**, 36, 3656.

150. G.C. Lee, "Phosphorus Ozone for Wood Pulp Bleaching," *Bechtel Corporation Report*, Bechtel Corp., **1994**.

151. A.C. Jenkins, "Laboratory Techniques for Handling High-Concentration Liquid Ozone," in *Ozone Chemistry and Technology*, Advances in Chemistry Series 21, American Chemical Society, Washington, D.C., **1959**, pp. 13–21.

152. OREC Product Line Brochure, General Literature Section, p. 22. Phoenix, AZ.

153. P. Lainé, *Ann. Physik*, **1935**, 3, 461.

154. G.A. Cook; A.D. Kiffer; C.V. Klumpp; A.H. Malik; L.A. Spence, "Separation of Ozone from Oxygen by a Sorption Process," in *Ozone Chemistry and Technology*, Advances in Chemistry Series 21, American Chemical Society, Washington, D.C., **1959**, pp. 44–52.

155. C.N. Hewitt; G.L. Kok; R. Fall, *Nature (Lond.)*, **1990**, 344, 56.

156. N. Hewitt; G. Terry, *Environ. Sci. Technol.*, **1992**, 26, 1891.

157. C. Nebel, "Ozone," in *Kirk-Othmer Encyclopedia of Chemical Technology*, Grayson, M., Ed., Vol. 16, Wiley, New York, **1978**, pp. 683–699.

158. Japan Air Cleaning Association, *Handbook*, Ohm Publishing Company, **1981**, p. 178.

159. R.J. Lewis, Jr., Ed., *Sax's Dangerous Properties of Industrial Materials*, Van Nostrand Reinhold, New York, **1993**.

160. U.S. Environmental Protection Agency, Safety Guidelines.

161. J.C. Hoff, *EPA Report 600/2–86/067*, U.S. Environmental Protection Agency, Research Triangle Park, NC, **1986**.

162. G.B. Wickramanayake; O.J. Sproul, *Ozone Sci. Eng.*, **1988**, 10, 123.

163. Q. Zhu; C. Liu; Z. Xu, *Ozone Sci. Eng.*, **1989**, 11, 169.

164. G. Davis, *Proc. 2nd Int. Symp. Ozone Technol., International Ozone Association*, Pan American Group, Stamford, CT, **1975**, p. 421.

165. M.D. Gurol; W.M. Bremen, "Kinetics and Mechanism of Ozonation of Free Cyanide Species in Water" *Environ. Sci. Technol.*, **1985**, 19, 804.

166. C. Nebel, *Industrial Odor Technology Assessment*, Ann Arbor Science Publishers, Ann Arbor, MI, **1975**, Ch. 23.

167. R.J. Strittmatter, et al., *Ozone Sci. Eng.*, **1993**, 15, 47.

168. L.F. Fieser; M. Fieser, *Reagents for Organic Synthesis*, John Wiley & Sons, 1967–1979, New York, Vols. 1–17.

169. R.G. Rice; M.E. Browning, Eds., Proc. *First Int. Symp. Ozone Water Wastewater Treatment*, International Ozone Institute, **1973**.

170. J. Hoigné, "The Chemistry of Ozone in Water," in *Process Technologies for Water Treatment*, S. Stucki, Ed., Plenum, New York, **1988**, pp. 121–143.

171. J. Hoigne; H. Bader, "The Role of Hydroxyl Radical Reactions in Ozonation Processes in Aqueous Solution," *Water Res.*, **1976**, 10, 377.

172. J. Hoigne; H. Bader, "Ozonation of Water: Role of Hydroxyl Radicals as Oxidizing Intermediates," *Science*, **1975**, 190, 782.

173. T.L. Theis; P.C. Singer, "Complexation of Iron (II) by Organic Matter and its Effect on Iron (II) Oxygenation," *Environ. Sci. Technol.*, **1974**, 8, 569.

174. W.R. Knocke, "Using Alternative Oxidation to Remove Dissolved Manganese from Waters Laden with Organics," *J. Am. Water Works Assoc.*, **1987**, 79, 75.

175. S.H. Lin, "Ozonation Kinetics of Iron and Manganese in Aqueous Solution," *J. Chem. Technol. Biotechnol.*, **1993**, 56, 163.

176. T. Loegager; J. Holcman; K. Sehested; T. Pedersen, "Oxidation of Ferrous Ions by Ozone in Acidic Solution," *Inorg. Chem.*, **1992**, 31, 3523.

177. D. Grasso, "Ozonation of Drinking Water: A Design Methodology, I. Background and Design Rationale," *Ozone Sci. Eng.*, **1987**, 9, 109.

178. P.W. Prendiville; D.G. McBride; G.F. Stolarik, "Ozonation–Direct Filtration of Los Angeles Drinking Water," paper presented at the 6th Ozone World Congress, International Ozone Association, Washington, D.C., May 22–26, **1983**.

179. M.R. Jekel, "The Benefits of Ozone Treatment Prior to Flocculation Processes," *Ozone Sci. Eng.*, **1983**, 5, 21.

180. D. Maier, "Microflockund Durch Ozone," in *Oxidationsverfahren in der Trinkwasseraufbereitung*, W. Kühn, H. Sontheimer, Eds., EPA Report 570/9-79-020, US EPA, Research Triangle Park, NC, **1979**, pp. 394–417.

181. B. Thorell; H. Borén; A. Grimvall; A. Nyström; R. Sävenhed, "Characterization and Identification of Odorous Compounds in Ozonated Waters," *Water Sci. Technol.*, **1992**, 25, 139.

182. M. Edwards; M. Boller; M.M. Benjamin, "Effect of Pre-Ozonation on Removal of Organic Matter During Water Treatment Plant Operations," *Water Sci. Technol.*, **1993**, 27, 37.

183. J.P. Duguet; C. Anselme; O. Wable; I. Baudin; J. Mallevialle, *Water Supply*, **1993**, 11, 141.

184. R.G. Rice, "Applications of Ozone for Industrial Wastewater Treatment — A Review," *Ozone Sci. Eng.*, **1997**, 18, 477.

185. W.K. Murphy, "The Use of Ozone in Recycled Oceanarium Water", in *Aquatic Applications of Ozone*, W.J. Blogoslawski; R.G. Rice, Eds., International Ozone Association, Pan American Group, Stamford, CT, **1975**, pp. 87–95.

186. S. Collins, "Ozone Applications at the Indianapolis Zoo," in *Proceedings of the 3rd International Symposium on the Use of Ozone in Aquatic Systems*, W.J. Blogoslawski, Ed., International Ozone Association, Stamford, CT, **1992**, pp. 119–122.

187. R.P. Selm, "Ozone Oxidation of Aqueous Cyanide Waste Solutions in Stirred Batch Reactors and Packed Towers," in *Ozone Chemistry and Technology*, Advances in Chemistry Series 21, American Chemical Society, Washington, D.C., **1959**, pp. 66–77.

188. G. Klingsick, "Applications of Ozone at the Boeing Co., Wichita, Kansas," in *Proc. First Int. Symp. Ozone Water Wastewater Treatment*, W.J. Blogoslawski; R.G. Rice, Eds., International Ozone Association, Pan American Group, Stamford, CT, **1975**, pp. 587–590.

189. M. Teramoto; S. Imamura; N. Yatagai; Y. Nishikawa; H. Taranishi, "Kinetics of Self-Decomposition of Ozone and the Ozonation of Cyanide Ion and Dyes in Aqueous Solution," *J. Chem. Eng. Jpn.*, **1981**, 14, 383.

190. A.J. Gough, "The Use of Ozone to Remove Colour from Waste Water," in *Proc. Int. Symp. Ozone-Oxidation Methods Water Wastewater Treatment*, Berlin, Germany, International. Ozone Association, European-African Group, **1993**, pp. II.2.1–II.2.14.

191. J. Sarasa; P. Roche; A. Puig; P. Ormad; M.S. Mutuberria; J.L. Ovelleiro, "Determination of the Compounds Present in Previously Chlorinated Wastewater Resulting from the Production of Azoic Dyes by GC/MS and Treatment with Ozone," in *Ozone in Water and Wastewater Treatment*, Proc. 11th Ozone World Congress, San Francisco, California, **1993**, International. Ozone Association, Pan American Group, **1993**, Vol. 2, pp. S-7-91–S-7-105.

192. M.P. Roche; J. Sarasa; M.A. Martin; A. Puig; J.L. Ovelleiro, "Treatment for a Waste water of a Colorant Synthesis Industry [O_3 and $Ca(OH)_2$]," in *Proc. 12th World Congress of the International Ozone Association*, Lille France, International Ozone Association, European-African Group, Paris, France, **1995**, pp. 489–500.

193. J. Carriére; J.P. Jones; A.D. Broadbent, "Effect of Dyeing Aid on the Oxidation Reaction of Color from an Insoluble and a Soluble Dye in a Simulated Effluent," *Ozone Sci. Eng.*, **1993**, 15, 189.

194. C.R. Zmolek, "Ultra Pure Water for Integrated Circuits Processing," *Ind. Water Eng.*, **1977**, Dec., 6.

195. C. Nebel; W.W. Nesgod, "Purification of Deionized Water by Oxidation with Ozone," *Solid State Technol.*, Oct. **1984**, 185.

196. R.A. Hango; F. Doane; L.J. Bollyky, "Wastewater Treatment for Reuse in Integrated Circuit Manufacturing," in *Wasser Berlin, '81*, Colloquium Verlag Otto H. Hess, **1981**, pp. 303–313.

197. R. Stone, "Ozonation Leaves a Hole," *Science*, **1995**, 267, 1771.

198. J.C. Kruithof; J.C. Schippers, "The Formation and Removal of Bromate," *Water Supply*, **1993**, 11, 149.

199. M. Doré; N. Merlet; B. Legube; J.Ph. Croue, "Interactions between Ozone, Halogens, and Organic Compounds," *Ozone Sci. Eng.*, **1988**, 10, 153.

200. T. Eriksson; J. Gierer, *J. Wood Chem. Technol.*, **1985**, 5, 53.

201. M.P. Godsay; E.M. Pearce, *Proc. 1984 TAPPI Oxygen Delignification Symposium*, San Francisco, California, **1984**, p. 55.

202. S. Chandra; J. Gratzl, *Proc. 1985 Int. Pulp Bleaching Conf.*, Quebec City, Quebec, Canada, **1985**, p. 27.

203. L.B. Sonnenberg, "Overview of Ozone Bleaching", in *Extended Abstracts, Division of Environmental Chemistry*, American Chemical Society Fall National Meeting, Chicago, Illinois, **1995**, pp. 264–267.

204. S.C. Stinson, "Fine and Intermediate Chemicals Makers Emphasize New Products and Processes," *Chem. Eng. News*, July 17, **1995**, p. 10.

205. D.K. Parker; J.R. Purdon, *U.S. Patent 5,039,737*, Aug. 23, **1991**, Assigned to Good year Tire and Rubber, Co.

206. L. Raber, "Green Chemistry Honored," *Chem. Eng. News*, June 30, **1997**, p. 7.

207. A. Sawa; S. Kosaka; H. Obara; K. Aoki; J.S. Edmends, "Role of Ozone in the Oxidation of $YBa_2Cu_3O_{7-X}$ Thin Films," in *Adv. Supercond. IV. Proc. 4th Int. Symp. Supercond.*, H. Hayakawa; N. Koshizuka, Eds., Springer, Tokyo, **1992**, pp. 659–662.

208. R.R. Mathews, "Process and Apparatus for the Treatment of Semiconductor Wafers in a Fluid," *U.S. Patent 5,464,480*, **1995**, Assigned to Legacy Systems, Inc.

209. W.H. Thompson; Z. Yamani; L.H. Abu Hassan; J.E. Greene; M. Nayfeh; M.-A. Hasan, "Room Temperature Oxidation Enhancement of Porous Si(001) Using Ultraviolet-Ozone Exposure", *J. Appl. Phys.*, **1996**, 80, 5415.

210. A. Kurokawa; S. Ichimura; H.J. Kang; D.W. Moon, "Low Temperature Oxidation Processing with High Purity Ozone," *Mater. Res. Soc. Symp. Proc.*, **1996**, 429, 269.

211. A. Kurokawa; S. Ichimura, "High Purity Ozone Oxidation on Hydrogen Passivated Silicon Surface," *Appl. Surf. Sci.*, **1996**, 100/101, 436.

212. A. Kazor; I.W. Boyd, "Ozone Oxidation of Silicon," *Electron Lett.*, **1993**, 29, 115.

213. M. Niwano; M. Suemitsu; Y. Ishibashi; Y. Takeda; N. Miyamoto; K. Honma, "Ultraviolet Ozone Oxidation of Silicon Surface Studied by Photoemission and Surface Infrared Spectroscopy," *J. Vac. Sci Technol. A*, **1992**, 10, 3171.

214. I.W. Boyd; V. Cracium; A. Kazor, "Vacuum-Ultra-Violet and Ozone-Induced Oxidation of Silicon and Silicon-Germanium," *Jpn. J. Appl. Phys., Part 1*, **1993**, 32, 6141.

215. Z.H. Lu; B. Bryskiewicz; J. McCaffrey; Z. Wasilewski; M.J. Graham, "Ultraviolet-Ozone Oxidation of Gallium Arsenide(100) and Indium Phosphide(100)," *J. Vac. Sci. Technol. B*, **1993**, 11, 2033.

216. M.P. Besland; P. Louis; Y. Robach; J. Joseph; G. Hollinger; D. Gallet; P. Viktorovitch, "Growth of Passivating UV/Ozone Oxides on Indium Phosphide. Correlations between Chemical Compositions and Interfacial Electrical Properties," *Appl. Surf. Sci.*, **1992**, 56–58, 846.

217. T. Nakanishi; S. Ohkubo; Y. Tamura, "Oxidation in Ozone," *Fujitsu Sci. Technol. J.*, **1996**, 32, 128.

218. J. Schwarz; J.W. McMillan; H.E. Fischer, "Partial Oxidation of Olefins by Molecular Oxygen Catalyzed by (Alumina)Rh(O_2). Part 2," *J. Mol. Catal.*, **1992**, 74, 443.

219. D.H. Parker; B.E. Koel, "Chemisorption of High Coverages of Atomic Oxygen on the Pt(111), Pd(111), and Au(111) Surfaces," *J. Vac. Sci. Technol. A*, **1990**, 8, 2585.

220. "Methods of Controlling Odors," in *Odors from Stationary and Mobile Sources*, National Academy of Sciences, Washington, D.C., **1979**, Ch. 6, p. 179. Committee on Odour from Stationary and Mobile sources.

221. L.J. Bollyky, "Ozone Systems for Odor Control," in *Proceedings of the First International Symposium on Ozone for Water & Wastewater Treatment*, R.G. Rice; M.E. Browning, Eds., International Ozone Institute, **1973**, p. 436.

222. R.S. Sheinson; N.S. Smyth, "Hydrocarbon Decomposition and Oxidation in a Plasma Discharge," in *Proc. 8th Int. Symp. Plasma Chem.*, (ISPC-8), Tokyo, Japan, Aug. **1987**, pp. 648–653.

223. A. Gervasini; G.C. Vezzoli; V. Ragaini, "VOC Removal by Synergic Effect of Combustion Catalyst and Ozone," *Catal. Today*, **1996**, 29, 449.

224. J. Aso; M. Chikuni, "Household Ozone-Type Sterilization Device", *Japan Kokai Tokkyo Koho 07-16163*, Jan. **1995**, Assigned to Toto, Ltd.

225. C.D. Garbutt; D.A. Moxley, "Method and Apparatus for Disinfecting a Closed Space," *US Patent 5,514,345*, May **1996**, Assigned to Ozact, Inc.

226. M. Yoshimoto; T. Nakatsuji; K. Nagano; K. Yoshida; M. Isono, "Decomposition of Ethylene with Ozone and Metal Oxide Catalysts", *Japan Kokai Tokkyo Koho 04-89402*, March, **1992**, Assigned to Sakai Chemical Industry, Co.

227. I.K. Puri, "The Removal of NO by Low-Temperature O_3 Oxidation," *Combust. Flame*, **1995**, 102, 512.

228. R.Kh. Amirov; É.I. Asinovskii; I.S. Samailov; A.V. Shepelin, "Ozone Synthesis Initiated by a Nanosecond Corona in Air," *Khimiya Vysokikh Énergii* (English translation) **1992**, 26, 61.

229. S. Masuda, "Pulse Corona Induced Plasma Chemical Process; A Horizon of New Plasma Chemical Technologies," *Pure Appl. Chem.*, **1988**, 60, 727.

230. L. Civitano; G. Dinelli, *La Termotenical*, March **1988**, p. 43.

231. I. Gallimberti, "Impulse Corona Stimulation for Flue Gas Treatment," *Pure Appl. Chem.*, **1988**, 60, 663.

232. S. Jordan; H.R. Paur, "Recent Development in Flue Gas Treatment by Electron Irradiation," *Beta-Gamma*, **1988**, 1, 19.

233. J.S. Clemens; A. Mizuno; W.C. Finney; R.H. Davis, "Combined Removal of SO_2, NO_x, and Fly Ash from Simulated Flue Gas Using Pulsed Streamer Corona," *IEEE Trans. Ind. Appl.*, **1989**, 25, 62.

234. V.Yu. Simachev, et al., "An Investigation of the Ozone-Ammonia Method of Simultaneous Desulphurization and Denitrification of Flue Gases When Burning Donetsk Coals," *Therm. Eng.*, **1988**, 35, 171.

235. *ENEL Research on Electrostatic Technologies for Pollution and Emission Control: Results and Perspectives.*, Italian Nat. Electricity Board, Thermal and Nuclear Res. Center, **1988**, p. 38.

236. W.H. Glaze, *Ozone News*, **1994**, 22, 42.

237. W.H. Glaze; J.-W. Kang; D.H. Chapin, "Chemistry of Water Treatment and Processes Involving Ozone, Hydrogen Peroxide, and Ultraviolet Radiation," *Ozone Sci. Eng.*, **1987**, 9, 335.

238. T.C. Huang; D.H. Chen, "Kinetics of Ozone Decomposition in Aqueous Solution with and without Ultraviolet Radiation," *J. Chin. Inst. Chem. Eng.*, **1993**, 24, 207.

239. W.D. McGrath; R.G.W. Norrish, "Studies of the Reactions of Excited Oxygen Atoms and Molecules Produced in the Flash Photolysis of Ozone," *Proc. R. Soc. A*, **1960**, 254, 317.

240. G.R. Peyton; W.H. Glaze, "New Developments in the Ozone/UV Process," *Proc. 6th Ozone World Congress*, Washington, DC, **1983**.

241. K. Ikemizu; M. Orita; M. Sagiike; S. Morooka; Y. Kato, "Ozonation of Organic Refractory Compounds in Water in Combination with UV Radiation," *J. Chem. Eng. Jpn.*, **1987**, 20, 369.

242. Y. Ku; W.-J. Su; Y.-S. Shen, "Decomposition Kinetics of Ozone in Aqueous Solution," *Ind. Eng. Chem. Res.*, **1996**, 35, 3369.

243. H.W. Prengle, Jr., "Evolution of the Ozone/UV Process for Wastewater Treatment," Paper presented at the IOI/EPA Colloquium on Wastewater Treatment and Disinfection with Ozone, Cincinnatti, Ohio, Sep. **1977**.

244. W. McPhee; L. Wagg; P. Martin, "Advanced Oxidation Process for the Destruction of Ordnance and Propellant Compounds Using Rayox," *Chem. Oxid.*, **1994**, 3, 249.

245. F.J. Benitez; J. Beltran-Heredia; J.L. Acero; T. Gonzalez, "Degradation of Protocatechuic Acid by Two Advanced Oxidation Processes: Ozone/UV Radiation and H_2O_2/UV Radiation," *Water Res.*, **1996**, 30, 1597.

246. M. Trapido; Y. Veressinina; R. Munter, "Ozonation and Advanced Oxidation Processes of Polycyclic Aromatic Hydrocarbons in Aqueous Solutions — A Kinetic Study," *Environ. Technol.*, **1995**, 16, 729.

247. S.C. Xu; H.D. Zhou; X.Y. Wei; J. Lu, "The pH Dependence and Effects of the Oxidation Products of Some Aromatic Compounds in Ozonation under UV Irradiation," *Ozone Sci. Eng.*, **1989**, 11, 281.

248. P.L. Yue, "Degradation of Organic Pollutants by Advanced Oxidation," *Process. Saf. Environ. Prot.*, **1992**, 70, 145.

249. J.R. Vig, "UV/Ozone Cleaning of Surfaces," *J. Vac. Sci. Technol. A*, **1985**, 3, 1027.

250. W.C. Neely; W.F. Welch; R.F. Askew, "Low-Temperature Oxidation at Surfaces Using Ozone Decomposition Products Formed by Microwave Discharge," *US Patent 5,443,863*, August **1995**, Assigned to Auburn University, Aubum, AL.

251. W.H. Glaze; J.W. Kang, "Advanced Oxidation Process. Test of a Kinetic Model for the Oxidation of Organic Compounds with Ozone and Hydrogen Peroxide in a Semibatch Reactor," *Ind. Eng. Chem. Res.*, **1989**, 28, 1850.

252. D.W. Ferguson; M.J. McGuire; B. Koch; R.L. Wolfe; E.M. Aieta, "Comparing PEROXONE and Ozone for Controlling Taste and Odor Compounds, Disinfection By-Products, and Microorganisms," *J. Am. Water Works Assoc.*, **1990**, 82, 181.

253. P.A. Hacker; C. Paszko-Kolva; M.H. Stewart; R.L. Wolfe; E.G. Means, III, "Production and Removal of Assimilable Organic Carbon under Pilot-Plant Conditions Through the Use of Ozone and PEROXONE," *Ozone Sci. Eng.*, **1994**, 16, 197.

254. T.M. Olson; P.F. Barbier, "Oxidation Kinetics of Natural Organic Matter by Sonolysis and Ozone," *Water Res.*, **1994**, 28, 1383.

255. K. Mizumoto; M. Horie, "Dyeing Wastewater Treatment by Combination of Ozone and Activated Carbon," *Jpn. Textile News*, **1974**, 89, 238.

256. M. Dore; B. Sohm; B. Legube; R. Brunet, "Effect of Ozone on Adsorbability and Biodegradability of a Few Organic Compounds over Granulated Activated Carbon," *Environ. Tech. Lett.*, **1982**, 3, 375.

257. Electric Power Research Institute, "Ozonation of Granulated Activated Carbon Beds," *EPRI TechApplication*, **1993**, 5 (3).

258. R. Munster; J. Kallas; L. Pikkov; S. Preis; S. Kamanev, "Ozonation and Advanced Oxidation Process Treatment of the Wastewater From Oil Shale and Pulp and Paper Industries," in *Ozone in Water and Wastewater Treatment*, Proc. 11th Ozone World Congress, San Francisco, California, 1993, Internat. Ozone Assoc., Pan American Group, **1993**, Vol. 2, pp. S-10-38–S-10-53.

12

SURFACE REACTIONS AND CATALYSIS

12.1 INTRODUCTION

The noncatalytic chemistry of ozone is very rich (Chapter 11), but the diversity of possible reactions also means that selectivity will be a problem. Catalysts offer the ability of controlling reaction pathways for precise chemical synthesis. Because of the cost of ozone (Chapter 11), however, application for production of chemicals is restricted to high-value–added substances. The most promising application of ozone is in the treatment of polluted streams containing VOCs [1]. The concentration of contaminants is generally low so that only small amounts of ozone are needed. Importantly, the reactivity of ozone is very high so that oxidation can occur close to room temperature for both liquid and gaseous streams. This results in substantial energy savings, as these streams do not need to be heated up and then cooled down. In conventional catalytic incinerators, the VOC concentration is too low for the reactions to be self-sustaining, and auxiliary fuels must be employed.

In the destruction of VOCs with ozone, selectivity may not seem to be important, as the desired reaction is the complete oxidation of the compounds. Many times when catalysts are not used, however, partial oxidation products are formed that are not completely mineralized. For example, oxalic, glyoxalic, and acetic acids in solution and acetaldehyde in the gas phase cannot be completely oxidized by ozone alone. In such cases, the use of catalysts is essential to promote complete oxidation.

A critical step in catalytic oxidation is adsorption and activation of the reactants on the surface of the catalyst. This chapter begins with a brief coverage of this topic. The catalytic chemistry of ozone can be categorized as homogeneous or heterogeneous depending on whether reactions involve reactive species in a fluid medium or solid substances that are phase separated from the fluid. Both types of processes are described. In some cases, a hybrid mechanism may operate that combines elements of both. For example, a solid material used in solution may be in equilibrium with a dissolved component, or a surface reaction may give rise to gas-phase radicals that react through chain mechanisms. Examples of these situations are cited.

In some cases, oxidation of organic compounds produces CO. If VOC concentrations in contaminant streams are low, some release of CO below regulated levels is acceptable. Otherwise, catalysts need to be designed for the further oxidation of CO to CO_2. Some work in this area is described.

Finally, an issue that must be considered in the use of ozone is that it is a toxic substance, and any releases must be avoided. A catalytic system for ozone oxidation must incorporate provision for its destruction after use. This can be accomplished by placing a decomposition catalyst at the end of the catalyst bed or by reacting the ozone with activated carbon. Both options are reviewed.

12.2 OZONE ADSORPTION AND REACTIONS WITH ADSORBATES

The adsorption of organic substances on solid surfaces is not covered here, as it is a well-studied area that has been thoroughly investigated in connection with selective oxidation [2–7] and catalytic combustion [8–11]. The emphasis in this section is on the interactions of ozone with individual elements and surfaces. Two approaches have been taken in this area: the investigation of matrix-isolated ozone species and the study of adsorption complexes on oxides.

Matrix isolation studies have concentrated on reactions between ozone and alkali or alkaline earth metals and the study of products by infrared and resonance Raman spectroscopy [12–15]. The infrared spectra showed very intense bands near 800 cm^{-1} and weaker bands near 600 cm^{-1} that were assigned to the ν_3 and ν_2 bands of the ozonide ion according to their oxygen isotopic shifts [12]. The ozonide species was proposed to be symmetrically bound to the metal cations in a C_{2v} structure. The resonance Raman spectra of the matrix reaction products consisted of a strong fundamental near 1010 cm^{-1} for the alkali metals [13] and 1020 cm^{-1} for the alkaline earth metals [15], both with a regular progression of overtones with decreasing intensity. This fundamental was assigned to the ν_1 band of the ozonide ion in the $M^+O_3^-$ species. Similar results with polycrystalline KO_3 [16] and CsO_3 samples [17] were also reported.

Adsorbed ozone species have been studied by electron paramagnetic resonance (EPR), infrared (IR) spectroscopy, and kinetic measurements. In some cases, ozone does not remain in molecular form and dissociates to atomic or diatomic oxygen species. A surface ozonide ion (O_3^-) has been identified by its EPR signal [18–21]. The spectrum shows characteristic features at $g_1 = 2.0147$, $g_2 = 2.0120$, and $g_3 = 2.0018$.

Ozone adsorption at 77 K on a series of oxides was studied with Fourier transform infrared (FTIR) spectroscopy [22–25]. On SiO_2 [22], TiO_2 [23],

Al_2O_3, ZrO_2, MgO, and CeO_2 [24], ozone was found to form hydrogen bonds with the surface OH groups via one of the terminal oxygen atoms. In these studies, it was suggested that the ozone molecule acted as a Lewis base, adsorbing on weaker Lewis acid sites to form a coordination complex and decomposing on stronger sites on TiO_2, Al_2O_3, or ZnO to form atomic oxygen. This interpretation was supported by the finding that preadsorbed acetonitrile or pyridine considerably decreased the amount of strongly bonded O_3 species and significantly inhibited ozone decomposition. This interpretation of ozone as a Lewis base is in conflict with the common view that ozone is an electrophile and hence more like a Lewis acid. A scale of basicity has, however, been established for various adsorbents interacting with the acidic OH groups of HY and MFI zeolites [26]. Based on shifts of the IR band for OH, basicity was deduced to increase in the order $O_2 < N_2O < COS < O_3 < CO < C_2H_4$. The results here are strongly influenced by the interactions of the molecules with the zeolites. In fact, studies with gas-phase species using Fourier transform ion cyclotron resonance mass spectroscopy indicate a different order of basicity: $COS > O_3 > CO$ [27]. The measured gas-phase proton affinity was 35.4 ± 0.7 kJ mol^{-1} (148 ± 3 kcal mol^{-1}), in agreement with the results of two high-level theoretical studies that gave values of 35.7 kJ mol^{-1} (149.4 kcal mol^{-1}, CCSDT) [28] and 35.4 kJ mol^{-1} (148.0 kcal mol^{-1}, LCGTO-DF) [29].

The interaction of ozone with a MnO_2 surface has been examined in depth by *in situ* laser Raman spectroscopy and theoretical calculations [30]. The Raman measurements were carried out at reaction conditions as the ozone underwent continuous decomposition. A sharp feature at 884 cm^{-1} that increased with ozone concentration but decreased with increasing temperature or time was shown to be due to the reactive intermediate. It was established by transient and isotopic experiments that the intermediate was a peroxide species, O_2^{2-}. For example, coadsorption of $^{16}O_3$ and $^{18}O_3$ produced three Raman peaks due to the three possible surface species $^{16}O^{16}O$, $^{16}O^{18}O$, and $^{18}O^{18}O$. In a sequential experiment in which $^{18}O_3$ was first admitted and removed followed by $^{16}O_3$ introduction, however, only two bands due to $^{16}O^{18}O$ and $^{16}O^{16}O$ were produced. This is consistent with the reaction between an adsorbed oxygen atom and an ozone molecule and rules out the recombination of oxygen atoms. Theoretical calculations with an *ab initio* Hartree-Fock method gave excellent agreement for the isotopic shift in the vibrational frequency of the peroxide species. As discussed later, this peroxide species also plays a key role in the decomposition reaction and in the oxidation of ethanol on MnO_2.

The reactions of ozone with molecules preadsorbed on a surface has been exploited in two ways: the synthesis of ozonolysis products and the degradation of contaminant wastes. In these cases, the reactions occur between species physically adsorbed on the surface of a solid, and the processes are not catalytic.

An example of a synthesis application is the preparation of nitro compounds from the corresponding amines [31, 32]. The technique consists of adsorbing an amine onto silica gel, cooling to 195 K, and then passing 3 equivalents of ozone through the solid. The product is obtained by warming to room temperature and eluting the product. At these conditions, there is preferential attack of ozone on N centers rather than CH bonds and apparently the formation of dipolar amine complexes, $R{-}NH_2^{+}{-}O_3^{-}$, which go on to form the nitro compounds.

An example of a waste treatment application is the use of ozone to oxidize chloroaromatic compounds from flue gas adsorbed on a zeolite (Wessalith DAY). The reaction occurs at 323–363 K (50–90°C) and is enhanced by reaction time and water vapor [33]. The only product reported is CO_2, and the fate of chlorine is not reported.

12.3 LIQUID PHASE CATALYTIC OXIDATIONS

Ozonolysis of alkyl aromatics in the absence of catalysts proceeds by ozone attack on the sp^2 ring resulting in destruction of the aromatic ring [34]. In aqueous phase in the presence of cobalt or other metal salts [35], catalytic oxidation of methyl homologs of benzene occurs, resulting in transformation of the alkyl group to carboxylic acids [36, 37]. With multiple substitution such as in xylenes and mesitylene, oxidation of the methyl groups occurs stepwise [38, 39]. Ozone appears to function by oxidizing Co(II) to Co(III). Co(III) oxidizes the aromatic ring first to an aryl radical cation ($Ar^{+\bullet}$) [39], which probably rearranges to a benzyl cation in a subsequent process [38]. The critical steps are depicted below:

$$\mathrm{Ar} + \mathrm{Co(III)} \longrightarrow \mathrm{Ar}^{+\bullet} + \mathrm{Co(II)} \tag{12.1}$$

$$\mathrm{O_3} + \mathrm{Co(II)} \longrightarrow \mathrm{O_3}^{-} + \mathrm{Co(III)} \tag{12.2}$$

The ozonide ion reacts further by the following reactions:

$$\mathrm{O_3}^{-} + \mathrm{H}^{+} \longrightarrow \mathrm{O_2} + \mathrm{OH}\bullet \tag{12.3}$$

$$\mathrm{OH}\bullet + \mathrm{Co(II)} \longrightarrow \mathrm{OH}^{-} + \mathrm{Co(III)} \tag{12.4}$$

$$\mathrm{OH}^{-} + \mathrm{H}^{+} \longrightarrow \mathrm{H_2O} \tag{12.5}$$

The further reactions of $Ar^{+\bullet}$ occur with molecular oxygen to yield oxidation products and generate H^+, which balance Eqs. 3 and 4. A similar mechanism has been suggested for the cobalt-catalyzed oxidation of nitrotoluene in which a benzyl cation is the initial intermediate [40].

In some cases when solid catalysts are employed in liquid phase oxidations, the role of the catalyst may be to decompose the ozone to generate HO• hydroxyl radicals. This is similar to the role of H_2O_2 and UV radiation in advanced oxidation processes described in Chapter 11. The HO• radicals have rate constants for reaction with organic compounds on the order of 10^9–10^{10} L $mol^{-1}s^{-1}$ [41], which are 10–100 times larger than those of ozone itself [42]. The role of the catalyst in decomposing ozone has been cited in a number of studies using Fe oxide particles [43], Fe(III) supported on Al_2O_3 [44], TiO_2 [45], CuO, Fe_2O_3, Cr_2O_3, and NiO [46].

In other cases, solid catalysts work in a truly catalytic manner through the activation of both the organic substrate and ozone. In solution, the solid catalyst may be in equilibrium with a solubilized form that also participates in the reaction. Such a situation has been reported for the MnO_2-catalyzed oxidation of oxalic acid in aqueous solution [47], which is suggested to involve the following steps:

$$\rangle Mn^{III}OH_2{}^+ + C_2O_4{}^{2-} \rightleftharpoons \rangle Mn^{III}C_2O_4{}^- + H_2O \tag{12.6}$$

$$\rangle Mn^{III}C_2O_4{}^- + O_3 + 2H^+ \longrightarrow \rangle Mn^{III}OH_2{}^+ + 2CO_2 + O_2 \tag{12.7}$$

The last reaction is likely to proceed through the following steps:

$$\rangle Mn^{III}C_2O_4{}^- \longrightarrow \rangle Mn^{II} + 2CO_2 \tag{12.8}$$

$$\rangle Mn^{II} + O_3 \longrightarrow \rangle Mn^{III}O^- + O_2 \tag{12.9}$$

$$\rangle Mn^{III}O^- + 2H^+ \longrightarrow \rangle Mn^{III}OH_2{}^+ \tag{12.10}$$

Ozone has been used in the oxidation of dye effluents with ferric oxide/alumina catalysts [48] and the oxidation of chlorobenzene [43]. Catalytic ozonation of wastewaters by transition metals [49], TiO_2-SiO_2, Pt, Rh, lanthanum oxide catalysts [50], TiO_2-ZrO_2, and Pd [51] has been described. Ozone has also been used to oxidize propionitrile to cyanoacetic acid using heteropoly acid catalysts [52], to oxidize trifluoroethanol to fluoroethanediol using V_2O_5/TiO_2 [53], to oxidize hydrogen chloride to chlorine with cupric chloride and manganous sulfate [54], and to oxidize anisaldehyde to anisic acid using cobalt, copper, or vanadium catalysts [55]. Ozone has also been used as a single oxygen atom donor to generate high-valent oxometalloporphyrin complexes, which can be used for various catalytic oxidations [56, 57]. In this case, perbromo-substituted Mn and Fe porphyrins were used with ozone to epoxidize a number of olefins like cyclooctene and styrene [58]. The oxidation of the complexes was suggested to occur by a simple atom transfer:

$$(Por)M^{III} + O_3 \longrightarrow (Por)M^{V}{=}O + O_2 \tag{12.11}$$

Manganese oxide has been used extensively because of its high catalytic activity. The kinetics of the homogeneously catalyzed oxidation of formic acid in acetic acid by ozone with a Mn(IV) catalyst [59] and the oxidation of dioxin pollutants by transition metal oxides like MnO_2 and TiO_2 [60, 61] have been presented elsewhere.

12.4 GAS PHASE CATALYTIC OXIDATIONS

The heterogenous catalytic oxidation reactions of ozone have been applied to a variety of systems, some of which have been commercialized. Most of the studies have been empirical, and only a few investigations have addressed the mechanism of reaction. Table 12.1 summarizes the research that has been done on various systems. An analysis of the contents follows, with an emphasis on what is known about the critical steps in the oxidation reactions with ozone. Then, a summary of applications is presented.

Alcohols are an important constituent of VOC streams, and several studies have addressed their reactions. The low temperature oxidation of a number of lower aliphatic alcohols by ozone was studied [62] over two materials, SiO_2 (214 m^2g^{-1}) and γ-Al_2O_3 (261 m^2g^{-1}), which have relatively low activity for ozone decomposition but which have good adsorption properties for the alcohols. It was found that the selectivity, S, depended on the alcohol. Ethanol produced

TABLE 12.1 Summary of Heterogeneous Reactions of Ozone

Substrates	Catalysts	Reaction Conditions	Products	Rate[a] (mol $m^{-2}s^{-1}$)	Ref.
Ethanol, propanol, isopropanol	γ-Al_2O_3, SiO_2	293–363 K	Aldehydes, ketone, CO_2	1.7×10^{-9b}	62
Ethanol	MoO_3/Al_2O_3	300–550 K	Acetaldehyde	0.9×10^{-9b}	63
	MnO_2/Al_2O_3		CO_2	1.9×10^{-9b}	
Methane	MgO Li/MgO	350–720 K	CO_2	1×10^{-5}	64, 65
n-Heptane *n*-Hexane	3 CO_3O_4: 1 MnO_2	393–493 K	CO_2	—	66
Benzene	MnO_2	283–353 K	CO, CO_2	1.5×10^{-9b}	67
Styrene, benzene, toluene, cyclohexane	$BaCuCrO_x$	323–723 K	Not specified	—	68
Dichloromethane, tetrachloro-ethylene, *p*-chlorotoluene	Pt/Al_2O_3 Pd/Al_2O_3 $BaCuCrO_x$	323–473 K	Mostly CO_2	—	69
CO	CeO_2	322–357 K O_3 0.3 mol m^{-3} CO 0.08–0.2 mol m^{-3}	CO_2	4.6×10^{-10b}	70
CO	MnO_2	322–357 K O_3 0.3 mol m^{-3} CO 0.08–0.2 mol m^{-3}	CO_2	9.4×10^{-8b}	71
Ozone	MnO_2	283–353 K	O_2	1.1×10^{-7b}	67
Ozone	MnO_2	343 K	O_2	2.3×10^{-7b}	67
Ozone	MnO_2/Al_2O_3	313 K	O_2	8.5×10^{-9}	72
Ozone	Carbon	298 K	CO/CO_2	1.9×10^{-10}	73
Ozone	Carbon	298 K	CO/CO_2	4.5×10^{-9}	74
Ozone	Carbon	298 K	CO/CO_2	5.6×10^{-9}	75
Ozone	Carbon	298 K	CO/CO_2	6.8×10^{-11}	76
SO_2	Al_2O_3, SiO_2 AC	292 K 290 ppm SO_2	SO_3	—	77

[a]Typical rates, normalized to the surface area of the support for the supported samples (assumes high dispersion). Total pressure is 1 bar, except where specified. AC, activated carbon.
[b]Temperature = 343 K, or middle of the range.

mostly acetaldehyde (S = 50–90%), *n*-propanol produced mostly propionaldehyde (S = 30–40%), and *i*-propanol produced acetone (S = 20–40%), with the balance being CO_2 in all cases. As expected, ozone was essential for any reaction to occur at the low temperatures (293–358 K, 20°–90°C) employed in the study, and larger amounts of ozone produced more CO_2. Typical rates on the Al_2O_3 catalyst (1.5 g) at concentrations of $[ROH] = 8.5 \times 10^{-4}$ M and $[O_3] = 0.5 \times 10^{-4}$ M were relatively invariant with flow rate (100–700 L/h) and were $r_{ROH} = 1.7 \times 10^{-9}$ mol $m^{-2}s^{-1}$ and $r_{O3} = 0.5 \times 10^{-9}$ mol $m^{-2}s^{-1}$. The rates were approximately first order with ozone and were first order in alcohol at low concentrations and zero order at high concentrations. These results were consistent with a reversible adsorption of alcohol followed by a rate determining reaction with ozone:

$$ROH + S \underset{k_{-1}}{\overset{k_1}{\rightleftharpoons}} ROH \cdot S \tag{12.12}$$

$$ROH \cdot S + O_3 \xrightarrow{k_2} S + Products \tag{12.13}$$

The kinetics could be described by the equation

$$r = \frac{k_1 k_2 [S[ROH]O_3]}{k_1[ROH] + k_{-1} + k_2[O_3]} \tag{12.14}$$

with $k_2 \ll k_{-1} < k_1$.

In related studies [63], the oxidation of ethanol was compared over MoO_3/Al_2O_3 and MnO_2/Al_2O_3, which are respectively poor and excellent ozone decomposition catalysts. As in the previous study, it was found that the use of ozone as an oxidant opened up new low-temperature reaction channels with much higher rate and lower activation energy than the oxidation with oxygen (Fig. 12.1). With ozone, the rate obtained for the MnO_2/Al_2O_3 catalyst at 363 K was 1.9×10^{-9} mol $m^{-2}s^{-1}$, while the rate on the MoO_3/Al_2O_3 catalyst at 438 K was 0.9×10^{-9} mol $m^{-2}s^{-1}$, both similar to the rates reported earlier [62] for the Al_2O_3 support. The selectivity with MoO_3, however, was about 80% to acetaldehyde and 20% to CO_2, while with MnO_2 it was 100% to CO_2, demonstrating the strong effect of the catalyst. Laser Raman spectroscopic measurements carried out simultaneously with reaction [63] allowed quantitation of the surface concentrations of the major reactive species, which were surface peroxo ($O_2{}^{2-}$) and ethoxy groups. The coverage in these species was smaller than those measured at the steady state with the separate reactants in ozone decomposition and ethanol oxidation with O_2. This indicated that a fast surface reaction was consuming the reactants when they were both present on the surface.

$$O_3 + 2^* \longrightarrow O_2{}^* + O^* \tag{12.15}$$

$$EtOH + 2^* \longrightarrow EtO^* + H^* \tag{12.16}$$

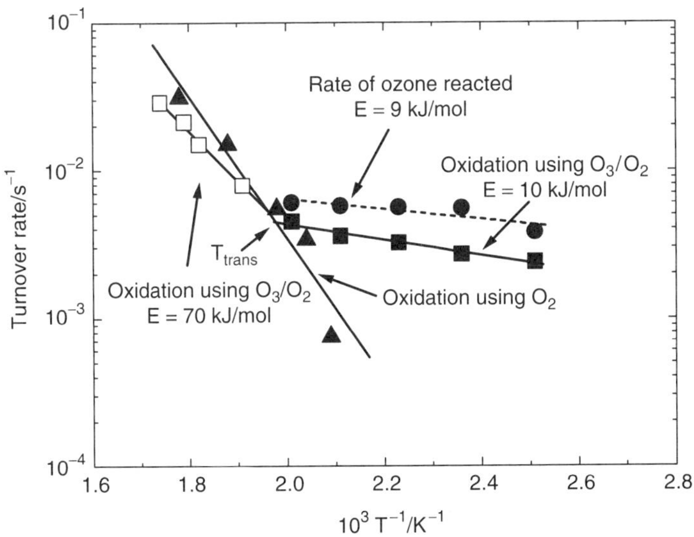

Figure 12.1 Arrhenius plots of ethanol oxidation with oxygen and ozone.

$$EtO^* + O_2{}^* \longrightarrow CO_2 \quad (12.17)$$

$$2O^* \text{ (or } O_2{}^*) \longrightarrow O_2 + {}^* \text{ (or } 2^*) \quad (12.18)$$

$$O^* \text{ (or } O_2{}^*) + 2H^* \longrightarrow H_2O \quad (12.19)$$

In these equations, * represents surface sites. The decomposition of ozone is closely associated with the oxidation of ethanol, as both occur simultaneously on the surface. The decomposition reaction provides active oxygen species that allow the oxidation to proceed at lower temperatures.

The conversion of aliphatic compounds poses a special challenge, as their reactivity in direct reactions with ozone is low (see Chapter 10). Ozone can cause reaction of such compounds by initiating free radical chain reactions, but in environmental cleanup applications the concentration of hydrocarbons is low and at such conditions chains will not propagate. The use of catalysts offers an alternative pathway, the possibility of producing active surface oxygen species from ozone that can interact with the hydrocarbons at low temperatures. Relatively little work has, however, been carried out in this area.

The reactions of methane [78] and ethane [79] with ozone in the absence of catalysts have been studied, but mostly from the standpoint of producing oxygenated products. Pure hydrocarbons were contacted with 5–10% oxygen containing 5% ozone in empty tubes at long residence times (300–600 s). It was surmised that ozone decomposed thermally to form oxygen atoms that initiated free radical chain reactions. The main products (S = 30–90%) were oxygenates (methanol, formaldehyde, ethanol, and acetaldehyde) and were obtained mostly at low temperatures (<523 K, 250°C) at conversions <10%.

The reactions of methane in the presence of MgO and Li/MgO catalysts have been the subject of a number of investigations [64, 65, 80]. In oxygen at high temperatures (>973 K), these catalysts are known for their activity in methane coupling [81, 82], and this behavior was duly confirmed [64]. At low temperatures (<523 K), activity in oxygen was virtually nil, but was substantial with ozone. For both catalysts at intermediate contact times (5 s) and low temperatures with an ozone in oxygen concentration of 5%, the main products were H_2 and CO_2, with no oxygenates formed [64]. At high temperatures the ozone decomposed, and the same products as with oxygen were obtained. At short contact times (0.05 s) and the same low temperatures with an ozone in oxygen concentration of 2%, CO and CO_2 were observed (H_2 was not analyzed) [65]. At higher temperatures as much as 80–90% formaldehyde was formed. These results are consistent with a combined surface/gas phase mechanism where ozone forms an active oxygen species (probably the radical O^-) on the MgO or Li/MgO surface. This species reacts with methane to release methyl radicals to the gas phase [81, 82] which then initiate chain reactions.

$$O_3 + {}^* \longrightarrow {}^*O^- + O_2 \quad (12.20)$$

$$CH_4 + {}^*O^- \longrightarrow {}^*OH + CH_3\bullet \quad (12.21)$$

$$CH_3\bullet + O_2 \longrightarrow CH_2O + HO\bullet \quad (12.22)$$

$$CO + HO\bullet \longrightarrow CO_2 + H\bullet \quad (12.23)$$

$$2H\bullet \longrightarrow H_2 \quad (12.24)$$

In support of the above mechanism, the activation energy for the methane oxidation reaction (74 kJ mol^{-1}) was found to be close to that of ozone decomposition (80 kJ mol^{-1}) [65]. This would be the expected result if the first step that forms O^- were the rate-determining step. The rate on both MgO and Li/MgO was high, close to 1×10^{-5} mol $m^{-2}s^{-1}$ (turnover rate $\sim$0.6 s^{-1}) at room temperature and decreased with increasing temperature, as ozone decomposed [65]. More work with other materials is needed to find catalysts that operate by an entirely surface mechanism.

Next to aliphatic compounds, aromatics are the most unreactive molecules in direct reactions with ozone (Chapter 10). The catalytic oxidation of benzene on an unsupported MnO_2 of surface area 74 m^2g^{-1} was studied in the temperature range 283–353 K (10°–80°C) with an ozone concentration of 0.20 mol m^{-3} and a benzene concentration of 0.013 mol m^{-3} [67]. A benzene oxidation rate of 1.5×10^{-9} mol $m^{-2}s^{-1}$ was found at 343 K. The activation energy was 30 kJ mol^{-1}, and the order in ozone was 1.0 and in benzene 0.4. In the same oxidation with O_2, the reaction order in benzene was 0.9 and the activation energy was 88 kJ mol^{-1}. The rate of ozone decomposition at the same conditions as the benzene oxidation was 1.1×10^{-7} mol $m^{-2}s^{-1}$, and the activation energy was 32 kJ mol^{-1}. The similarity in the activation energy with that of the oxidation suggested that ozone decomposition was involved in the mechanism of benzene oxidation.

The oxidation of VOCs such as styrene, toluene, benzene, and cyclohexane has been studied with a commercial copper chromite catalytic combustion catalyst in laboratory scale and pilot plant units [68]. Considerably better performance was obtained with ozone than with oxygen, but the results depended on VOC concentration. For example, using air in the oxidation of toluene at concentrations of 1700, 550, and 250 ppm, the temperatures necessary for reduction of the exit concentration to a value lower than the threshhold limit value (TLV) of 100 ppm [83] were 723, 653, and 463 K (450°, 380°, and 190°C). These were lowered by 10%, 30%, and 70%, respectively, with ozone. These results can be understood from the ozone concentration employed, which was 1–2 wt% in air or 60–120 ppm. Thus, the concentration of ozone has to be on the order of the VOC concentration for catalytic oxidation to proceed. It was also found in the study that the oxidation of a compound in a mixture became more difficult than with the pure compound alone [68]. This is probably due to competitive adsorption.

In a related VOC oxidation study [69], platinum and palladium catalysts supported on alumina at high (PTH and PDH) and low (PTL and PDL) concentrations were compared with a barium-promoted copper chromite (BCC) catalyst. It was found that for typical VOCs (acetone, toluene) activity with ozone was highest in the temperature range 323–473 K (50°–200°C) and that the order in activity was PTH > PDH > PDL > PTH > BCC. Noble metal combustion catalysts are known to be poisoned by halogens [84]. The BCC catalyst was, however, also effective in the ozone oxidation of chlorinated hydrocarbons (e.g., dichloromethane, tetrachloroethylene, *p*-chlorotoluene), and produced mostly HCl and HClO. Importantly, in mixed VOC feeds it produced extremely low levels of hazardous products such as chlorinated dibenzodioxins and dibenzofurans (three orders of magnitude below permissible levels).

The oxidation of *n*-heptane and *n*-octane in the presence of ozone has been carried out with various catalysts in the temperature range 393–573 K (120°–300°C) [66]. It was found that a mixture of Co_3O_4 and MnO_2 in a weight ratio of 3 to 1 supported on kieselguhr (6 wt%) was the most active composition, superior to Al_2O_3 and considerably better than a 0.4-nm molecular sieve. The primary product was CO_2. At temperatures below 333 K (160°C), the hydrocarbon concentration had an effect on the conversion, as in the case of the toluene oxidation above [68], because the hydrocarbon concentration was higher than the ozone concentration. Above that temperature, there was little effect because the ozone decomposed [85, 86].

The oxidation of hydrazine on Al_2O_3 or SiO_2 has also been reported [87]. Cobalt catalysts in the form of supported cobalt oxide on alumina have been used for the gas phase oxidation of xylene [88]. Ozone (640 ppm, 0.064%) has been employed for the oxidation of SO_2 (290 ppm) to SO_3 at 292 K [77]. It was reported that Al_2O_3, SiO_2, and activated carbon were the most active catalysts, while V_2O_5, Pt/Al_2O_3, and Pd/Al_2O_3 were poorly active. These results are probably equivocal, however, because SO_3 in the effluent was not analyzed directly, but instead SO_4^{2-} remaining adsorbed on the catalysts was quantitated.

The oxidation of carbon monoxide is an important final step in the destruction of VOCs. Several studies have addressed this problem using CaO/Al_2O_3 [89], CeO_2 [70], and MnO_2 [67]. Cerium dioxide is a component of automobile three-way catalysts and acts to store oxygen at lean conditions [90–93]. The oxidation was carried out with O_3 at a concentration of 0.33 mol m^{-3} and CO in the range 0.08–0.16 mol m^{-3} at 322–357 K (49–84°C). At 343 K, the rate was 4.6×10^{-10} mol $m^{-2}s^{-1}$. Electron spin resonace spectra reveal the presence of a species with anisotropic signals ($g_1 = 2.0124$, $g_2 = 2.0048$, $g_3 = 2.0003$, $g_{ave} = 2.0058$) that was attributed to O^{3-}. This species was inactive, however, and its signal remained unchanged with time of treatment in ozone. Although not directly observed, the active species was suggested to be an O^- radical that reacted in the following manner:

$$Ce^{4+}O^- + CO \longrightarrow Ce^{3+} + CO_2 \qquad (12.25)$$

A similar scheme involving O^- was suggested for CO oxidation on MnO_2 [67]. At the same conditions as in the previous study, the oxidation rate on MnO_2 was substantially higher, 9.4×10^{-8} mol $m^{-2}s^{-1}$. With O_2 instead of O_3 as the oxidant, the rate was only 5.8×10^{-10} mol $m^{-2}s^{-1}$. The rate of decomposition of O_3 was 2.3×10^{-7} mol $m^{-2}s^{-1}$ and was slightly higher than the rate of oxidation, suggesting that the two reactions were linked.

12.5 GAS PHASE CATALYTIC OZONE DECOMPOSITION

As was noted in the previous sections, the catalytic decomposition of ozone is intimately tied to the oxidation chemistry. The catalytic decomposition has been extensively studied because of ozone's toxic properties and the need to remove it from human surroundings. The subject of catalytic decomposition of ozone has been recently reviewed [72]. Ozone decomposition is utilized in a number of situations. For example, ozone generated by photocopiers or drawn into aircraft cabins [94, 95] from the lower stratosphere is decomposed by catalytic filters. Similarly, discharges from sterilization [96], deodorization [97], and waste treatment units [98, 99] must have their ozone levels reduced. A recent development that is undergoing evaluation is the PremAir system [100, 101] for the large-scale cleanup of outdoor air. The concept is to catalytically destroy ozone in the air that enters through the radiator section of vehicles by having the radiator coated with a Pd catalyst.

Most of the work on ozone decomposition catalysts is reported in the patent literature [72]. The catalytic materials selected are usually noble metals like Pd or Pt or oxides of transition metals such as Mn, Co, Ni, and Ag. These components are often supported on materials such as γ-Al_2O_3, TiO_2, SiO_2, zeolites, activated carbon (AC), or a combination of these. For example, a filter to remove ozone from aircraft cabin air consists of Pd/Al_2O_3 [102, 103], while one to treat the effluent of photocopiers is MnO_2/AC [104, 105]. There are numerous challenges to be met in the development of new catalysts because the conditions for their use vary substantially. For instance, the air stream to be treated may contain particulates like dust or toner particles, humidity may fluctuate over wide values, and operating temperatures may vary over considerable ranges (typically from 313 K for photocopiers to 423 K for aircraft compressor effluents). Presently, some catalysts experience a long-term decline in activity. Development of catalysts of lighter weight, higher activity, and lack of deactivation are important goals in catalyst improvement.

Considerable information is now available on the mechanism of the decomposition. In an early investigation on Ag_2O [106], it was found that *p*-type oxides were active in the decomposition and that the mechanism probably involved a charged species such as superoxide (O_2^-). The correlation between activity and *p*-type conductivity in oxides was confirmed [72] through Hall effect measurements (Fig. 12.2).

The kinetics and mechanism have been thoroughly investigated on MnO_2, the most active of the transition metal oxides. The rate has been found to be independent of O_2 and H_2O partial pressure and to be fractional order in O_3 [107]. Transient and steady-state measurements of the rate coupled with *in situ* laser Raman spectroscopy and isotopic substitution experiments have confirmed that the active intermediate is a peroxide species, O_2^{2-} [30, 108].

$$O_3 + {}^* \longrightarrow O^* + O_2 \tag{12.26}$$

$$O_3 + O^* \longrightarrow O_2^* + O_2 \tag{12.27}$$

$$O_2^* \longrightarrow {}^* + O_2 \tag{12.28}$$

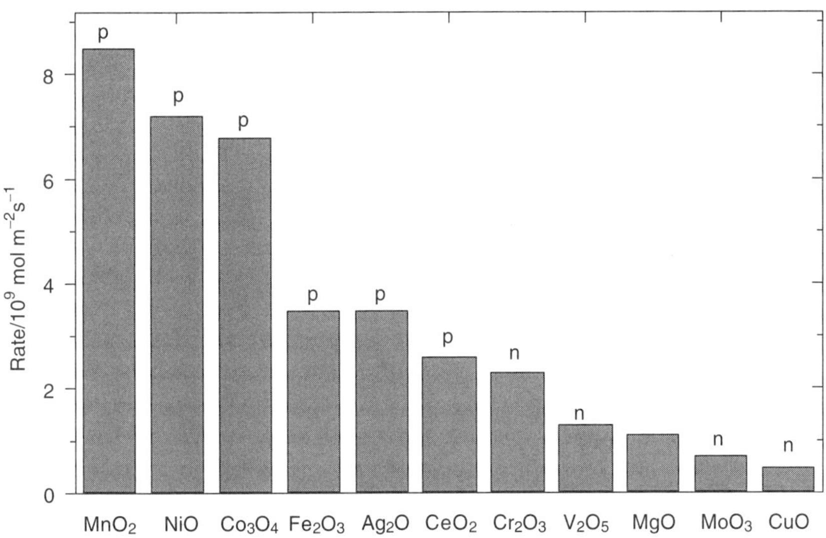

Figure 12.2 Catalytic activity of *p*- and *n*-type oxides in ozone decomposition. (Data from Dhandapani and Oyama [72].)

In these reactions * represents surface sites, and O_2^* is the peroxide species. The first reaction is fast so that there are few empty sites on the surface. The two other reactions are slow, and together they determine the coverage in O_2^* [108].

12.6 REACTIONS WITH CARBON

Activated carbons have traditionally been used to purify aqueous and gaseous streams by adsorption [109, 110]. This is primarily due to the very high surface areas and porous structures that give activated carbons a large adsorptive capacity. In fact, activated carbons with BET surface areas of over 3000 m^2g^{-1} have been processed and are commercially available [111]. A number of studies have been carried out on the interaction of ozone with carbon [73, 112, 113]. There is an indication that ozone oxidizes carbon at room temperature to CO and CO_2 [74, 75, 114]. The surface of graphite has been studied by scanning tunneling microscopy following oxidation by oxygen and ozone to determine the nature of reactive sites [115].

The interaction of ozone and carbon has been described in a number of investigations. A set of studies was carried out with fibrous carbons with microporosities in the range of 0.3–0.4 cm^3 g^{-1} and mesoporosities in the range of 0.1–0.2 cm^3 g^{-1} [112, 116, 117]. Good decomposition activity was claimed, but surface areas were not reported and areal rates cannot be calculated. In another study [73], commercially available reticulated vitrious carbon in the form of low-surface area, free standing foams were found to have low activity for ozone decomposition. Emulsion-derived foams made from resorcinol-formaldehyde polymers were, however, found to display high efficiencies in ozone decomposition. A limitation of these materials was that they produced high back pressures. Attempts to introduce straight channels by stamping and molding techniques resulted in the collapse of the emulsion microstructure and were only partly successful. An areal rate at room temperature of 1.9×10^{-10} mol $m^{-2}s^{-1}$ can be calculated at a flow rate of 8 l min^{-1}, assuming that the 5 mm thick filter used weighed 1 g.

In an investigation of coconut- and coal-based charcoals, it was found that samples exposed to ozone initially underwent weight increases and then decreases [75]. This was attributed to the formation of surface compounds that subsequently decomposed to O_2, CO, and CO_2. An areal rate at room temperature of 5.6×10^{-9} mol $m^{-2}s^{-1}$ can be calculated from the reported data [75]. These findings were largely confirmed by an investigation of the kinetics of the reaction on commercial charcoal [74]. It was found that the ozone loss rate reached a steady-state value after an amount corresponding to a monolayer of adsorbed ozone molecules was adsorbed on the carbon sample. A mass balance showed that for every ozone molecule reacted on the surface, an oxygen molecule was formed. The odd oxygen went on to form CO and CO_2. This sequence was accompanied by a slower process involving surface diffusion/reaction of an active oxygen species. It was concluded that no catalytic reaction occurred. An areal rate at room temperature of 4.5×10^{-9} mol m^{-2}s^{-1} can be calculated from this investigation.

The ozone decomposition rate at 313 K on pure carbons has been measured on two carbons of different surface areas [76]. Rates were determined to be 3.5×10^{-11} mol m^{-2}s^{-1} for Maxsorb (3100 m^{2}g^{-1}) and 6.8×10^{-11} mol m^{-2}s^{-1} for Darco (1600 m^{2}g^{-1}). Differences in areal rate with those obtained from the other studies cited above can be attributed to slight differences in conditions (pressure, ozone concentration, flow rate) as well as to differences between carbon samples. Nevertheless, the rates are on the same order of magnitude and lower than what is found on oxide catalysts.

12.7 TECHNOLOGY FOR CATALYTIC OZONE TREATMENT OF POLLUTED AIR STREAMS

In the catalytic treatment of air streams containing small concentrations (<2000 ppm) of volatile organic compounds, the pollutants are activated chemically on the surface of the catalyst. At the same time, in practical applications large volumes of air are generally treated, and this requires high flow rates. For this reason, the geometric form of the catalysts needs to be designed so that there is good contact between the fluid phase and the solid while generating minimal pressure drop.

The available physical forms of catalysts are pellets, fibers, corrugated structures, monoliths, and foams. Typical examples of monoliths, foams, and pellets are shown in Figure 12.3. Pellets are usually obtained by extrusion and come in a variety of forms, including spherical, cylindrical, and ring shapes. Fibers are made from filamentous materials and can be woven into mats. Corrugated structures are made of deformable materials in sheet form (cardboard or metal) that are bent to form parallel passages. Monoliths are unitary structures resembling honeycombs that provide parallel straight-through pore shapes. Foams are unitary structures composed of open cells that form a three-dimensional network of passages. The active catalytic material can be prepared separately and then made into any of the above-mentioned forms, or it can be impregnated on these substrates.

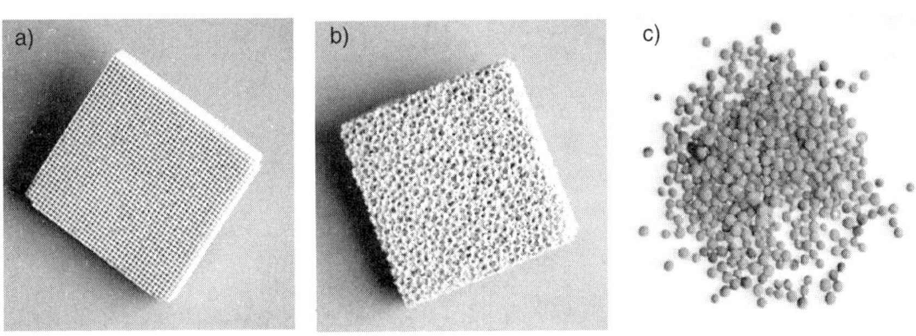

Figure 12.3 Typical geometric shapes for catalysts: **a**, monolith, **b**, foam, **c**, pellets.

The pellets and fibers result in substantial pressure buildup at high flow rates, while the corrugated structures, monoliths, and foams are able to accommodate high gas throughput with little resistance. Corrugated cardboard suffers from problems of distortion caused by the high moisture content of ambient air. Because of these reasons, monolithic forms have been the predominant geometry employed in commercial products. The foam geometry is an emerging technology that may offer advantages in gas–solid contact. Metal substrates (Al, Ti, and stainless steel) are also used as catalytic filters in some cases [118–120].

In a comparative study of pellets and monolith structures [121], it was found that important parameters in the comparison were mass transfer, heat transfer, and pressure drop effects. It was found that for the same bed height and flow rate, the pressure drop for a monolith structure was two to three orders of magnitude lower than pellets. A simple procedure for calculating mass transfer in monoliths has been described [95]. It was found that a reactant molecule will experience tens to hundreds of collisions with the monolith walls at typical residence times employed in air treatment applications. In certain laminar flow regimes, however, contact between gas and substrate may be limited by diffusion to the walls. An alternate form that addresses this mass-transfer problem is the open cell foam.

A comparison of pressure drop in pellets, monoliths, and foams is illustrated in Figure 12.4 [72]. These were measured while the geometric surface area was kept constant at 620 cm^2 for the three substrates. As expected, the pressure drop for the pellets was much higher than for the monolith or foam for the same flow rates. The monolith and foam, however, have very similar gas resistances. Thus, the use of these substrates can allow the handling of large gas flow rates.

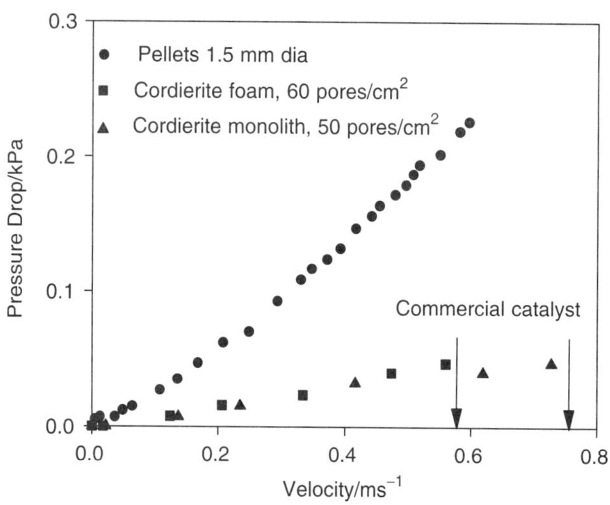

Figure 12.4 Comparison of pressure drop for pellets, foam, and monolith geometry (equal surface area basis).

A technology has been developed for the deodorization of smells in toilets or refrigerators involving the catalytic oxidation of waste gases with ozone [122–124]. Units containing this technology were developed by Toto, Inc., in Japan and are widely used. The technology is incorporated into the appliance itself as a unit that draws air in with a fan, generates ozone by corona discharge [125, 126], and passes the contaminated air stream over a catalyst [127]. A variety of catalysts have been reported, including naphthohydroquinones and metal catalysts dissolved in alkaline liquids [128], manganese oxide [129, 130] or nickel oxide in conjunction with aluminum oxide [131], manganese oxide with nickel hydroxide or carbonate [132], manganese oxide with copper oxide [133], manganese oxide with iron and platinum on monoliths [134], alkaline earth oxides or hydroxides in conjunction with transition metal oxides [135], metal peroxides on activated carbon [136], and metal sulfates [137].

Another technology commercialized in Japan by Hitachi [138] is an air cleaner purifier called Dashpure. The device draws in air and passes it through a corona discharge-type ozone generator and then directly to a catalytic bed for oxidation and ozone decomposition (Fig. 12.5). Because the impure stream passes directly through the ionizing region, radical and ionic gas-phase reactions are likely to assist the destruction of organic compounds. The device is highly effective not only for deodorization, but also for disinfection (destruction of bioorganisms). It incorporates a filter for particulate removal. Catalytic reaction systems that incorporate a discharge have been the subject of limited studies because of their complicated nature [139, 140]. For example, the hydrogenation of CO_2 in an electric discharge has been investigated in the presence and absence of a catalyst [141, 142]. Both experiments and numerical simulation showed that CH_4 and CO are formed without a catalyst, while with a CuO/ZnO catalyst small amounts of methanol are produced.

The combustion of VOCs has been studied in a pilot plant unit designed for catalytic combustion that could be employed with or without ozone [68]. The plant had a capacity of 100 Nm^3h^{-1} and was operated at temperatures of 323–723 K (50°–450°C) with space velocity of 20,000 h^{-1}. The design (Fig. 12.6) incorporated features such as heat exchangers that could be useful in a purely ozone application, using a modest amount of heating (50–100 K). An

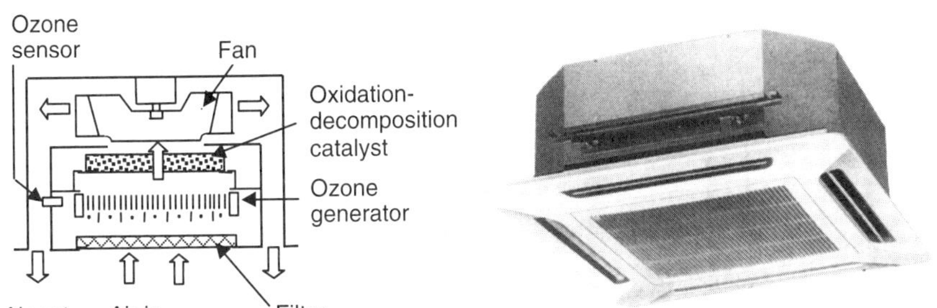

Figure 12.5 Dashpure ozone air cleaner.

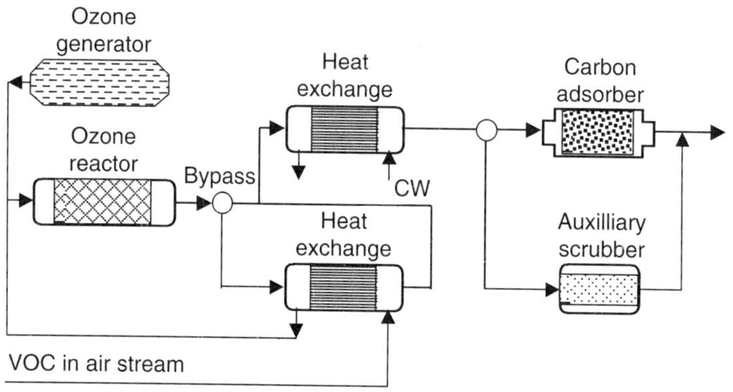

Figure 12.6 Schematic of an ozone VOC oxidation plant (Adapted from Ragaini, et al, [68].)

auxilliary scrubber could beused to remove acidic halogen species. Substantially better performance was obtained with ozone over oxygen in the conversion of such VOCs as styrene, toluene, benzene, and cyclohexane at temperatures below 523 K (250°C).

A complete cost analysis for the system has not been published. Comparing the energy needed to generate ozone (5–10 Wh/m^3 air) with the energy savings resulting from lower reaction temperatures, however, it has been estimated that an amount of 10–15 Wh/m^3 of energy could be saved depending on the type of VOC oxidized [69].

The treatment of indoor air streams containing ppb levels of VOCs has been studied in a (50 cfm) pilot plant and a (3500 cfm) demonstration unit [143]. The units were operated at room temperature and atmospheric pressure and normal relative humidity (25–75%) using ozone in the concentration range 1–10 ppm. The systems had two catalyst beds in series, the first designed for hydrocarbon

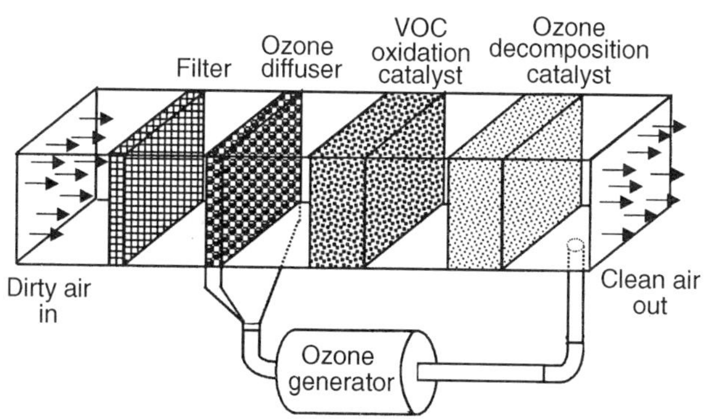

Figure 12.7 Schematic of the UCAIR indoor air pollution abatement process.

oxidation and the second for ozone decomposition. The technology was developed by Union Carbide and was named UCAIR, and it was reported to be effective in the reduction of concentration of a wide variety of VOCs, CO, and ozone, as well as microbiological aerosols.

In conclusion, the use of ozone in conjunction with catalysts opens up new reaction pathways that result in complete oxidation of organic compounds. The catalytic activity is high, and temperatures are considerable lower than those without catalysts. New technologies are being developed for the purification of contaminated air streams.

REFERENCES

1. S.J. Masten; S.H.R. Davies, "The Use of Ozonization to Degrade Organic Contaminants in Wastewaters," *Environ. Sci. Technol.*, **1994**, 28, 181A.
2. M. Che; E. Giamello, "Electron Paramagnetic Resonance-Principles and Applications to Catalysis" in *Catalyst Characterization: Physical Techniques for Solid Materials*, B. Imelik, J.C. Vedrine, Eds., Plenum, New York, 1994 Chp 6, pp. 131–180.
3. A.A. Davydov, *Infrared Spectroscopy of Adsorbed Species on the Surface of Transition Metal Oxides*, Wiley, New York, 1984.
4. H.H. Kung, *Transition Metal Oxides: Surface Chemistry and Catalysis*, Elsevier, Amsterdam, 1990.
5. R.K. Grasselli; S.T. Oyama; A.M. Gaffney; J.E. Lyons, Eds., *3rd World Congress on Oxidation Catalysis, Studies in Surface Science and Catalysis*, Vol. 110, Elsevier, Amsterdam, 1997.
6. B.K. Warren; S.T. Oyama, Eds., *Heterogeneous Hydrocarbon Oxidation*, ACS Symposium Series 638, American Chemical Society, Washington, D.C., 1996.
7. S.T. Oyama; J.W. Hightower, *Catalytic Selective Oxidation*, ACS Symposium Series 523, American Chemical Society, Washington, D.C., 1993.
8. J.J. Spivey, "Complete Catalytic Oxidation of Volatile Organics," *Ind. Eng. Chem. Res.*, **1987**, 26, 2165.
9. M.F.M. Zwinkels; S.G. Järås; P.G. Menon; T.A. Griffin, "Catalytic Materials for High-temperature Combustion," *Catal. Rev. Sci. Eng.*, **1993**, 35, 319.
10. L.D. Pfefferle; W.C. Pfefferle, "Catalysis in Combustion," *Catal. Rev. Sci. Eng.*, **1987**, 29, 219.
11. R. Prasad; L.A. Kennedy; E. Ruckenstein, "Catalytic Combustion," *Catal. Rev. Sci. Eng.*, **1984**, 26, 1.
12. R.C. Spiker; L. Andrews, "Matrix Reactions of Alkali Metal Atoms with Ozone: Infrared Spectra of the Alkali Metal Ozonide Molecules," *J. Chem. Phys.*, **1973**, 59, 1851.
13. L. Andrews; R.C. Spiker, *J. Chem. Phys.*, **1973**, 59, 1863.
14. L. Andrews; R.C. Spiker, "Resonance Raman Spectrum and Vibrational Analysis of the Ozonide Ion in the Argon Matrix-Isolated $M^+O_3^-$ Species," *J. Phys. Chem.*, **1972**, 76, 3208.

15. B. Ault; L. Andrews, "Argon Matrix Raman and Infrared Spectra and Vibrational Analysis of Ozone and the Oxygen-18 Substituted Ozone Molecules," *J. Mol. Spectrosc.*, **1977**, 62, 437.
16. A.I. Karalin; V.Y. Rosolovsky; C.A. Tokareva, "Visible Absorption and Resonance Raman Spectra of Alkaline Earth Metal Ozonides in Argon and Nitrogen Matrices at 10 K," *Volnov Dokl. Akad. Nauk. SSSR Ser. Fiz. Khim.*, **1972**, 206, 641.
17. J.B. Bates; M.H. Brooler; G.E. Boyd, *Chem. Phys. Lett.*, **1972**, 76, 3208.
18. M. Che; A.J. Tench, "Characterization and Reactivity of Mononuclear Oxygen Species on Oxide Surfaces," *Adv. Catal.*, **1982**, 31, 77.
19. A.J. Tench; T. Lawson, "The Formation of O^- and O_3^- Adsorbed on an Oxide Surface," *Chem. Phys. Lett.*, **1970**, 7, 459.
20. A.J. Tench; T. Lawson; J.F.J. Kibblewhite, "Oxygen Species Adsorbed on Oxides Part 1. Formation of (O^-) on MgO," *J. Chem. Soc. Faraday Soc. I*, **1972**, 68, 1169.
21. C. Naccache; M. Che, *Proc. Int. Cong. Catal.*, **1973**, 2, 1389.
22. K.M. Bulanin; A.V. Alexeev; D.S. Bystrov; J.C. Lavalley; A.A. Tsyganenko, "IR Study of Ozone Adsorption on SiO_2," *J. Phys. Chem.*, **1994**, 98, 5100.
23. K.M. Bulanin; J.C. Lavalley; A.A. Tsyganenko, "Infrared Study of Ozone Adsorption on TiO_2 (Anatase)," *J. Phys. Chem.*, **1995**, 99, 10294.
24. K.M. Bulanin; J.C. Lavalley; A.A. Tsyganenko, "IR Spectra of Adsorbed Ozone," *Coll. Surf. A*, **1995**, 101, 153.
25. K.M. Bulanin; J.C. Lavalley; A.A. Tsyganenko, "Infrared Study of Ozone Adsorption on CaO," *J. Phys. Chem. B*, **1997**, 101, 2917.
26. L. Mariey; J. Lamotte; P. Hoggan; J.-C. Lavalley; K. Boulanine; A. Tsyganenko, "On the Basicity of Ozone", *Chem. Lett.*, **1997**, 835.
27. F. Cacace; M. Speranza, "Concerning the Basicity of Ozone", *Chem. Lett.*, **1998**, 419.
28. C. Meredith; G.E. Quelch; H.F. Schaefer III, "Open-chain and Cyclic Protonated Ozone: The Ground-State Potential-Energy Hypersurface," *J. Am. Chem. Soc.*, **1991**, 113, 1187.
29. M. Toscano; N. Russo; J. Rubio, "Calculation of Proton Affinities and Absolute Gas Basicities of X_3 Group VI Triatomics: A Density Functional Study," *J. Chem. Soc. Faraday Trans.*, **1996**, 92, 2681.
30. W. Li; S.T. Oyama, "Mechanism of Ozone Decomposition on a Manganese Oxide Catalyst. 1. In Situ Raman Spectroscopy and Ab Initio Molecular Orbital Calculations," *J. Am. Chem. Soc.*, **1998**, 120, 9041.
31. Z. Cohen; E. Keinan; Y. Mazur; T.H. Varkony, "Dry Ozonation. A Method for Stereo selective Hydroxylation of Saturated Compounds on Silica Gel," *J. Org. Chem.*, **1975**, 40, 2141.
32. E. Keinan; Y. Mazur, "Dry Ozonation of Amines. Conversion of Primary Amines to Nitro Compounds," *J. Org. Chem.*, **1977**, 42, 844.
33. S. Leichsenring; D. Lenoir; H.G. May; A. Kettrup, "Catalytic Oxidation of Chloroaromatic Trace Contaminants Adsorbed on Wessalith DAY by Ozone", *Chemosphere*, **1996**, 33, 343.
34. P.S. Bailey, *Ozonation in Chemistry, Vol. II, Nonolefinic Compounds*, Academic Press, New York, **1982**, Ch. III.

35. A.S. Hay; J.W. Eustance; H.S. Blanchard, "Catalytic Oxidation of Hydrocarbons. Initiation by Ozone," *J. Org. Chem.*, **1960**, 25, 616.

36. G.A. Galstyan; M.M. Dvortsevoi; S.M. Sokolova, *Izv. Vyssh. Uchebn. Zaved. Khim. Khim. Tekhnol.*, **1978**, 21, 192.

37. G.A. Galstyan; V.A. Yakobi; M.M. Dvortsevoi; T.M. Galstyan, "Liquid-Phase Catalytic Oxidation of *p*-Nitrotoluene by Ozone-Air Mixture," *J. Appl. Chem. USSR*, [English Translation], **1978**, 51, 123.

38. G.A. Galstyan; V.A. Yakobi; M.M. Dvortsevoi, "Liquid-Phase Oxidation of Alkylbenzenes Initiated with Ozone," *Pet. Chem. USSR (Neftekhimiya)*, [Engl. Translation], **1976**, 16, 100.

39. G.A. Galstyan; V.A. Yakobi; A.B. Tarasenko; T.M. Galstyan, "Oxidation of Mesitylene by an Ozone-Oxygen Mixture," *Pet. Chem. USSR (Neftekhimiya)*, [Engl. Translation], **1975**, 15, 146.

40. G.A. Galstyan; V.A. Yakobi; M.M. Dvortsevoi; T.M. Galstyan, "Liquid-Phase Catalytic Oxidation of Nitrotoluene by Ozone-Air Mixture," *Zh. Prikladnoi Khimii*, [Engl. Translation], **1978**, 51, 133.

41. M. Ambar; Farhataziz; A.B. Ross, "Selected Specific Rates of Reactions of Transients from Water in Aqueous Solution," *National Bureau of Standards*, NSRDS-NBS59, Washington, DC, 1977.

42. J. Hoigné; H. Bader, "Rate Constants of Reactions of Ozone with Organic and Inorganic Compounds in Water I, Non-Dissociating Organic Compounds," *Water Res.*, **1983**, 17, 173.

43. N.N. Bhat; M.D. Gurol, "Oxidation of Chlorobenzene by Ozone and Heterogeneous Catalytic Ozonation," *Hazard. Ind. Wastes*, **1995**, 27, 371.

44. N. Al-Hayek; B. Legube; M. Doré, "Ozonation Catalytique (FeIII/Al_2O_3) du Phénol et de ses Produits d'Ozonation," *Environ. Technol. Lett.*, **1989**, 10, 415.

45. H. Allemane; B. Delouane; H. Paillard; B. Legube, "Comparative Efficiency of Three Systems (O_3, O_3/H_2O_2, O_3/TiO_2) for the Oxidation of Natural Organic Matter in Water," *Ozone Sci. Eng.*, **1993**, 15, 419.

46. R.R. Munter; S.B. Kamenev, et al., "Catalytic Treatment of Wastewater with Ozone," *Khim. Technol. Vody*, **1985**, 7, 17.

47. R. Andreozzi; A. Insola; V. Caprio; R. Marotta; V. Tufano, "The Use of Manganese Dioxide as a Heterogeneous Catalyst for Oxalic Acid Ozonation in Aqueous Solution," *Appl. Catal. A Gen.*, **1996**, 138, 75.

48. F. Ferrero; M. Iennaco; G. Rovero, "Oxidative Degradation of Dyes with Ozone," *Inquinamento*, **1992**, 34, 66.

49. T. Sekine; O. Hamamoto, *Japan Kokai Tokkyo Koho 07-80482*, March **1995**, Assigned to Hazama Gumi, Mitsui Shipbuilding Eng.

50. K. Mitsui; T. Ishii; K. Sano; A. Inoue, "Treatment of Wastewater by Catalytic Wet Oxidation," *Japan Kokai Tokkyo Koho 01-218685*, Aug. **1989**, Assigned to Nippon Shokubai Kagaku Kogyo, Co.

51. K. Mitsui; T. Ishii; K. Sano; A. Inoue, "Treatment of Wastewater by Catalytic Wet Oxidation," *Japan Kokai Tokkyo Koho 01-218684*, Aug. **1989**, Assigned to Nippon Shokubai Kagaku Kogyo, Co.

52. A.R. Bulls; J.D. Fellmann; R.A. Periana, "Process for the Production of Cyanoacetic Acid," *World Patent WO 9212962*, Aug. **1992**, Assigned to Catalytica, Inc.

53. V.G. Kharchuk, R.E. Ilatovskii; T.A. Sinitsina; V.I. Saloutin, "Preparation of 2,2,2-Trifluoroethane-1,1-diol from 2,2,2-Trifluoroethanol," *Soviet Patent SU 1715799*, Feb. **1992**, Assigned to the Ural Institute of Chemistry.

54. O.Y. Netsvetaev; A.B. Solomonov; V.I. Skudaev; Y.P. Stafeichuk, "Oxidation of Hydrogen Chloride by Ozone in Aqueous Solution," *Zh. Prikl. Khim. (S.-Petersburg)*, **1991**, 64, 2169.

55. H. Mitsui; M. Mukoyama; Y. Masaharu; H. Yatani; O. Kaieda, "Preparation of Anisic Acid as Material for Drugs and Photosensitive Agents," *Japan Kokai Tokkyo Koho 01-316342*, Dec. **1989**, Assigned to Nippon Shokubai Kagaku Kogyo, Co.

56. T.J. McMurry; T.J. Groves, in *Cytochrome P-450: Structure, Mechanism and Biochemistry*, O. de Montellano, Ed., Plenum, New York, 1986, Ch. 1.

57. I. Tabushi, *Coord. Chem. Rev.*, **1988**, 86, 1.

58. S. Campestrini; A. Robert; B. Meunier, "Ozone Epoxidation of Olefins Catalyzed by Highly Robust Manganese and Iron Porphyrin Complexes," *J. Org. Chem.*, **1991**, 56, 3725.

59. B.I. Tarunin; V.N. Tarunina; E.V. Talina, "Kinetics of Homogeneously Catalyzed Oxidation of Formic Acid in Acetic Acid Medium," *Kinet. Katal.*, **1994**, 35, 724.

60. F.E. Thompson; P.N. Sharratt; J. Hutchinson, "Heterogeneous Catalytic Oxidation of Organic Pollutants in Aqueous Solutions," *IChemE Res. Event, 1st Eur. Conf. Young Res. Chem. Eng.*, **1996**, 1, 297.

61. F.E. Thompson; P.N. Sharratt; J. Hutchinson, "Oxidation of the Organic Pollutant 1,4-Dioxan in Aqueous Solution in the Presence of Ozone and Transition Metal Oxide Catalysts," *IChemE Res. Event, 2nd Eur. Conf. Young Res. Chem. Eng.*, **1996**, 2, 651.

62. M.N. Klimova; B.I. Tarunin; Y.A. Aleksandrov, "Oxidation of Lower Aliphatic Alcohols by Ozone in the Presence of Silicon and Aluminum Oxides," *Kinet. Katal.*, **1985**, 26, 1143 [Engl. Translation], Plenum, New York, 1986, p. 987.

63. W. Li; S.T. Oyama, "Ethanol Oxidation Using Ozone over Supported Manganese Oxide Catalysts: An in situ Laser Raman Study," in *3rd World Congress on Oxidation Catalysis*, R.K. Grasselli, S.T. Oyama, A.M. Gaffney, J.E. Lyons, Eds., Studies in Surface Science and Catalysis, Vol. 110, p. 873, Elsevier, Amsterdam, 1997.

64. G.J. Hutchings; M.S. Scurrell; J.R. Woodhouse, "Direct Partial Oxidation of Methane: Effect of the Oxidant on the Reaction," *Appl. Catal.*, **1988**, 38, 157.

65. W. Li; S.T. Oyama, "Catalytic Methane Oxidation at Low Temperatures Using Ozone," in *Heterogeneous Hydrocarbon Oxidation*, B.K. Warren, S.T. Oyama, Eds., ACS Symposiurm Series 638, American Chemical Society, Washington DC, 1966, Ch. 27, p. 364.

66. K. Hauffe; Y. Ishikawa, "The Catalytic Oxidation of Hydrocarbons in the Presence of Ozone," *Chemie Ing. Techn.*, **1974**, 46, 1053.

67. A. Naydenov; D. Mehandjiev, "Complete Oxidation of Benzene on Manganese Dioxide by Ozone," *Appl. Catal. A Gen.*, **1993**, 97, 17.

68. V. Ragaini; C.L. Bianchi; G. Forcella; A. Gervasini, "A Comparison Between the Catalytic Combustion of Volatile Organic Compounds (VOC) with and without Ozone to Reduce the Industrial Air Pollution," in *Trends in Ecological Physical Chemistry*, L. Bonati, et al., Eds., Elsevier, Amsterdam, 1993, p. 275.

69. A. Gervasini; C.L. Bianchi; V. Ragaini, "Low-Temperature Catalytic Combustion of Volatile Organic Compounds Using Ozone," in *Environmental Catalysis*,

J.N. Armor, Ed., ACS Symp. Series 552, American Chemical Society, Washington, D.C., 1994, Ch. 29, p. 353.

70. A. Naydenov; R. Stoyanova; D. Mehandjiev, "Ozone Decomposition and CO Oxidation on CeO_2," *J. Mol. Catal.*, **1995**, 98, 9.
71. A. Naydenov; D. Mehandjiev, "CO Oxidation on MnO_2 by Ozone," C.R. (Doklady Acad. Bulgare des Sciences), **1993**, 46, 49.
72. B. Dhandapani; S.T. Oyama, "Gas Phase Ozone Decomposition Catalysts," *Appl. Catal. B Env.*, **1997**, 11, 129.
73. L. Whinnery; D. Coutts; C. Shen; R. Adams; C. Quintana; S. Showalter, *Mat. Res. Soc. Symp. Proc., Vol. 371, Advances in Porous Materials*, S. Komarneni; D.M. Smith; J.S. Beck, Eds., Materials Research Society, **1995**, p. 425.
74. S. Stephens; M.J. Rossi; D.M. Golden, "The Heterogeneous Reaction of Ozone on Carbonaceous Surfaces," *Int. J. Chem. Kinet.*, **1986**, 18, 1133.
75. V.R. Deitz; J.L. Bitner, "Interaction of Ozone with Adsorbent Charcoals," *Carbon*, **1973**, 11, 393.
76. C. Heisig; W. Zhang; S.T. Oyama, "Decomposition of Ozone Using Carbon-Supported Metal Oxide Catalysts," *Appl. Catal. Bkeep Environ.*, **1997**, 14, 117.
77. J.A. Halstead; R. Amstrong; B. Pohlman; S. Sibley; R. Maier, "Nonaqueous Heterogeneous Oxidation of Sulfur Dioxide," *J. Phys. Chem.*, **1990**, 94, 3261.
78. H.D. Gesser; N.R. Hunter; P.A. Das, "The Ozone Sensitized Oxidative Conversion of Methane to Methanol and Ethane to Ethanol," *Catal. Lett.*, **1992**, 16, 217.
79. H.D. Gesser; G. Zhu; N.R. Hunter, "Conversion of Ethane to Methanol and Ethanol by Ozone Sensitized Partial Oxidation at Near Atmospheric Pressure," *Catal. Today*, **1995**, 24, 321.
80. G.J. Hutchings; M.S. Scurrell; J.R. Woodhouse, "Effect of O_3 Versus O_2 as Oxidants for Methane," in *Methane Conversion*, D.M. Dibby, C.D. Chang, R.F. Howe, S. Yurchak, Eds., p. 415, Elsevier, Amsterdam, 1988.
81. T. Ito; J.H. Lunsford, "Synthesis of Ethylene and Ethane by Partial Oxidation of Methane over Lithium-Doped Magnesium Oxide," *Nature (Lond.)*, **1985**, 314, 721.
82. D.J. Driscoll; W. Martir; J.-X. Wang; J.H. Lunsford, "Formation of Gas-Phase Methyl Radicals over MgO," *J. Am. Chem. Soc.*, **1985**, 107, 58.
83. N.I. Sax, *Dangerous Properties of Industrial Materials*, 6th Ed., Van Nostrand Reinhold, New York, 1983.
84. S.L. Hung; L.D. Pfefferle, "Methyl Chloride and Methylene Chloride Inciniration in a Catalytically Stabilized Thermal Combustor," *Environ. Sci. Technol.*, **1989**, 23, 1085.
85. C.A. Clausen; C.D. Cooper; M. Hewett; A. Martinez, "Enhancement of Organic Vapor Incineration by Using Ozone," *J. Hazard. Mat.*, **1992**, 31, 75.
86. A. Gervasini; G.C. Vezzoli; V. Ragaini, "VOC Removal by Synergic Effect of Combustion Catalyst and Ozone," *Catal. Today*, **1996**, 29, 449.
87. T.S. Lazareva; L.F. Atyaksheva; G.I. Emel'yanova; B.V. Strakhov; N.A. Strigunenko, "Heterogeneous Catalytic Oxidation of Hydrazine by Ozone," *Vestn. Mosk Univ., Ser. 2 Khim*, **1990**, 31, 348.
88. V.M. Zlotopol'skii; T.Yu. Volkova; T.E. Slavikovskaya, "Oxidation of *o*-Xylene by Ozone on a Supported Cobalt Oxide Catalyst," *Khim. Prom-st. (Moscow)*, **1993**, 12, 620.

89. L.A. Kireeva; S.N. Tkachenko; M.P. Popovich; L.A. Lozhenko; L.V. Lyapunova, "Aluminum-Calcium Oxide Systems. Decomposition of Ozone and Oxidation of Carbon Monoxide," *Vestn. Mosk. Univ., Ser. 2 Khim.*, **1992**, 33, 342.

90. C.E. Hori, H. Permana, K.Y.S. Ng, A. Brenner, K. More, K.M. Rahmoeller, D. Belton, "Thermal Stability of Oxygen Storage Properties in a Mixed CeO_2-ZrO_2 System," *Appl. Catal. B Environ.*, **1998**, 16, 105

91. T. Shido; Y. Iwasawa, "Regulation of Reaction Intermediate by Reactant in the Water-Gas Shift Reaction on CeO_2 in Relation to Reactant-Promoted Mechanism," *J. Catal.*, **1992**, 136, 493.

92. J.M. Schwartz; L.D. Schmidt, "Microstructures of Pt-Ce and Rh-Ce Particles on Alumina and Silica," *J. Catal.*, **1992**, 138, 283.

93. G.S. Zafiris; R.J. Gorte, "A Study of CO, NO, and H_2 Adsorption on Model Pt/CeO_2 Catalysts," *Surf. Sci.*, **1992**, 276, 86.

94. R.M. Heck; R.J. Farrauto; H.C. Lee, "Commercial Development and Experience with Catalytic Ozone Abatement in Jet Aircraft," *Catal. Today*, **1992**, 13, 43.

95. R.M. Heck; R.J. Farrauto, *Catalytic Air Pollution Control, Commercial Technology*, Van Nostrand Reinhold, New York, **1995**.

96. S. Tsuruta, "Apparatus for Sterilization with Ozone and Ultraviolet-Ray Radiation," *Jpn. Patent 62227358*, **1988**.

97. R.D. S. Stevens; P.M. Brown, "Method for Removing Malodorous or Toxic Gases from an Air Stream," *Eur. Pat. EP 261987*, Assigned to Solarchem Research, March **1988**.

98. J.P. Gould; W.J. Weber, Jr., *J. Water Pollut. Control*, **1976**, 98, 471.

99. M. Teramoto; S. Imamura; N. Yatagai; Y. Nishikawa; H. Teranishi, "Kinetics of the Self-Decomposition of Ozone and the Ozonation of Cyanide Ion and Dyes in Aqueous Solutions," *J. Chem. Eng. Jpn.*, **1981**, 14, 383.

100. J. Hoke; D. Anderson; R. Heck; T. Poles; J. Steger, SAE Technical Paper Series 96-0800, in *Cold-Start Emission Control and Catalyst Technologies (SP-1173)*, SAE International, Warrendale, PA, **1996**.

101. W. Rudy; R. Ober; M. Durilla; D. Anderson; J. Hoke; R. Heck, Paper 96-MP4A.08, presented at the *Air & Waste Management Association 89th National Meeting*, June **1996**.

102. W.F. Carr; M. Junction; H.M. Chen, *US. Patent 4,343,776*, Assigned to Engelhard Corporation, Aug. **1982**.

103. M.P. Galligan; J.C. Dettling, *U.S. Patent 5,422,331*, Assigned to Engelhard Corporation, June **1995**.

104. A. Akira, *Jpn. Patent 52,133,894*, Assigned to Fuji-Xerox Co., Ltd., 9 Nov. **1977**.

105. A. Inoue; M. Kobayashi, *Jpn. Patent 62,176,541*, Assigned to Nippon Shokubai Kagaku Kogyo Co., Ltd., 3 Aug. **1987**.

106. S. Imamura; M. Ikebata; T. Ito; T. Ogita, "Decomposition of Ozone on a Silver Catalyst," *Ind. Eng. Chem. Res.*, **1991**, 30, 217.

107. B. Dhandapani; S.T. Oyama, "Kinetics and Mechanism of Decomposition of Ozone on Manganese Oxide," *Chem. Lett.*, **1995**, 413.

108. W. Li; S.T. Oyama, "Mechanism of Ozone Decomposition on a Manganese Oxide Catalyst. 2. Steady-State and Transient Kinetic Studies," *J. Am. Chem. Soc.*, **1998**, 120, 9047.

109. J.S. Mattson; H.B. Mark, *Activated Carbon: Surface Chemistry and Adsorption from Solution*, Marcel Dekker, New York, **1971**, p. 129.

110. K.K. Robinson, *Granulated Activated Carbon for Water Treatment, U.S. Patent 4,954,469*, Assigned to Amoco Oil, **1990**.

111. T. Otowa, *Proceedings of Seiken Symposium, Tokyo*, **1991**, 7, 273–280.

112. T.L. Rakit-Skaya; A.Y. Bandurko; V.V. Litvinskaya, *Russian J. Appl. Chem.*, **1991**, 66, 1993 [Translated from *Zhurnal Prikladnoi Khimii*, **1993**, 66, 2141].

113. E. Papirer; J.-B. Donnet; A. Schutz, "Etude Cinetique de l'Oxydation des Noirs de Carbone par l'Ozone," *Carbon*, **1967**, 5, 113.

114. V.R. Deitz; J.L. Bitner, "The Reaction of Ozone with Adsorbent Charcoal," *Carbon*, **1972**, 10, 145.

115. M. Balooch; W.J. Siekhaus; R.J. Tench, "Scanning Tunneling Microscopy of Reactive Sites: Gasification of Graphite by Oxygen and Hydrogen," *AIP Conf. Proc.*, **1992**, 241, 269.

116. L.F. Atyaksheva; G.I. Emel'yanova; T.S. Lazareva, et al., *Mekh. Kompoz Mater.*, **1988**, 1, 166.

117. L.F. Atyaksheva; G.I. Emel'yanova, *Zh. Fiz. Khim.*, **1989**, 63, 2606.

118. M. Yoshimoto; T. Nakatsuji; K. Nagano; K. Yoshida, *Eur. Pat. 90,302,545*, **1990**, Assigned to Sakai Chemical Industry, Co., Ltd.

119. G.R. Lester; S.T. Homeyer, "Catalytic Converter with Metal Monolith Having and Integral Catalyst," *World Patent WO 9409903*, May **1994**, Assigned to Allied-Signal, Inc.

120. M.P. Galligan, J.C. Dettling, *US 5,422,331*, June 6, **1995**, Assigned to Engelhard Corporation.

121. J.P. DeLuca; L.E. Campbell, "Monolithic Catalyst Supports," in *Advanced Materials in Catalysis*, J.J. Burton, R.L. Garten, Eds., Academic Press, New York, **1977**.

122. T. Watabe; A. Kitamura, "Deodorization in Toilet with Ozone," *Japan Kokai Tokkyo Koho 05-300935*, Nov. **1993**, Assigned to Toto, Ltd.

123. T. Watabe, "Apparatus for Air Deodorization," *Japan Kokai Tokkyo Koho 03-242149*, Nov. **1991**, Assigned to Toto, Ltd.

124. K. Ogura; H. Miyazaki, "Catalytic Deodorizing Apparatus," *Japan Kokai Tokkyo Koho 04-285563*, Nov. **1992**, Assigned to Toto, Ltd.

125. I. Kikuchi; H. Takagiwa, "Corona Discharge Apparatus with Integrated Piezoelectric Transformer for Ozonizer," *Japan Kokai Tokkyo Koho 08-91807*, April **1996**, Assigned to Toto, Ltd.

126. T. Watabe, "Ozonizers Using Corona Discharge for Deodorization Apparatus," *Japan Kokai Tokkyo Koho 07-309605*, Nov. **1995**, Assigned to Toto, Ltd.

127. H. Haraga; H. Miyazaki; K. Akamatsu; A. Hirano, "Apparatus for Deodorization of Indoor Air using Ozone," *Japan Kokai Tokkyo Koho 04-200467*, July **1992**, Assigned to Toto, Ltd.

128. S. Terui; S. Kitaguchi; A. Nose; T. Hayashi; T. Kuhara; T. Sonoda, "Apparatus and Method for Catalytic Deodorization of Waste Gases or Polluted Air," *Japan Kokai Tokkyo Koho 06-254347*, Sep. **1994**, Assigned to Nippon Catalytic Chem. Ind.

129. S. Terui; K. Sano; K. Nishikawa; A. Inoue; Y. Yokota, "Catalyst for Oxidation and Ozone Decomposition in Gas Treatment," *Eur. Patent Appl. EP 399302 A2*, Nov. **1990**, Assigned to Nippon Shokubai Kagaku Kogyo, Co.

130. S. Terui; K. Sano; K. Nishikawa; A. Inoue; Y. Yokota, "Catalyst for Oxidation and Ozone Decomposition in Gas Treatment," *Eur. Patent Application EP 399302 A2*, Nov. **1990**, Assigned to Nippon Shokubai Kagaku Kogyo, Co.

131. M. Yoshimoto; T. Nakatsuji; K. Nagano; K. Yoshida, "Deodorization with Ozone and Catalysts," *Japan Kokai Tokkyo Koho 06-142455*, May **1994**, Assigned to Sakai Chemical Industry, Co.

132. T. Nakatsuji; K. Yoshida; M. Yoshimoto, "Deodorization," *Japan Kokai Tokkyo Koho 06-15140*, Jan. **1994**, Assigned to Sakai Chemical Industry, Co.

133. T. Watabe, "Apparatus for Air Deodorization," *Japan Kokai Tokkyo Koho 03-242148*, Oct. **1991**, Assigned to Toto, Ltd.

134. H. Kobayashi; H. Myazaki; A. Hirano; K. Akamatsu, "Apparatus for Deodorizing Toilet Air Using Ozone," *Japan Kokai Tokkyo Koho 05-300935*, Aug. **1993**, Assigned to Toto, Ltd.

135. T. Nakatsuji; K. Yoshida; M. Yoshimoto, "Deodorization of Gases," *Japan Kokai Tokkyo Koho 05-49863*, March **1993**, Assigned to Sakai Chemical Industry, Co.

136. S. Seo; S. Matsumoto; Y. Kimura; N. Yoshida, "Deodorization of Gases and Agents for It," *Japan Kokai Tokkyo Koho 05-49862*, March **1993**, Assigned to Nippon Kayaku Kabushiki Kaisha.

137. M. Yoshimoto; T. Nakatsuji; K. Yoshida, "Deodorization of Waste Gases by Catalytic Oxidation with Ozone," *Japan Kokai Tokkyo Koho 04-231059*, Aug. **1992**, Assigned to Sakai Chemical Industry, Co.

138. S. Ichikawa, Personal communication, Hitachi, Dashpure (Model DE-10TU) Air Cleaner Specifications, 1998.

139. U. Kogelschatz, "Dielectric-Barrier Discharges Principle and Applications," invited Plenary Lecture, Int. Conf. Phen. Ioniz. Gases (ICPIG XXIII), July 17–22, Toulousse, France.

140. B. Eliasson; U. Kogelschatz, "Nonequilibrium Volume Plasma Processing," *IEEE Trans. Plasma Sci.*, **1991**, 19, 1063.

141. B. Eliasson; F.G. Simon; W. Egli, "Hydrogenation of CO_2 in a Silent Discharge," in *Non-Thermal Plasma Techniques for Pollution Control*, B.M. Penetrante, S.E. Schultheis, Eds., *NATO Adv. Sci. Series, Vol. G34, Part B*, Springer-Verlag, Berlin, 1993.

142. B. Eliasson; W. Egli; U. Kogelschatz, "Modelling of Dielectric Barrier Discharge Chemistry," *Pure Appl. Chem.*, **1994**, 66, 1275.

143. J.A. Rodberg; J.F. Miller; G.E. Keller; J.E. Woods, "A Novel Technique to Permanently Remove Indoor Air Pollutants," *CIB/ASHRAE Healthy Buildings*, **1991**.

APPENDIX 1.1

INTERNET LINKS TO INTERNATIONAL EFFORTS IN POLLUTION PREVENTION*

Air and Waste Management Association (http://www.awma.org/)
One of the premier international environmental organizations promoting global environmental responsibility

Amazing Environmental Organization Web Directory (http://www.webdirectory.com/)
Links to thousands of environment-related sites from over 100 countries

BCN Environment Center (http://bcn.boulder.co.us/environment/Global/EnvTopics.html)
Boulder Community Network's clearing house of links to web resources related to worldwide environmental issues

BNA Environment Information Products (http://www.bna.com/hub/bna/envir/index.html)
Environmental reference material from the Bureau of National Affairs, Inc.

Brown's Environmental Studies Department (http://www.brown.edu/Departments/Environmental_Studies/espgm.html)
Brown University's list of links to Environmental Studies Programs and Departments at colleges and universities in the United States and worldwide

Canadian Environmental Assessment Agency (http://www.ceaa.gc.ca/)
A Canadian government organization that facilitates the integration of environmental factors into federal planning and decision making

EINET Galaxy (http://galaxy.einet.net/)

* Reproduced from the EPA Office of Air and Radiation home page (http://www.epa.gov/docs/oar/oarhome.html).

A guide to worldwide information and services and is provided as a public service by EINet

EnviroLink Network (http://envirolink.org/)

A non-profit organization that is known as an on-line source for environmental information

Environment Canada's Atlantic Region (http://atlenv.bed.ns.doe.ca:80/)

Links to Canadian weather and climatological information, key publications, major initiatives and programs, and other worldwide environmental servers

Environmental Research Tools (http://www.princeton.edu/~mlevy/)

Princeton University site with links to Internet resources useful for research on global environmental policy

Environmental Technology Gateway (http://www.nttc.edu/environmental.html)

Links to other Web sites with environmental information

Global Network of Environment and Technology (GNET) (http://www.gnet.org/)

Provides links to the information, people, and resources that make up the environment and technology community

National Institute for the Environment (http://www.cnie.org/)

The Committee for the National Institute for the Environment (CNIE) is a national, non-profit organization working to improve the scientific basis for making decisions on environmental issues

Seed Web List of Environmental Servers
(http://www.studorg.nwu.edu/seed/SEEDWeb.html)

Resource page for Northwestern University's Students for Environmental and Ecological Development. Many links to government agencies, environmental organizations, environmental topics, and indices to online environmental information

Yahoo Environment and Nature List
(http://www.yahoo.com/Society_and_Culture/Environment_and_Nature/)

From the Yahoo Internet resource directory, links relating to the environment and nature

APPENDIX 1.2

POLLUTANTS REGULATED UNDER THE CLEAN AIR ACT AMENDMENTS†

		Regulating Program					
CAS No.	Pollutant Name	Criteria[a]	NSPS	HAP[b]	Accidental[c]	ODS[d]	VOC[e]
75-07-0	Acetaldehyde			✓	F [10,000]		✓
60-35-5	Acetamide			✓			✓✓
141-78-6	Acetic acid ethyl ester						✓
67-64-1	Acetone						×
75-05-8	Acetonitrile			✓			✓
98-86-2	Acetophenone			✓			✓
53-96-3	2-Acetylaminofluorene			✓			✓✓
74-86-2	Acetylene				F [10,000]		✓
2278-22-0	Acetyl nitroperoxide						✓
107-02-8	Acrolein			✓	T [5,000]		✓
79-06-1	Acrylamide			✓			✓✓
79-10-7	Acrylic acid			✓			✓✓
107-13-1	Acrylonitrile			✓	T [20,000]		✓
814-68-6	Acrylyl chloride				T [5,000]		✓✓
1071-88-6	Allyl alcohol				T [15,000]		✓✓
107-05-1	Allyl chloride			✓			✓
107-11-9	Allylamine				T [10,000]		✓✓
92-67-1	4-Aminobiphenyl			NR			✓✓
7664-41-7	Ammonia (anhydrous)			NR	T [10,000]		
766-41-7	Ammonia (conc. 20% or greater)				T [20,000]		

(continued)

† (Adapted from a list originally compiled by D.K. Berry and S.R. Sabatini.)

CAS No.	Pollutant Name	Criteria[a]	NSPS	HAP[b]	Accidental[c]	ODS[d]	VOC[e]
		Regulating Program					
62-53-3	Aniline			✓			✓
90-04-0	*o*-Anisidine			✓			✓✓
	Antimony compounds			NR			
	Arsenic compounds (inorganic, including arsine)			✓			
7784-34-1	Arsenous trichloride				T [15,000]		
7784-42-1	Arsine				T [1,000]		
1332-21-4	Asbestos			✓			
100-52-7	Benzaldehyde						✓
71-43-2	Benzene			✓			✓
30025-33-3	Benzenedicarboxaldehyde						✓
92-87-5	Benzidine			✓			✓✓
100-47-0	Benzonitrile						✓
98-07-7	Benzotrichloride			✓			✓✓
32368-69-7	Benzoyl nitro peroxide						✓
100-44-7	Benzyl chloride			✓			✓
	Beryllium compounds			✓			
92-52-4	Biphenyl			✓			✓✓
117-81-7	Bis(2-ethylhexyl)phthalate (DEHP)			NR			✓✓
542-88-1	Bis(chloromethyl)ether			✓			✓✓
10294-34-5	Boron trichloride				T [5,000]		
7637-07-2	Boron trifluoride				T [5,000]		
353-42-4	Boron trifluoride compound w/ methyl ether (1 : 1)				T [15,000]		
7726-95-6	Bromine				T [10,000]		
108-86-1	Bromobenzene						✓
107-04-0	1-Bromo-2-chloroethane						✓
34652-54-5	Bromochloropropane						✓
109-70-6	1-Bromo-3-chloropropane						✓
3737-00-6	3-Bromo-1-chloro-1-propene						✓
75-27-4	Bromodichloromethane						✓
75-25-2	Bromoform			✓			✓
74-83-9	Bromomethane (methyl bromide)			✓			✓
28807-97-8	Bromomethylbenzene						✓
79504-02-2	Bromopentachloroethane						✓
26446-77-5	Bromopropane						✓
106-95-6	3-Bromo-1-propene						✓
596-73-2	Bromotrifluorethylene				F [10,000]		✓✓
75-63-8	Bromotrifluoromethane (Halon 1301)					I	✓✓
106-99-0	1,3-Butadiene			✓	F [10,000]		✓
123-72-8	Butanal						✓

CAS No.	Pollutant Name	Regulating Program				ODS[d]	VOC[e]
		Criteria[a]	NSPS	HAP[b]	Accidental[c]		
106-97-8	Butane				F [10,000]		✓
25167-67-3	Butene				F [10,000]		✓
106-98-9	1-Butene				F [10,000]		✓
107-01-7	2-Butene				F [10,000]		
590-18-1	2-Butene(cis)				F [10,000]		✓
624-64-6	2-Butene(trans)				F [10,000]		✓
111-76-2	2-Butoxyethanol						✓
123-86-4	Butyl acetate						×
104-51-8	Butylbenzene						✓
	Butyne						✓
	Cadmium compounds			NR			
156-62-7	Calcium cyanamide						
105-60-2	Caprolactam			✓			✓✓
133-06-2	Captan			NR			✓✓
63-25-2	Carbaryl			NR			✓✓
75-15-0	Carbon disulfide			✓	T [20,000]		✓
630-08-0	Carbon monoxide	✓	✓				
463-58-1	Carbon oxysulfide				F [10,000]		✓
56-23-5	Carbon tetrachloride			✓		I	✓
120-80-9	Catechol			NR			✓✓
133-90-4	Chloramben			NR			✓✓
57-74-9	Chlordane			NR			✓✓
7782-50-5	Chlorine			NR	T [2,500]		
10049-04-4	Chlorine dioxide				T [1,000]		
7791-21-1	Chlorine monoxide				F [10,000]		
79-11-8	Chloroacetic acid			✓			✓✓
532-27-4	2-Chloroacetophenone			✓			✓✓
587-04-2	3-Chlorobenzaldehyde						✓
108-90-7	Chlorobenzene			✓			✓
27134-26-5	Chlorobenzeneamine						✓
510-15-6	Chlorobenzilate			NR			✓
75-45-6	Chlorodifluoromethane						×
75-69-4	Chlorofluorocarbon-11 (CFC-11)	E				I	×
954-56-3	Chlorofluorocarbon-111 (CFC-111)					I	
76-12-0	Chlorofluorocarbon-112 (CFC-112)					I	
76-13-1	Chlorofluorocarbon-113 (CFC-113)	E				I	×
76-14-2	Chlorofluorocarbon-114 (CFC-114)	E				I	×
76-15-3	Chlorofluorocarbon-115 (CFC-115)	E				I	
75-71-8	Chlorofluorocarbon-12 (CFC-12)	E				I	×
75-72-9	Chlorofluorocarbon-13 (CFC-13)					I	

(continued)

		Regulating Program					
CAS No.	Pollutant Name	Criteria[a]	NSPS	HAP[b]	Accidental[c]	ODS[d]	VOC[e]
422-78-6	Chlorofluorocarbon-211 (CFC-211)					I	
3182-26-1	Chlorofluorocarbon-212 (CFC-212)					I	
2354-06-5	Chlorofluorocarbon-213 (CFC-213)					I	
29255-31-0	Chlorofluorocarbon-214 (CFC-214)					I	
4259-43-2	Chlorofluorocarbon-215 (CFC-215)					I	
661-97-2	Chlorofluorocarbon-216 (CFC-216)					I	
422-86-6	Chlorofluorocarbon-217 (CFC-217)					I	
67-66-3	Chloroform			✓	T [20,000]		✓
627-42-9	1-Chloro-2-methoxyethane						✓
542-88-1	Chloromethyl ether				T [1,000]		✓✓
107-30-2	Chloromethyl methyl ether			NR	T [5,000]		✓✓
106-43-4	1-Chloro-4-methylbenzene						✓
25167-93-5	Chloronitrobenzene						✓
126-99-8	Chloroprene			✓			✓
590-21-6	1-Chloropropylene				F [10,000]		✓✓
557-98-2	2-Chloropropylene				F [10,000]		✓✓
95-49-8	2-Chlorotoluene						✓
	Chromium compounds			✓			
	Cobalt compounds			NR			
	Coke Oven Emissions			✓			
108-39-4	*m*-Cresol			✓			✓
95-48-7	*o*-Cresol			✓			✓
106-44-5	*p*-Cresol			✓			✓
1319-77-3	Cresols/cresylic acid (isomers and mixture)			✓			✓✓
4170-30-3	Crotonaldehyde				T [20,000]		✓✓
123-73-9	(E)-Crotonaldehyde				T [20,000]		✓✓
98-82-8	Cumene			✓			✓
	Cyanide Compounds			NR			
460-19-5	Cyanogen				F [10,000]		
506-77-4	Cyanogen chloride				T [10,000]		
291-64-5	Cycloheptane						✓
110-82-7	Cyclohexane						✓
108-94-1	Cyclohexanone						✓
110-83-3	Cyclohexene						✓
108-91-8	Cyclohexylamine				T [15,000]		✓✓
287-92-3	Cyclopentane						✓
142-29-0	Cyclopentene						✓
75-19-4	Cyclopropane				F [10,000]		✓✓

CAS No.	Pollutant Name	Regulating Program Criteria[a]	NSPS	HAP[b]	Accidental[c]	ODS[d]	VOC[e]
541-02-6	Decamethylcyclopentasiloxane						×
124-18-5	Decane						✓
334-88-3	Diazomethane			NR			✓✓
132-64-9	Dibenzofuran		✓	NR			✓✓
19287-45-7	Diborane				T [2,500]		
594-18-3	Dibromodichloromethane						✓
74-95-3	Dibromomethane						✓
124-48-1	Dibromochloromethane						✓
67708-83-2	Dibromochloropropane						✓
55162-35-1	1,1-Dibromo-2-chloropropane						✓
96-12-8	1,2-Dibromo-3-chloropropane			✓			✓
106-93-4	1,2-Dibromoethane						✓
74-95-3	Dibromomethane						✓
78-75-1	1,2-Dibromopropane						✓
84-74-2	Dibutylphthalate			NR			✓✓
25321-22-6	Dichlorobenzene						✓
106-46-7	1,4-Dichlorobenzene			✓			✓
541-73-1	1,3-Dichlorobenzene						✓
95-50-1	1,2-Dichlorobenzene						✓
91-94-1	3,3-Dichlorobenzidine			✓			✓✓
72-55-9	1,1-Dichloro-2,2-bis (*p*-chlorophenyl) ethylene			NR			✓✓
110-56-5	1,4-Dichlorobutane						✓
107-06-2	1,2-Dichloroethane						✓
25323-30-2	Dichloroethene						✓
156-59-2	(Z)-1,2-Dichloroethene						✓
159-60-5	(E)-1,2-Dichloroethene						✓
111-44-4	Dichloroethyl ether			✓			✓
75-09-2	Dichloromethane (methylene chloride)			✓			✓✓
4885-02-3	Dichloromethoxymethane						✓
29797-40-8	Dichloromethylbenzene						✓
28699-88-9	Dichloronaphthalene						✓
27900-75-0	Dichloronitrobenzene						✓
78-87-5	1,2-Dichloropropane (propylene dichloride)			✓			✓
26952-23-8	Dichloro-1-propene						✓
542-75-6	1,3-Dichloropropene			✓			✓✓
10061-02-6	(E)-1,3-Dichloro-1-propene						✓
10061-01-5	(Z)-1,3-Dichloro-1-propene						✓
4109-96-0	Dichlorosilane				F [10,000]		
1320-37-2	Dichlorotetrafluoroethane						✓
62-73-7	Dichlorvos			NR			✓✓

(continued)

		Regulating Program					
CAS No.	Pollutant Name	Criteria[a]	NSPS	HAP[b]	Accidental[c]	ODS[d]	VOC[e]
111-42-2	Diethanolamine			✓			✓✓
135-01-3	1,2-Diethylbenzene						✓
141-93-5	1,3-Diethylbenzene						✓
105-05-5	1,4-Diethylbenzene						✓
1762-28-03	Diethyldimethyl lead						✓
64-67-5	Diethyl sulfate			NR			✓
75-37-6	Difluoroethane				F [10,000]		
119-90-4	3,3-Dimethoxybenzidine			NR			✓✓
124-40-3	Dimethylamine				F [10,000]		✓✓
60-11-7	Dimethyl aminoazobenzene			NR			✓✓
121-69-7	*N*,*N*-Dimethylaniline			✓			✓✓
119-93-7	3,3-Dimethylbenzidine			✓			✓✓
75-83-2	2,2-Dimethylbutane						✓
79-29-8	2,3-Dimethylbutane						✓
75-97-8	3,3-Dimethyl-2-butanone						✓
79-44-7	Dimethylcarbamoyl chloride			NR			✓✓
27195-67-1	Dimethylcyclohexane						✓
28729-52-4	Dimethylcyclopentane						✓
2532-58-3	c-1,3-Dimethylcyclopentane						✓
1759-58-6	t-1,3-Dimethylcyclopentane						✓
75-78-5	Dimethyldichlorosilane				T [5,000]		✓✓
624-92-0	Dimethyldisulfide						✓
68-12-2	*N*,*N*-Dimethylformamide			✓			✓✓
2213-23-2	2,4-Dimethylheptane						✓
2216-30-0	2,5-Dimethylheptane						✓
584-94-1	2,3-Dimethylhexane						✓
589-43-5	2,4-Dimethylhexane						✓
592-13-2	2,5-Dimethylhexane						✓
57-14-7	1,1-Dimethylhydrazine			✓	T [15,000]		✓✓
79-92-5	2,2-Dimethyl-3-methylene-bicyclo[2.2.1]-heptane						✓
127-91-3	6,6-Dimethyl-3-methylene-bicyclo[3.1.1]-heptane						✓
565-59-3	2,3-Dimethylpentane						✓
108-08-7	2,4-Dimethylpentane						✓
562-49-2	3,3-Dimethylpentane						✓
575-26-1	2,6-Dimethylphenol						✓
108-68-9	3,5-Dimethylphenol						✓
131-11-3	Dimethyl phthalate			✓			✓✓
463-82-1	2,2-Dimethylpropane				F [10,000]		✓✓
583-61-9	2,3-Dimethylpyridine						✓

		Regulating Program					
CAS No.	Pollutant Name	Criteria[a]	NSPS	HAP[b]	Accidental[c]	ODS[d]	VOC[e]
108-47-4	2,4-Dimethylpyridine						✓
108-48-5	2,6-Dimethylpyridine						✓
77-78-1	Dimethyl sulfate			✓			✓
534-52-1	4,6-Dinitro-*o*-cresol, and salts			NR			✓✓
51-28-5	2,4-Dinitrophenol			✓			✓✓
121-14-2	2,4-Dinitrotoluene			✓			✓✓
123-91-1	1,4-Dioxane (1,4-diethyleneoxide)			✓			✓
122-66-7	1,2-Diphenylhydrazine			✓			✓✓
	Dioxin/furan (total tetra through octa chlorinated dibenzo-*p*-dioxins and dibenzofuran)		✓				✓✓
	Diphenyl ether						✓
119-61-9	Diphenylmethanone						✓
540-97-6	Dodecamethylcyclo-hexasiloxane						×
112-40-3	Dodecane						✓
123-01-3	Dodecylbenzene						✓
106-89-8	Epichlorohydrin (1-chloro-2,3-epoxypropane)			✓	T [20,000]		✓
106-88-7	1,2-Epoxybutane			✓			✓✓
74-84-0	Ethane	E			F [10,000]		×
107-00-6	Ethyl acetylene				F [10,000]		✓✓
140-88-5	Ethyl acrylate			✓			✓✓
75-04-7	Ethylamine				F [10,000]		✓✓
100-41-4	Ethylbenzene			✓			✓
51-79-6	Ethyl carbamate (urethane)			NR		I	✓✓
75-00-3	Ethyl chloride (chloroethane)			✓	F [10,000]		✓
1678-91-7	Ethylcyclohexane						✓
1640-89-7	Ethylcyclopentane						✓
29224-55-3	Ethyldimethylbenzene						✓
934-80-5	4-Ethyl-1,2-dimethylbenzene						✓
933-98-2	1-Ethyl-2,3-dimethylbenzene						✓
1068-87-7	3-Ethyl-2,4-dimethylpentane						✓✓
74-85-1	Ethylene				F [10,000]		✓
107-15-3	Ethylenediamine				T [20,000]		✓✓
106-93-4	Ethylene dibromide (dibromoethane)			✓			✓✓

(continued)

CAS No.	Pollutant Name	Regulating Program Criteria[a]	NSPS	HAP[b]	Accidental[c]	ODS[d]	VOC[e]
107-06-2	Ethylene dichloride (1,2-dichloroethane)			✓			✓✓
151-56-4	Ethyleneimine (aziridine)			NR	T [10,000]		✓✓
60-29-7	Ethyl ether				F [10,000]		✓✓
107-21-1	Ethylene glycol			✓			✓✓
619-99-8	3-Ethylhexane			✓			✓
75-08-1	Ethyl mercaptan			✓	F [10,000]		✓✓
611-14-3	1-Ethyl-2-methylbenzene						✓
620-14-4	1-Ethyl-3-methylbenzene						✓
622-96-8	1-Ethyl-4-methylbenzene						✓
30677-34-0	Ethylmethylcyclohexane						✓
109-95-5	Ethyl nitrite				F [10,000]		✓✓
75-21-8	Ethylene oxide			✓	T [10,000]		✓
96-45-7	Ethylene thiourea			NR			✓✓
75-34-3	Ethylidene chloride (1,1-dichloroethane)			✓			✓
1762-26-1	Ethyltrimethyllead						✓
	Fine mineral fibers			NR			
	Fluorides		✓				
7782-41-4	Fluorine				T [1,000]		
50-00-0	Formaldehyde			✓			✓
50-00-0	Formaldehyde (solution)				T [15,000]		✓
64-18-6	Formic acid						✓
110-00-9	Furan				T [5,000]		✓✓
	Glycol ethers			✓			✓✓
421-01-2	Halon-1211					I	
75-63-8	Halon-1301					I	
124-73-2	Halon-2402					I	
76-44-8	Heptachlor			NR			✓✓
629-78-7	Heptadecane						✓
142-82-5	Heptane						✓
14686-13-6	(E)-2-Heptene						✓
6443-92-1	(Z)-2-Heptene						✓
118-74-1	Hexachlorobenzene			✓			✓✓
87-68-3	Hexachlorobutadiene			✓			✓
58-89-9	1,2,3,4,5,6-Hexachlorocyclohexane (all stereo isomers, including lindane)			NR			
77-47-4	Hexachlorocyclopentadiene			NR			✓
67-72-1	Hexachloroethane			✓			✓
556-88-3	Hexadecamethylcyclooctasiloxane						×
544-76-3	Hexadecane						✓
541-05-9	Hexamethylcyclotrisiloxane						×
822-06-0	Hexamethylene-1,6-diisocyanate			NR			✓✓
680-31-9	Hexamethylphosphoramide			NR			✓✓

		Regulating Program					
CAS No.	Pollutant Name	Criteria[a]	NSPS	HAP[b]	Accidental[c]	ODS[d]	VOC[e]
100-54-3	Hexane			✓			✓
592-41-6	1-Hexene						✓
4050-45-7	(E)-2-Hexene						✓
7688-21-3	(Z)-2-Hexene						✓
13269-52-8	(E)-3-Hexene						✓
7642-09-3	(Z)-3-Hexene						✓
1077-16-3	Hexylbenzene						✓
302-01-2	Hydrazine			NR	T [15,000]		
7647-01-0	Hydrochloric acid		✓	NR	T (15,000)		
75-43-4	Hydrochlorofluorocarbon-21 (HCFC-21)					II	
75-45-6	Hydrochlorofluorocarbon-22 (HCFC-22)					II	
593-70-4	Hydrochlorofluorocarbon-31 (HCFC-31)					II	
134237-32-4	Hydrochlorofluorocarbon-121 (HCFC-121)					II	
354-21-2	Hydrochlorofluorocarbon-122 (HCFC-122)					II	
306-83-2	Hydrochlorofluorocarbon-123 (HCFC-123)	E				II	
2837-89-0	Hydrochlorofluorocarbon-124 (HCFC-124)	E				II	
134237-34-6	Hydrochlorofluorocarbon-131 (HCFC-131)					II	
33579-37-2	Hydrochlorofluorocarbon-132 (HCFC-132)					II	
431-07-2	Hydrochlorofluorocarbon-133 (HCFC-133)					II	
2516-78-8	Hydrochlorofluorocarbon-141 (HCFC-141)					II	
338-65-8	Hydrochlorofluorocarbon-142 (HCFC-142)					II	
134237-35-7	Hydrochlorofluorocarbon-221 (HCFC-221)					II	
134237-36-8	Hydrochlorofluorocarbon-222 (HCFC-222)					II	
134237-37-9	Hydrochlorofluorocarbon-223 (HCFC-223)					II	
134237-38-0	Hydrochlorofluorocarbon-224 (HCFC-224)					II	
127564-92-5	Hydrochlorofluorocarbon-225 (HCFC-225)					II	
134308-72-8	Hydrochlorofluorocarbon-226 (HCFC-226)					II	
134190-48-0	Hydrochlorofluorocarbon-231 (HCFC-231)					II	

(continued)

		Regulating Program					
CAS No.	Pollutant Name	Criteria[a]	NSPS	HAP[b]	Accidental[c]	ODS[d]	VOC[e]
134237-39-1	Hydrochlorofluorocarbon-232 (HCFC-232)					II	
134237-40-4	Hydrochlorofluorocarbon-233 (HCFC-233)					II	
127564-83-4	Hydrochlorofluorocarbon-234 (HCFC-234)					II	
134237-83-5	Hydrochlorofluorocarbon-235 (HCFC-235)					II	
134190-49-1	Hydrochlorofluorocarbon-241 (HCFC-241)					II	
134237-42-6	Hydrochlorofluorocarbon-242 (HCFC-242)					II	
134237-43-7	Hydrochlorofluorocarbon-243 (HCFC-243)					II	
134190-50-4	Hydrochlorofluorocarbon-244 (HCFC-244)					II	
134190-51-5	Hydrochlorofluorocarbon-251 (HCFC-251)					II	
134190-52-6	Hydrochlorofluorocarbon-252 (HCFC-252)					II	
134237-44-8	Hydrochlorofluorocarbon-253 (HCFC-253)					II	
134237-45-9	Hydrochlorofluorocarbon-261 (HCFC-261)					II	
134190-53-7	Hydrochlorofluorocarbon-262 (HCFC-262)					II	
134190-54-8	Hydrochlorofluorocarbon-271 (HCFC-271)					II	
74-90-8	Hydrocyanic acid				T (2,500)		
1333-74-0	Hydrogen				F [10,000]		
7647-01-0	Hydrogen chloride (hydrochloric acid)		✓	NR	T (5,000)		
7647-01-0	Hydrogen chloride (anhydrous)			NR	T (5,000)		
7664-39-3	Hydrogen fluoride (hydrofluoric acid)			✓			
7664-39-3	Hydrogen fluoride (conc. 50% or greater)				T (1,000)		
7783-07-5	Hydrogen selenide				T (500)		
7783-06-4	Hydrogen sulfide		✓	NR	T (10,000)		
123-31-9	Hydroquinone			✓			✓✓
13463-40-6	Iron, pentacarbonyl-				T (2,500)		
75-28-5	Isobutane				F [10,000]		✓
78-82-0	Isobutyronitrile				T [20,000]		✓✓
931-54-5	Isocyanobenzene						✓
78-78-4	Isopentane				F [10,000]		✓
78-59-1	Isophorone			✓			✓✓

		Regulating Program					
CAS No.	Pollutant Name	Criteria[(a)]	NSPS	HAP[(b)]	Accidental[(c)]	ODS[(d)]	VOC[(e)]
78-79-5	Isoprene				F [10,000]		✓
75-29-6	Isopropyl chloride				F [10,000]		✓✓
108-23-6	Isopropyl chloroformate				T (15,000)		✓✓
75-31-0	Isopropylamine				F [10,000]		✓✓
119-65-3	Isoquinoline						✓
	Lead compounds	✓	✓	NR			
108-31-6	Maleic anhydride			✓			✓
	Manganese compounds			NR			
	Mercury compounds			✓			
126-98-7	Methacrylonitrile				T (10,000)		✓✓
74-82-8	Methane	E			F [10,000]		✓
67-56-1	Methanol			✓			✓✓
72-43-5	Methoxychlor			NR			✓✓
74-89-5	Methylamine				F [10,000]		✓✓
1334-78-7	Methylbenzaldehyde						✓
74-83-9	Methyl bromide (bromomethane)			✓			✓
563-46-2	2-Methyl-1-butene				F [10,000]		✓
563-45-1	3-Methyl-1-butene				F [10,000]		✓
513-35-9	2-Methyl-2-butene						✓
2049-94-7	(3-Methylbutyl)-benzene						✓
74-87-3	Methyl chloride (chloromethane)			✓	T (10,000)		✓
71-55-6	Methyl chloroform (1,1,1,-trichloroethane)	E		NR		I	✓
79-22-1	Methyl chloroformate				T (5,000)		✓✓
108-87-2	Methylcyclohexane						✓
1502-38-1	Methylcyclooctane						✓
96-37-7	Methylcyclopentane						✓
693-89-0	1-Methylcyclopentene						✓
75-09-2	Methylene chloride (dichloromethane)			✓			✓
101-77-9	4,4-Methylenedianiline			✓			✓✓
101-68-8	Methylenediphenyl diisocynate (MDI)			✓			✓✓
98-83-9	1-(Methylethenyl)-benzene						✓
115-10-6	Methyl ether				F [10,000]		✓✓
98-82-8	1-(Methyl)-ethylbenzene						✓
78-93-3	Methyl ethyl ketone (2-butanone)			✓			✓
107-31-3	Methyl formate				F [10,000]		✓✓
135-98-8	1-(Methylpropyl)benzene						✓
592-27-8	2-Methylheptane						✓
589-81-1	3-Methylheptane						✓
591-76-4	2-Methylhexane						✓
589-34-4	3-Methylhexane						✓

(continued)

CAS No.	Pollutant Name	Regulating Program					
		Criteria[a]	NSPS	HAP[b]	Accidental[c]	ODS[d]	VOC[e]
60-34-4	Methylhydrazine			✓	T (15,000)		✓✓
74-88-4	Methyl iodide (iodomethane)			NR			✓
108-10-1	Methyl isobutyl ketone (hexone)			✓			✓✓
624-83-9	Methyl isocyanate			✓	T (10,000)		✓✓
74-93-1	Methyl mercaptan				T (10,000)		✓✓
80-62-6	Methyl methacrylate			✓			✓✓
	7-Methyl-3-methylene-1,6-octadiene						✓
5989-27-5	(R)-1-Methyl-4-(1-methylethenyl)-cyclohexene						✓
5989-54-8	(S)-1-Methyl-4-(1-methylethenyl)-cyclohexene						✓
99-87-6	1-Methyl-4-(1-methyl-ethyl)-benzene						✓
90-12-2	1-Methylnaphthalene						✓
91-57-6	2-Methylnaphthalene						✓
871-83-0	2-Methylnonane						✓
17301-94-9	4-Methylnonane						✓
2216-33-3	3-Methyloctane						✓
2216-34-4	4-Methyloctane						✓
96-14-0	3-Methylpentane						✓
763-29-1	2-Methyl-1-pentene						✓
625-27-4	2-Methyl-2-pentene						✓
616-12-6	(E)-3-Methyl-2-pentene						✓
674-76-0	(E)-4-Methyl-2-pentene						✓
691-38-3	(Z)-4-Methyl-2-pentene						✓
115-11-7	2-Methylpropene				F [10,000]		✓
1333-41-1	Methylpyridine						✓
109-06-8	2-Methylpyridine						✓
108-99-6	3-Methylpyridine						✓
108-89-4	4-Methylpyridine						✓
636-41-9	2-Methyl-1H-pyrrole						✓
616-43-3	3-Methyl-1H-pyrrole						✓
1634-04-4	Methyl tert-butyl ether			✓			✓✓
554-14-3	2-Methylthiophene						✓
616-44-4	3-Methylthiophene						✓
75-79-6	Methyltrichlorosilane				T (5,000)		✓✓
75-93-4	Monomethylester sulfuric acid						✓
91-20-3	Naphthalene			✓			✓
55-18-5	*N*-ethyl-*N*-nitrosoethanamine						✓
13463-39-3	Nickel carbonyl				T (1,000)		
	Nickel compounds			NR			

CAS No.	Pollutant Name	Regulating Program Criteria[a]	NSPS	HAP[b]	Accidental[c]	ODS[d]	VOC[e]
7697-37-2	Nitric acid (conc. 80% or greater)				T (15,000)		
10102-43-9	Nitric oxide				T (10,000)		
98-95-3	Nitrobenzene			✓			✓
92-93-3	4-Nitrobiphenyl			NR			✓✓
10102-44-0	Nitrogen dioxide	✓					
	Nitrogen oxides	✓	✓				
5796-89-4	Nitro-1-oxopropylperoxide						✓
100-02-7	4-Nitrophenol			✓			✓✓
79-46-9	2-Nitropropane			✓			✓
62-75-9	*n*-Nitrosodimethylamine			NR			✓
684-93-5	*n*-Nitroso-*N*-methylurea			NR			✓✓
59-89-2	*n*-Nitroso morpholine			NR			✓
111-84-2	Nonane						✓
124-11-8	1-Nonene						✓
593-45-3	Octadecane						✓
558-87-2	Octamethylcyclotetrasiloxane						×
124-13-0	Octanal						✓
111-65-9	Octane						✓
111-66-0	1-Octene						✓
7642-04-8	(Z)-2-octene						✓
8014-95-7	Oleum (fuming sulfuric acid)			✓	T (10,000)		
10028-15-6	Ozone	✓					
56-38-2	Parathion			NR			
	Particulate matter (PM10)	✓					
608-93-5	Pentachlorobenzene						✓
55880-77-8	Pentachloro-1,3-butadiene						✓
76-01-7	Pentachloroethane						✓
82-68-8	Pentachloronitrobenzene (quintobenzene)			NR			✓✓
87-86-5	Pentachlorophenol			NR			✓✓
629-62-9	Pentadecane						✓
504-60-9	1,3-Pentadiene				F [10,000]		✓✓
109-66-0	Pentane				F [10,000]		✓
107-87-9	2-Pentanone						✓
96-22-0	3-Pentanone						✓
109-67-1	1-Pentene				F [10,000]		✓
646-04-8	(E)-2-Pentene				F [10,000]		✓
627-20-3	(Z)-2-Pentene				F [10,000]		✓
79-21-0	Peracetic acid				T (10,000)		✓
127-18-4	Perchloroethylene (tetrachloroethylene)			✓			✓
594-42-3	Perchloromethylmercaptan				T (10,000)		✓✓
108-95-2	Phenol			✓			✓
106-50-3	p-Phenylenediamine			✓			✓✓
93-55-0	1-Phenyl-1-propanone						✓

(continued)

		Regulating Program					
CAS No.	Pollutant Name	Criteria[a]	NSPS	HAP[b]	Accidental[c]	ODS[d]	VOC[e]
75-44-5	Phosgene			✓	T (500)		✓
7803-51-2	Phosphine			NR	T (5,000)		
723-14-0	Phosphorus			NR			
10025-87-3	Phosphorus oxychloride				T (5,000)		
7719-12-2	Phosphorus trichloride				T (15,000)		
85-44-9	Phthalic anhydride			✓			
110-89-4	Piperidine				T (15,000)		✓✓
1336-36-3	Polychlorinated biphenyls (aroclors)			NR			✓✓
	Polycyclic organic matter			✓			✓✓
463-49-0	Propadiene				F [10,000]		✓
74-98-6	Propane				F [10,000]		✓
1120-71-4	1,3-Propane sulfone			✓			✓✓
57-57-8	beta-Propiolactone			✓			✓✓
123-38-6	Propionaldehyde			✓			✓
107-12-0	Propionitrile				T (10,000)		✓
114-26-1	Propoxur (baygon)			NR			✓✓
2807-30-9	2-Propoxy ethanol						✓
109-61-5	Propyl chloroformate				T (15,000)		✓✓
103-65-1	Propylbenzene						✓
115-07-1	Propylene				F [10,000]		✓
78-87-5	Propylene dichloride (1,2-dichloropropane)			✓			✓✓
75-56-9	Propylene oxide			✓	T (10,000)		✓
75-55-8	Propyleneimine				T (10,000)		✓✓
75-55-8	1,2-Propylenimine (2-methyl aziridine)			✓			✓✓
74-99-7	Propyne				F [10,000]		✓
110-86-1	Pyridine						✓
109-97-7	1H-pyrrole						✓
91-22-5	Quinoline			✓			✓
106-51-4	Quinone			✓			✓
	Radionuclides (including radon)			NR			
	Reduced sulfur compounds		✓				
	Selenium compounds			NR			
7803-62-5	Silane				F [10,000]		
100-42-5	Styrene			✓			✓
96-09-3	Styrene oxide			NR			✓✓
7446-09-5	Sulfur dioxide	✓					
7446-09-5	Sulfur dioxide (anhydrous)				T (5,000)		
7783-60-0	Sulfur tetrafluoride				T (2,500)		
7446-11-9	Sulfur trioxide				T (10,000)		
7664-93-9	Sulfuric acid mist		✓				
17700-09-3	Tetrachlorobenzene						✓

		Regulating Program					
CAS No.	Pollutant Name	Criteria[a]	NSPS	HAP[b]	Accidental[c]	ODS[d]	VOC[e]
1746-01-6	2,3,7,8-Tetrachlorodibenzo-*p*-dioxin			NR			✓✓
630-20-6	1,1,1,2-Tetrachloroethane						✓
79-34-5	1,1,2,2-Tetrachloroethane			✓			✓
127-18-4	Tetrachloroethylene (perchloroethylene)			✓			✓
29797-40-8	Tetrachloromethylbenzene						✓
107-50-6	Tetradecamethylcycloheptasiloxane						×
629-59-4	Tetradecane						✓
78-00-2	Tetraethyllead						✓
116-14-3	Tetrafluoroethylene				F [10,000]		✓
75-73-0	Tetrafluoromethane						
109-99-9	Tetrahydrofuran						✓
25619-60-7	Tetramethylbenzene						✓
458-23-3	1,2,3,4-Tetramethylbenzene						✓
527-53-7	1,2,3,5-Tetramethylbenzene						✓
95-93-2	1,2,4,5-Tetramethylbenzene						✓
75-74-1	Tetramethyllead				T (10,000)		✓
75-76-3	Tetramethylsilane				F [10,000]		✓✓
509-14-8	Tetranitromethane				T (10,000)		✓✓
75-18-3	Thiobismethane						✓
110-02-1	Thiophene						✓
7550-45-0	Titanium tetrachloride			NR	T (2,500)		
108-88-3	Toluene			✓			✓
95-80-7	2,4-Toluenediamine			✓			✓✓
26471-62-5	Toluene diisocynate (unspecified isomers)			NR	T (10,000)		✓✓
584-84-9	Toluene-2,4-diisocynate				T (10,000)		✓✓
94-75-7	2,4-D, salts and esters			NR			✓✓
91-08-7	Toluene-2,6-diisocynate				T (10,000)		✓✓
95-53-4	*o*-Toluidine			✓			✓✓
	Total reduced sulfur (TRS)		✓				
	Total suspended particulate		✓				
8001-35-2	Toxaphene (chlorinated camphene)			NR			✓✓
12002-48-1	Trichlorobenzene						✓
87-61-6	1,2,3-Trichlorobenzene						✓
120-82-1	1,2,4-Trichlorobenzene			✓			✓
108-70-3	1,3,5-Trichlorobenzene						✓
71-55-6	1,1,1-Trichloroethane (methyl chloroform)	E		NR		I	×
79-00-5	1,1,2-Trichloroethane			✓			✓

(continued)

CAS No.	Pollutant Name	Regulating Program					
		Criteria[a]	NSPS	HAP[b]	Accidental[c]	ODS[d]	VOC[e]
79-01-6	Trichloroethylene			✓			✓
30583-33-3	Trichloromethylbenzene						✓
95-95-4	2,4,5-Trichlorophenol			✓			✓✓
88-06-2	2,4,6-Trichlorophenol			✓			✓✓
10025-78-2	Trichlorosilane				F [10,000]		
629-50-5	Tridecane						✓
121-44-8	Triethylamine			✓			✓✓
1762-28-3	Triethylmethyl lead						✓
79-38-9	Trifluorochloroethylene				F [10,000]		
1582-09-8	Trifluralin			NR			
75-50-3	Trimethylamine				F [10,000]		✓✓
576-73-8	1,2,3-Trimethylbenzene						✓
95-63-6	1,2,4-Trimethylbenzene						✓
108-067-8	1,3,5-Trimethylbenzene						✓
80-56-8	2,6,6-Trimethyl-bicyclo[3.1.1]-hept-2-ene						✓
13466-78-9	3,7,7-Trimethyl-bicyclo[4.1.0]-hept-3-ene						✓
464-06-2	2,2,3-Trimethylbutane						✓
755-77-4	Trimethylchlorosilane				T (10,000)		✓✓
30498-63-6	Trimethylcyclohexane						✓
1795-27-3	1A,3A,5A-Trimethylcyclohexane						✓
3522-94-9	2,2,5-Trimethylhexane						✓
564-02-3	2,2,3-Trimethylpentane						✓
540-84-1	2,2,4-Trimethylpentane			✓			✓
565-75-3	2,3,4-Trimethylpentane			✓			✓
144-19-4	2,2,4-Trimethyl-1,3-pentanediol			✓			✓
496-78-6	2,4,5-Trimethylphenol						✓
527-60-6	2,4,6-Trimethylphenol						✓
108-75-8	2,4,6-Trimethylpyridine						✓
1120-21-4	Undecane						✓
51-79-6	Urethane (ethyl carbamate)						✓✓
108-05-4	Vinyl acetate			✓			✓✓
108-05-4	Vinyl acetate monomer				T (15,000)		✓✓
689-97-4	Vinyl acetylene				F [10,000]		✓✓
593-60-2	Vinyl bromide			NR			✓✓
75-01-4	Vinyl chloride			✓	F [10,000]		✓✓
109-92-2	Vinyl ethyl ether				F [10,000]		✓✓
75-02-5	Vinyl fluoride				F [10,000]		✓✓
107-25-5	Vinyl methyl ether				F [10,000]		✓✓
75-35-4	Vinylidene chloride (1,1-dichloroethylene)			✓	F[10,000]		✓

CAS No.	Pollutant Name	Regulating Program Criteria[a]	NSPS	HAP[b]	Accidental[c]	ODS[d]	VOC[e]
75-38-7	Vinylidene fluoride				F [10,000]		✓✓
	Volatile organic compounds (VOC)	✓	✓				✓
1330-20-7	Xylenes (isomers and mixture)			✓			✓✓
108-38-3	*m*-Xylene			✓			✓
95-47-6	*o*-Xylene			✓			✓
106-42-3	*p*-Xylene			✓			✓

[a] E indicates that the chemical is exempt as a VOC (Ozone Precursor). See Table 1.3.
[b] A checkmark designates a regulated hazardous air pollutant (HAP); NR indicates that it is listed as a HAP but not regulated under 40 CFR PARTS 61 or 63.
[c] F and T designate flammable (F) and toxic (T) ACCIDENTAL Release Substances as given in Tables 1 and 3 of 40 CFR PART 68 (January 31,1994). Threshold quantities shown in parentheses are expressed in pounds.
[d] I and II designate class I and class II ODS. The list of Ozone Depleting Chemicals (ODS) in FR Vol. 58 No. 236 also includes methyl bromide and 34 other HBFCs not on this list.
[d] A checkmark indicates a compound from the EPA VOC database reported in J.T. Shaw, H.B. Singh, *Environ. Sci. Technol.* **1988**, 22, 1381. A double checkmark indicates that the compound is a likely VOC. A cross indicates that the chemical appears in the original EPA database, but is exempt as a VOC (see Table 1.3).
This table was compiled by D.K. Berry and S.R. Sabatini of Environmental Quality Management, Inc., Cincinnati, Ohio. It is adapted and reproduced with permission of the authors.

INDEX